# Research Methods in Geography

# Critical Introductions to Geography

Critical Introductions to Geography is a series of textbooks for undergraduate courses covering the key geographical subdisciplines and providing broad and introductory treatment with a critical edge. They are designed for the North American and international market and take a lively and engaging approach with a distinct geographical voice that distinguishes them from more traditional and out-dated texts.

Prospective authors interested in the series should contact the series editor:
John Paul Jones III
Department of Geography and Regional Development
University of Arizona
jpjones@email.arizona.edu

## Published

Cultural Geography
*Don Mitchell*

Political Ecology
*Paul Robbins*

Geographies of Globalization
*Andrew Herod*

Geographies of Media and Communication
*Paul C. Adams*

Social Geography
*Vincent J. Del Casino Jr*

Mapping
*Jeremy W. Crampton*

Environment and Society
*Paul Robbins, Sarah Moore and John Hintz*

Research Methods in Geography
*Basil Gomez and John Paul Jones III*

## Forthcoming

Geographic Thought
*Tim Cresswell*

Cultural Landscape
*Donald Mitchell and Carolyn Breitbach*

# Research Methods in Geography

## A Critical Introduction

## Edited by Basil Gomez and John Paul Jones III

WILEY-BLACKWELL

A John Wiley & Sons, Ltd., Publication

This edition first published 2010
© 2010 Blackwell Publishing Ltd

Blackwell Publishing was acquired by John Wiley & Sons in February 2007. Blackwell's publishing program has been merged with Wiley's global Scientific, Technical, and Medical business to form Wiley-Blackwell.

*Registered Office*
John Wiley & Sons Ltd, The Atrium, Southern Gate, Chichester, West Sussex, PO19 8SQ, United Kingdom

*Editorial Offices*
350 Main Street, Malden, MA 02148-5020, USA
9600 Garsington Road, Oxford, OX4 2DQ, UK
The Atrium, Southern Gate, Chichester, West Sussex, PO19 8SQ, UK

For details of our global editorial offices, for customer services, and for information about how to apply for permission to reuse the copyright material in this book please see our website at www.wiley.com/wiley-blackwell.

The right of Basil Gomez and John Paul Jones III to be identified as the author of the editorial material in this work has been asserted in accordance with the UK Copyright, Designs and Patents Act 1988.

*Library of Congress Cataloging-in-Publication Data*

Research methods in geography / edited by Basil Gomez and John Paul Jones III.
    p. cm. – (Critical introductions to geography)
  Includes bibliographical references and index.
  ISBN 978-1-4051-0710-5 (hardcover : alk. paper) – ISBN 978-1-4051-0711-2 (pbk. : alk. paper)  1. Geography–Research–Methodology.  I. Gomez, Basil.  II. Jones, John Paul, 1955–
  G73.R45   2010
  910.7′2–dc22

                                                                        2009040193

A catalogue record for this book is available from the British Library.

Set in 10/12.5pt Minion by Toppan Best-set Premedia Limited

01   2010

# Contents

# List of Figures

# List of Tables

# List of Boxes

# List of Exercises

# Notes on Contributors

**Hannah Allen** (BA, Geography, 2006, University of Birmingham) is an independent scholar.

**Debbie Allsop** (BA, Geography, 2006, University of Birmingham) is an auditor for Price Waterhouse Cooper.

**Sandra Lach Arlinghaus** is Adjunct Professor, School of Natural Resources and Environment, University of Michigan, and Founding Director, Institute of Mathematical Geography, in Ann Arbor, Michigan, USA.

**Hayley Raxter** (BA, Geography, 2006, University of Birmingham) works at Waterstone's Bookstore, in Birmingham, UK.

**Stefania Bertazzon** is Associate Professor, Department of Geography, University of Calgary, in Calgary, Alberta, Canada.

**Helen Clare** (BA, Geography, 2006, University of Birmingham) teaches English at Bournville College, in Birmingham, UK.

**Ian Cook** is Associate Professor, Department of Geography, University of Exeter, in Exeter, England, UK.

**Michael J. Crozier** is Emeritus Professor, School of Geography, Environment and Earth Sciences, Victoria University of Wellington, in Wellington, New Zealand.

**Dydia DeLyser** is Associate Professor, Department of Geography and Anthropology, Louisiana State University, in Baton Rouge, Louisiana, USA.

**Deborah P. Dixon** is Senior Lecturer, Institute of Geography and Earth Sciences, Aberystwyth University, in Aberystwyth, Wales, UK.

**Thomas W. Gillespie** is Associate Professor, Department of Geography, University of California, Los Angeles, in Los Angeles, California, USA.

**Basil Gomez** is Professor, Geomorphology Laboratory, Indiana State University, in Terre Haute, Indiana, USA.

**Michael F. Goodchild** is Professor, Department of Geography, University of California, Santa Barbara, California, USA.

**Stephen P. Hanna** is Associate Professor and Chair, Department of Geography, University of Mary Washington, in Fredericksburg, Virginia, USA.

**Ulrike Hardenbicker** is Associate Professor, Department of Geography, University of Regina, in Regina, Saskatchewan, Canada.

**Andrew Herod** is Professor, Department of Geography, University of Georgia, in Athens, Georgia, USA.

**Ryan R. Jensen** is Associate Professor, Department of Geography, Brigham Young University, in Provo, Utah, USA.

**John Paul Jones III** is Professor of Geography and Dean, College of Social and Behavioral Sciences, University of Arizona, in Tucson, Arizona, USA.

**Glen M. MacDonald** is Professor, Department of Geography, University of California, Los Angeles, in Los Angeles, California, USA.

**Yvonne Martin** is Associate Professor, Department of Geography, University of Calgary, in Calgary, Alberta, Canada.

**Gordon F. Mulligan** is Professor Emeritus, School of Geography and Development, University of Arizona, in Tucson, Arizona, USA.

**Kathleen C. Parker** is Professor, Department of Geography, University of Georgia, in Athens, Georgia, USA.

**Marianna Pavlovskaya** is Associate Professor, Hunter College of the City University of New York, in New York City, New York, USA.

**Bruce L. Rhoads** is Professor and Head, Department of Geography, University of Illinois at Urban-Champaign, in Urbana, Illinois, USA.

**Paul F. Robbins** is Professor, School of Geography and Development, University of Arizona, in Tucson, Arizona, USA.

**Richard H. Schein** is Associate Professor, Department of Geography, University of Kentucky, in Lexington, Kentucky, USA.

**Anna J. Secor** is Associate Professor, Department of Geography, University of Kentucky, in Lexington, Kentucky, USA.

**Ian Graham Ronald Shaw** is a PhD student, School of Geography and Development, University of Arizona, in Tucson, Arizona, USA.

**J. Matthew Shumway** is Professor and Chair, Department of Geography, Brigham Young University, in Provo, Utah, USA.

**David M. Smith** is Professor Emeritus, Department of Geography, Queen Mary, University of London, in London, England, UK.

**Kevin St Martin** is Associate Professor, Department of Geography, Rutgers, the State University of New Jersey, in New Brunswick, New Jersey, USA.

**Douglas A. Stow** is Professor, Department of Geography, San Diego State University, California, USA.

**Christina Upton** (BA, Geography, 2006, University of Birmingham) is an independent scholar.

**Sent Visser** is retired from the Department of Geography, Texas State University, in San Marcos, Texas, USA.

**Alice Williams** (BA, Geography, 2006, University of Birmingham) is a broadcast journalist for BBC News in London, UK.

**David Wilson** is Professor, Department of Geography, University of Illinois at Urban-Champaign, in Urbana, Illinois, USA.

**Julie A. Winkler** is Professor, Department of Geography, Michigan State University, East Lansing, Michigan, USA.

# Acknowledgments

We are, first of all, indebted to the authors who agreed to offer their expertise in writing the chapters in this volume. They have been incredibly responsive and patient as the process unfolded, from initial request to final publication.

All the chapters in this volume were reviewed by the editors and by an independent reviewer. The editors would like to thank the following people for their assistance: Derek H. Alderman, Bernard O. Bauer, Paul D. Bates, C. Mark Cowell, Jeremy Crampton, Sandy Dall'erba, Vincent J. Del Casino Jr., Deborah P. Dixon, Mona Domosh, Sarah Elwood, C. Susan B. Grimmond, Julie Guthman, Robert J. Inkpen, Frank J. Magilligan, Sallie A. Marston, Sara McLafferty, Jonathan D. Phillips, E. Jeffrey Popke, Ian Graham Ronald Shaw, and Elizabeth A. Wentz.

The administrative, editorial and production advice and assistance of Barbara Duke, Ben Thatcher, Justin Vaughan, and Mervyn Thomas of Wiley-Blackwell and Robert Merideth is greatly appreciated. Finally, many thanks go to Ian Graham Ronald Shaw. His care and diligence as a research assistant during the months leading up to the volume's final production are warmly acknowledged.

# Chapter 1

# Introduction

*John Paul Jones III and Basil Gomez*

- Research Methodology and Methods
- Why *Research Methods in Geography?*
- Conclusion

For much of the twentieth century, as geographers' concerns ranged over contemporary physical and human space and into their past arrangements in so far as they could be documented, the methods they used to explain, model, and predict different aspects of the human and physical worlds became progressively more quantitative. But the new technologies and theoretical perspectives that emerged in the latter decades of the twentieth century helped to redefine the objects of geographers' inquiry and extend the methods in use for collecting and analyzing data and evaluating research. They also raised concern about the criteria, norms, and values for human action and conduct (ethics); how we position ourselves and are positioned by others (positionality); and the relationship we, as researchers, have with the world (reflexivity). In physical geography, data availability increased dramatically after Earth observation satellites were launched in the 1970s. At about the same time some human geographers began to seek alternatives to using spatial analytic (that is, quantitative, objective, and scientific) methods to explain, represent, and understand human actions and landscapes. Among the enduring approaches developed since the 1970s are humanistic, Marxist, feminist, and poststructuralist geography. Nevertheless, for many students research methods remain grounded in the traditional canons of spatial analysis and quantitative techniques, and methods instruction are, we feel, too often structured according to the prevailing divisions between human and physical geography.

This book is an introduction to research methods in geography. A research method is a way of collecting and analyzing data. This sounds very "nuts and bolts," but there is no way to properly engage in research – or in methods – without also tackling some of the fundamental theoretical questions facing both human and physical geographers. These "philosophical" questions concern the nature of reality (ontology) and how we go about understanding it (epistemology). Such philosophical concerns tend to get sorted out into "paradigms" – bodies of theory that groups of researchers follow as part of their everyday scientific practice. Nested within the theoretical coordinates of paradigms are a set of "middle level" decisions one has to make about methodology: the selection of research objects, the questions directed toward them, the design of a study, and the implications that our objectives have for carrying out research. Finally, at the most concrete and practical level we find research methods: the ways we go about collecting and analyzing data, and the conclusions we draw from these processes.

The intent of *Research Methods in Geography* is to provide a foundation for geography students, beginning with the big picture, moving through methodology, and finally introducing a number of commonly used methods in data collection and analysis. *Research Methods in Geography* therefore covers theory while providing a solid basis for engaging in concrete research activities. Schematically, the entire framework can be viewed like this:

Epistemology and Ontology

↓↑

Paradigms

↓↑

Research Methodology

↓↑

Research Methods

That the arrows work in both directions indicates that theory needs to be responsive – constantly amended and reworked – in light of the surprises and contradictions that emerge in concrete research activities.

Chapter 2 gives an overview of the essential philosophical issues surrounding ontology and epistemology. It also describes the broad contours of four important paradigms – or theoretical frameworks – operating in geography today (spatial science, humanism, critical realism, and poststructuralism). Before turning to that chapter, it might be helpful to further consider the questions of research methodology and their distinction to research methods. These are the two domains that constitute most of the content of the book.

## Research Methodology and Methods

Research methodologies are always situated within larger theories of the world. Many of the most creative aspects of the research process involve questions that translate those theories into more precise research objectives, questions and tasks. To these ends, it can be helpful to formulate a research methodology in terms of a series of questions, the most basic of which include the following:

- What objects or events should I select to analyze?
- How should I theorize their domain of operation?
- How should I theorize their relationships to other objects and events?
- What research questions are appropriate for explaining or understanding them?
- How should I collect data for answering those questions?
- What procedures should be used to analyze that data?
- What safeguards should I rely upon to ensure the validity and reliability of my account?
- What are the grounds for evaluating competing accounts?
- How are my findings influenced by the "theoretical priors" I brought to the research?
- What is the purpose of the research (e.g., the production of scientific knowledge, saving the Earth, transforming society, something else)?
- What ethical safeguards have been followed or need to be addressed?

Having posed these and other questions about the world and the intended research activity, a number of obvious connections to specific research methods may emerge, not least because some research methods tend to be tied very closely to specific objects of analysis (consider for example, the different methods required to analyze the physical landscape and travel writing). However, some methods are more utilitarian than others (for example, scientists, humanists, critical realists, and poststructuralists can all conduct interviews), and some approaches support multiple methods (consider, for example, the different methods used to describe and analyze physical landscapes). *Research Methods in Geography* covers a wide range of topics in data collection and analysis within the field of geography.

## Why *Research Methods in Geography*?

We developed the idea for this book after having taught many courses in our substantive areas in human and physical geography, respectively. What was lacking, we felt, was an introductory level textbook that spoke to theoretical issues but that also covered concrete aspects of methods as well as specific methods and techniques that geographers use to conduct research. *Research Methods in Geography* is intended primarily for second or third year undergraduates embarking on a more focused course of study in human or physical geography, in human–environment relations, or in geographic techniques. Most second and third year students won't yet have taken many substantive courses, and our thought is that surveying a book like this will help improve their ability to conduct the sort of research projects that they might be expected to undertake for a senior thesis or undergraduate dissertation and improve their understanding of the research papers they might encounter in courses on, say, population, economic, and urban geography, or on geomorphology, climatology, and biogeography, to name but a few. Some students will have already taken a technical course or two, or intend to specialize in geographical information systems, remote sensing, or spatial statistics, but even they probably haven't encountered the breadth of methods and techniques that geographers use, or that are represented in this book.

*Research Methods in Geography* is intended to assist students as they move forward in geography towards completion of their undergraduate degree. Its overarching objectives are to help them to understand and to begin to assess the research of others, and to assist them in the development and conduct of their own high quality research projects. If students find that research turns them on, then perhaps they will seek out more advanced training at the masters or doctoral level in geography. Some of the chapters in this volume might be profitably read by students at those levels, particularly those just starting out in the world of research, but our intention is, first and foremost, to assist undergraduates in the discipline.

Geography is a tremendously wide ranging field; most geographers have a difficult time just keeping up with their own area of research, much less staying abreast of developments in areas far from their domain of interest. Thus it would be almost impossible for a single author to have written with the level of depth and breadth represented in this book. Situations like this call for an edited volume. There are several excellent recent and competing texts that converge on the topic of research methods in geography, but several things set this book apart from those texts.

First, each chapter was authored by accomplished researchers, many of whom are leaders in their field. Second, while necessarily offering in-depth details in particular areas, *Research Methods in Geography* also includes several important cross-cutting chapters – including those on observing our world (Chapter 3), measurement and interpretation (Chapter 4), and operational decisions (Chapter 5) – which are co-authored by geographers of different theoretical and substantive backgrounds. The intention in these foundational chapters is to deliberately cross some of the divides that have emerged within the discipline of geography. Some divisions have arisen over theory, i.e., how geographers theorize how the subject works; some are over methods, for example, whether geographers should use qualitative or quantitative data; and sometimes a line of separation is drawn between human and physical geography. Such divisions may be inevitable within a discipline that is as diverse as geography, but that does not mean that there are no common denominators, or that geographers can't engage in dialog about those divisions. The fact that this book is co-edited by a physical geographer (BG) and a human geographer (JPJ) is a testament to our desire to peer over divides to see what lies on the other side, and to try to determine how they might be negotiated or bridged.

The fact that no single geographer would be likely to attempt to write a volume that includes such a range of material drawn from human and physical geography and closely related disciplines, as well as such a diversity of theory and methods, should suggest to students that the faculty instructor for their particular course on "research methods" will doubtless have more of a background in some of the chapters than in others. This means that individuals will have to step up to the plate, so to speak, and take some responsibility for their own education as a geographer. With that in mind, each chapter in *Research Methods in Geography* includes sections that contain "Additional Resources" and "Practical Exercises." These sections are provided to help students plant their feet and dive more deeply into the areas that most interest them, and provide instructors who come across unfamiliar content some direction and ideas about how to develop particular topics.

*Research Methods in Geography* is divided into four Parts. The first comprises five chapters that address overarching issues of theory and research methodology, including research design. In Part II the focus moves to the methods used to collect specific types of information, for example, about physical and cultural landscapes. Part III addresses the issues of how geographers represent and analyze different types of data. The final Part of the book concerns a researcher's obligations in terms of ethics and the important role of disseminating their work to others.

## Conclusion

Learning about research should be a rewarding experience that allows students to pursue their own interests, learn more about a chosen topic and, above all, examine a subject from different perspectives. The best reason for researching a topic in depth is that *you* find it stimulating and important. But students should also be encouraged to approach the eclecticism of their chosen discipline with a broad mind and an ecumenical spirit. Many prominent geographers have been attracted to the field precisely because of its wide remit, and some topics that are now considered mainstream were, as recently as a generation ago, not considered to be part of the discipline. So we encourage students to let their imagination run free as they select objects of analysis and ways to study them. Finally, while all research is constrained by such basic considerations as the amount of time available or the presence of supporting equipment, facilities, or funding, it is *your* curiosity about questions and *your* commitment to finding answers that are most important in influencing your success.

# Part I
# Theory and Methodology

# Chapter 2

# Theorizing Our World

*Ian Graham Ronald Shaw, Deborah P. Dixon, and*
*John Paul Jones III*

- Introduction
- Ontology
- Epistemology
- Research Paradigms
- Conclusion

Keywords

| | |
|---|---|
| Becoming | Marxist geography |
| Binary | Methodology |
| Critical realism | Monism |
| Dialectics | Ontology |
| Dualism | Paradigm |
| Epistemology | Poststructuralist geography |
| Humanistic geography | Spatial science |

## Introduction

Influential French philosopher Gilles Deleuze once wrote that all human existence, from the mundane to the exceptional, is founded upon a multiplicity of problems to be overcome. Writing against both ancient and modern theorists who sought to order the world around us into finely-cut, timeless categories, Deleuze insisted on the creative potential of life to challenge us and to "call us into being." The world constantly asks questions of us, he wrote, and we try to give answers back. Sometimes we succeed, sometimes we fail – but the world never stops unfolding. For Deleuze, it is the manner in which we reply to these different challenges that constitutes our subjectivity, or who we are. As human beings, or simply as researchers, then, we always find ourselves in the *middle of things*: in the middle of a confrontation with social difference, an economic crisis, or an ecological change. Without the world's never-ending chain of problems, existence would be, well, very boring indeed.

Reflecting on the nature of existence may seem overly philosophical for a chapter focused on explaining how theory and methodologies intersect and work off one another. But the important lesson of Deleuze is that our research methods and projects do not begin in a vacuum, and neither do we as subjects. We do not, for example, suddenly wake up in the morning – without reason and without a context – and decide to sample tree-rings to collect evidence about global warming. Instead, we find ourselves situated in the middle of a life in which ideas, opportunities, and problems are thrown at us on a daily basis. If we are to take Deleuze at his word we must, therefore, acknowledge that our research projects are never truly our own creation; rather, the world invites us to solve its never ending sequence of problems. Put another way, our research projects belong as much to the world as they do to us. We chart a research path *with* the world, and never against it or above it. What challenges have you chosen to investigate, for what reason, and how will you carry them out? To begin to formulate answers to these questions is to start squarely in the middle of things.

If the world is constantly presenting us with problems, then what better way to begin than to ask what exactly is the nature of the world in the first place? It is only from this foundational starting point that we can begin to imagine those methods and toolkits that are the best fit for the research job. "Ontology," which focuses on debates concerning the nature of the world, and "epistemology," which addresses how we come to know that world, lie at the heart of geographic research, and yet it is fair to say that it is only recently that geographers have begun to regularly reflect on the character of these two theoretical terms in the everyday practice of their research. Whether one counts herself as a humanist, a social scientist, or a natural scientist, the everyday life of a geographer is more likely to involve defining research questions, or collecting and analyzing data. These practical aspects of the research process often take precedent over abstract theoretical reflections.

And yet, we want to emphasize here, how one approaches and analyses one's objects of interest are deeply embedded in these philosophical issues. No geographer, therefore, should ignore such "big picture" concerns. Doing so limits a critical evaluation of research – one's own and that of others. In the following two sections we trace the broad definitional contours of ontology and epistemology, two terms that, however nebulous they may sound, are always present within the heart of any geography research project, physical as well as human.

## Ontology

If the literal translation of geography from its Greek roots means "writing about the world," then it follows that we need to think through the characteristics the world possesses before we can begin writing about it. **Ontology** is the largest branch of metaphysics in philosophy, and traditionally deals with questions of existence, or what it means "to be." As noted above, at the most general level we can define ontology as a set of assumptions and theories that explore "what the world is like." It has a rich philosophical lineage, from pre-Socratic philosophers who wrote about a world in constant flux, to German Idealists who stressed the relational character of phenomena. These and other philosophers have attempted to tap into the true nature of "being".

One of the major distinctions made in ontology, for example, is between **monism** and **dualism**. In the case of the former, it is understood that every phenomena on the planet and in the universe essentially belongs together as "one" insofar as they are all ultimately formed from the same material or are subject to one overriding principle that governs their existence. This worldview, advocated by philosophers as diverse as Spinoza and Deleuze, is contrasted with dualism, wherein a profound distinction is drawn between the realm of the mental – of human thought – and the realm of the material, or the "stuff" of things. Importantly, these philosophical positions are woven throughout the natural and social sciences and humanities, including geography.

By way of an example, imagine that you were attempting to describe the fundamental structure, composition, or character of the world to an alien from another planet. Indeed, this is how the philosopher discerns our world – nothing about existence is taken for granted. To undertake such a task you would first need a common language, one that would enable you to discuss sets of categories that describe the fundamental characteristics of the world. You might then attempt to describe the "stuff" of the world, perhaps distinguishing between natural phenomena and social phenomena. Each of these categories, in turn, would have their own subcomponents. Nature, for example, might be broken down into animal and plant life, and air, land, and water. In turn, belief systems, language, and social groups and the family – however defined – might be described as subcomponents of society or culture. In either case the researcher is recounting what are thought to be the essential properties of the world.

In the process, what tends to happen is that we establish a series of differences between things that are thought to be profound in the sense that they separate out incompatible characteristics. Because a particular phenomenon is natural, for example, it cannot by default be social; because a something is a plant, it cannot be an animal and so on. The result is that we end up ordering the world into a series of component parts, each of which is predicated upon mutual opposition, or incompatibility. In Table 2.1 we have summarized some of the categories that have been put forth over the centuries to describe the world in the form of such mutual opposites, or **binaries**, whereby things belong "here"

**Table 2.1**  Ontology: what is the world like?

| *Binary* | |
| --- | --- |
| Material | Ideational |
| Reality | Fantasy |
| Body | Mind |
| Nature | Culture |
| Individual | Society |
| Self | Other |
| Organic | Technological |
| Space | Time |

but not "there", in "this" group but not "that" group. For example, a recurring distinction within dualist thought, as we noted earlier, has been made between the brute reality of earth and bodies on the one hand, and the mental images or ideas that we have about the world on the other. Such a distinction has given rise to at least three binaries – *Material–Ideational*, *Reality–Fantasy*, and *Corporeal–Mental* – wherein each side of the binary is considered somehow separate from its counterpart.

A fuller account of the world's elements would require a researcher to explain additional, more obviously nebulous phenomena, such as space and time. For the philosopher and geographer Immanuel Kant, for example, space and time were tremendously important things insofar as they are the necessary means by which we leverage other phenomena into meaningful, readily identifiable objects. These are fundamental because it is impossible to think of some aspect of the world that does not have a spatial and temporal dimension. Time, for example, is generally conceived of as a linear progression – past, present, and future. At any moment time situates individuals, society, and elements of nature. It is through this sense of time that our understanding of causality among the world's elements is established, for it is inconceivable that something in the present or future could cause something in the past (though predictions in the past or the present about possible futures may cause us to act differently at the time).

Likewise, elements of the world are also generally considered to be locatable within two and three-dimensional space. This view of space enables us to describe things as being "near" or "far away" from each other; it also leads to a number of fundamental geographic concepts, such as absolute and relative location, distance and direction, scale and movement, as well as perspective. It is upon the binary of *Space–Time* that the distinctions between the disciplines of geography and history have been emphasized – geography taking "space" as its prize, whereas history becomes centered on "time." This distinction was fundamental to Kant, as well as to twentieth-century geographers such as Richard Hartshorne (1939).

Historically, geographers have tended to examine objects "grounded" in the left hand side – the material, the real, the corporeal – but since the 1960s, with increasing work in behavioral and humanistic geography, and especially as a result of an emerging interest in the symbolic realm of words and images, research in geography has expanded to include objects classified as being located on the right hand sides of these binaries. What is more, there has been a surge of interest in how phenomena actually crossover such binaries, existing, for example, in both the physical and the social realms. People, for example, are not simply individual human beings: we are also part of larger society, and, as living organisms, we are also part of nature. Likewise, pets such as cats and dogs are not simply animals, and hence part of nature – they have become socialized into a world of human beings and artifacts. Indeed, through their relationships with people, pets domesticate us as much as we domesticate them. In other words, and to borrow an additional Deleuzian concept, humans and animals are constantly **becoming** with each other. Consider for example, the amount of medical research that is carried out on animals – from mice to fruit flies – that directly impinges on the drugs and solutions for illnesses we experience. Or more directly, the transplantations of animal body parts into human bodies.

These are all "hybrid" phenomena, whose existence serves as a ready reminder that all such categorizations are merely attempts to impose order upon the world. And, we can see how in recent years geographers have attempted to overcome their own institutionalized differences as either "physical" or "human" in order to gain a more comprehensive picture of highly complex phenomena such as climate change, environmental degradation, urban sprawl and epidemic disease, none of which clearly fit into natural or social categories. Even geography's overwhelming emphasis upon space has been rethought, as some scholars such as Doreen Massey (2005) point out the fact that a sense of space and a sense of time are so entwined as to be indistinguishable.

These concerns over categories are not simply pedantic; they have implications for causality, for how we go about explaining the world. For when phenomena are considered to be hybrid, then it is no longer possible to isolate an $X_1$ or an $X_2$. The distinctions between the two become blurred and we have to proceed cautiously in explanation. Consider for example the ontological status of a building. It might be seen as existing in the minds of individuals (in its design), in nature (in its materials), and in society (in the coordinated labor required for its construction). Even these components are hybrids. For example, the mental labor that produced the building is both natural (operating through bio-chemical activity) and social (operating through architectural design networks) (see Box 2.1).

With hybrid objects, the task of describing a particular phenomenon can become impossible to contain. When thinking through the ontology of phenomena, then, it behooves the researcher to look beyond the obvious borders of these – the enveloping skin of an animal, the sharp edges of a tool – and to consider how matter, force, and causal power flows *across* these borders, such that an animal becomes *also* the food it eats and the land it inhabits, or the tool becomes *also* the hand that wields it, and the artifacts it creates. These phenomena extend far beyond the readily observable lifeform or object to encompass all that makes their very existence possible.

There exists a second set of binaries associated with ontology; these describe aspects of the regularity of relations among phenomena. For example, are phenomena related to one another in ways that are orderly and hence predictable? The pairs of binaries shown in Table 2.2 circulate around the extent to which we can make statements about regularity. For example, one can assess the degree to which a causal effect will repeat itself without any difference whatsoever across different social or natural contexts. If it is repeated to a high degree, then the causal pattern is orderly; if not, it is chaotic. If a high degree of order is present, then one can offer some prediction as to future patterns of cause and effect. If not, then future relations of cause and effect will produce more unpredictable configurations.

Concepts such as order, predictability, and determinacy have tended to be more apparent in the study of physical, rather than social, phenomena because it is often assumed that objects in the natural world obey, with a high degree of order, an invariant set of physical processes, from the laws of gravity and chemical reactions to those of thermodynamics. Individuals, by contrast, are usually held to be much more unpredictable, owing to the serendipitous character of individual interactions, personal whimsy, or the complex and contingent ways in which individuals are embedded within different societies or cultures (which can produce wide variations in actions and thoughts). And yet, physical scientists have become increasingly concerned in recent years that they are in fact studying objects whose behaviors are fundamentally random and chaotic (even though the larger systems seem to obey quite nicely the laws of physical processes). And, we often hear quite deterministic assessments about people that seem to suggest that some degree of neurological or even ideological order governs thoughts and action.

**Box 2.1**   Discrete and Embedded Causality

One of the most common binaries at work in geography is the distinction between discrete and embedded relations. On this distinction pivots a great deal of the difference between physical and human geography, and between some forms of human geography (e.g., spatial science vs. critical realism). One of the most common forms of discrete relationships, found across the natural and social sciences, is that of causality, whereby one phenomena is thought to exert some form of force over and against another, causing it to change in some way. We might graphically depict such a causal relation in the following terms:

$$X \rightarrow Y$$

A more complicated process can be introduced via a feedback or loop, in which the interaction between these phenomena is thought to produce change in both. For example, one could state that humans degrade nature by polluting the environment, but that such pollution in turn has a detrimental impact upon human health. In this case, we are suggesting something more like this:

$$X \leftrightarrow Y$$

More usually, causality is considered to be a much more complex relation, for in the real world two phenomena do not exist in isolation. Instead, we find that a multitude of phenomena interact with each other, so that our pattern becomes:

$$X_2 \leftarrow X_1$$

$$\updownarrow \uparrow$$

$$X_3 \leftrightarrow X_4 \rightarrow Y$$

Here, it becomes difficult to pinpoint one key causal phenomenon; instead, one can point to a number of influences and try to ascertain their relative impact or significance.

Embedded relations are also causal, but they tend to be theorized in ways that do not permit the drawing of arrows between phenomena. This is because, as embedded relations, they are intrinsically connected to other phenomena by virtue of the *internal* relations they have with them. Instead of discrete causality that places attention, first, upon the identification of separated phenomena (X and Y), and second on the *external* relations (i.e., →) that connect them, in embedded relations the phenomena under study can never be sorted out as "simply X and Y." These phenomena are said to "co-constitute" each other internally, and at the moment of the production of one, the other is itself transformed. A famous example is found in the dialectical materialism of Karl **Marx**. He felt that "human nature" changes as humans work *on* nature – that is, as people appropriate nature into shelter, food, clothing and other items, they not only change nature, but themselves. Thus through human labor, nature is socialized, while at the same time nature itself becomes internal to human social relations (Marx 1976).

**Table 2.2** Ontology: how orderly is the world?

| *Binary* | |
| --- | --- |
| Orderly | Chaotic |
| Predictable | Random |
| Deterministic | Indeterministic |
| Necessary | Contingent |
| General | Particular |
| Global | Local |

# Epistemology

The questions and concerns raised in regard to how we categorize the phenomena that make up the world, and how we understand the connections between phenomena, lead us into the realm of **epistemology**. Epistemology deals with our understanding of knowledge – that is, how we come to know the world as a site for research and analysis. Some of the concerns we outlined in the section on ontology can be found here, for example the tension between monism and dualism. The dualist would insist that knowledge is gathered by a transcendent mind that exists apart from the world and is able to thus achieve a sense of objectivity concerning it. A monist, by contrast, would insist that all knowledge is produced via the individual's immersion within that world. Here, knowledge is formed from the continual gathering of sensory impressions; what is more, these impressions work to reconfigure our sense of who we are, and our place within the world. Deleuze (1993) uses the concept of the "fold" to discuss this form of subjectivity, by which he means our everyday embeddedness in social and also ecological networks. The upshot of this line of thought is that any notion of objectivity, whereby the individual stands outside of the world they observe, is undermined.

In understanding how we as researchers produce knowledge, then, we need to bear in mind and consider the ontological issues outlined above, and particularly those inherited binaries that mark a distinction between *Material–Ideational, Reality–Fantasy, Body–Mind*, and *Individual–Society*. How do we as subjects – individuals with bodies and minds – exist in the world? Do our minds roam free, or are our thoughts and ideas derived from the milieus and contexts experienced as part and parcel of that world? Are there distinct structures to the mind in the form of *id* and *ego*? Do we exist as individuals endowed with free will, or as subjects bound through relational ties to other subjects, such that our "will" is shaped by the ideas and attitudes of those around us (by political systems, the economy, or cultural beliefs, for example)? Do we exist as a part of nature, or are we distinct in some fundamental way, perhaps by virtue of a unique ability to reflect upon our own existence? All of these questions revolve around our understanding of the stuff we are made of. Thus, the central but somewhat vague definition of epistemology – assumptions and theories about "how we know the world" – can be more usefully stated as an understanding of how we exist as knowing subjects, how we actively engage with the world, and how we reflect upon our experience of the world.

In regard to the first question, how we conceptualize ourselves as knowing subjects, many of our assumptions are fairly recent in historical terms in that they are drawn from that body of philosophical thought referred to as the "Enlightenment" (Foucault 1970). This was a time spanning the eighteenth century, when Western societies came under the predominant influence of "reason" as the source of authority in regard to scientific inquiry, but also political rule. One of the main Enlightenment principles was the inalienable rights of "Man," which underpinned such political movements as the French Revolution, guided by the philosophy of Jean-Jacques Rousseau. Here, individuals were described and celebrated as fully knowing agents. The mind, it was argued, was an external observer, able to dwell upon everything outside of itself, and also capable of penetrating and understanding its own undertakings.

This positioning of the mind as a knowing agent capable of fully comprehending that upon which it dwells, including itself, upholds a number of binaries noted in Table 2.3 – *Mind–World, Self–Other, Subject–Object* – that maintain a distinction between the mind-as-knowing agent and everything outside of it. Famously, this Enlightenment view was encapsulated by the figure of René Descartes, a philosopher who, through a series of thought experiments, deduced that while he could be deceived into thinking that everything in the world was a phantom projection of an "evil demon," he was nonetheless unable to doubt his own existence which was revealed through the very same thought experiments. From this he concluded that the thinking self must be the primary source of knowledge; all else, including the coarse materiality of the body, must then become the object to be thought upon. He wrote:

> But what, then, am I? A thinking thing, it has been said. But what is a thinking thing? It is a thing that doubts, understands, [conceives], affirms, denies, wills, refuses; that imagines also, and perceives. (Descartes, *Meditation* II)

The characteristics, or capacities, of the mind-as-knowing agency are also listed in Table 2.3. In order for the mind to be able to be positioned as an external observer of itself and others, there must be *rational* and hence *masculine* (as opposed to *emotional* and *feminine*) and *rigorous* (as opposed to *playful*) thought, terms that for many come under the larger

**Table 2.3**   Epistemology: how do we know the world?

| *The mind-as-knowing agent* | |
| --- | --- |
| Mind | World |
| Self | Other |
| Subject | Object |
| *The capacities of the mind-as-knowing agency* | |
| Rational | Emotional |
| Rigor | Play |
| Objective | Subjective |
| Science | Art |
| Complete | Partial |
| Coherent | Illogical |
| Unified | Disjointed |
| General | Particular |
| Masculinist | Feminist |
| Explanation | Understanding |

heading of *objective* (as opposed to *subjective*) thought. Traditionally, within academia, objective thought has also been considered a hallmark of *science* (as opposed to *art*).

Given these capacities, the mind-as-knowing agent can provide an explanation for the way in which the world works. Although no one would claim that a complete knowledge of the world has indeed been produced, many would hold to the belief that knowledge is rather like building a wall with bricks, wherein each explanation builds upon others, such that more and more truths about the world become apparent over time. There is a belief, then, that the potential for a *complete* (as opposed to *partial*) explanation exists. Within this mode of thought, each truth is compatible with the next, such that a *coherent* (as opposed to *illogical*) and *unified* (as opposed to *disjointed*) field of knowledge emerges.

While such a view is still very much predominant across the natural sciences, there has emerged a series of critiques of just such a stance within the social sciences, humanities and arts. Certainly within human geography those terms listed on the right hand side of the binaries set out in Table 2.3 have come to the forefront. Feminist geography, for example, has been at the forefront in questioning the dualist philosophy that underlies the notion of the objective self (Massey 1994). For these geographers, reference is made to the manner in which the self is necessarily shaped by its embeddedness within a highly complex, continually changing world. Feminists adopt stances that are *situated* (as opposed to *detached*), that require *participation* (as opposed to *observation*), and that views researchers as *political* agents (as opposed to *neutral*).

## Research Paradigms

Having stressed the importance of reflecting on the nature of the world, and how we come to gain knowledge of the world – a twin process that should run throughout the research process – we are now in a position to discuss in more detail how ontology and epistemology find their place within different **paradigms** in geographic research. A paradigm refers here to a body of literature that shares fundamental assumptions about what the world is like and how we should research it, but also, and more specifically, about what the key objects of analysis for geographers should be, and the role of geography within wider society. At the risk of stereotyping research within geography, it is useful to provide an outline of some of the contemporary paradigms in order to show how they commit to certain ontological and epistemological stances. We should also note here, however, that within particular projects both **methodology** and methods can overlap and, moreover, each paradigm contains within it some often contentious debates.

### Spatial science

**Spatial science** rests on the foundational pillars of objectivity and generality in searching for orderly causal processes; it also adopts what is said to be a realist approach to representation (Abler et al. 1971). Ontologically speaking, spatial scientists maintain a strict divide between space and time, and between space and society. In turn, both space and time are discerned as measurable, insofar as they can be divided into increments, and these increments can be empirically assessed by instruments such as clocks, chronometers, compasses and so on. Within this paradigm, geography is taken to be synonymous with the term "spatial," such that geographic inquiry becomes a matter of asking: (1) How do objects and practices vary and/or move across the earth's surface?; and (2), why do these variations take the spatial forms that they do? Descriptive accounts of spatial variation (1), then, should in turn lead to explanatory ones (2). The latter typically rely on the development of qualitatively- or quantitatively-stated conceptual models that set forth hypothetical explanations for the variations of interest. Research questions suggested by the models lead to data collection and analysis (again, these can be qualitative or quantitative). Importantly, this type of research is made possible by assuming that the objects of analysis are discrete; this enables questions that take the form of the following: "how does the spatial distribution of X compare with that of Y?"; "how does X cause Y"; and "how do X and Y cause each other?"

## Humanistic geography

**Humanistic geography** questions the emphasis upon explanation within scientific ways of knowing, offering in turn a hermeneutic concern for understanding and interpretation of the complex, singular actions and attitudes of people (see Chapter 4 on "Interpretation"). Most commonly, humanistic geography is associated with an interest in the dense lifeworld of individuals, which is a sensible and sensuous world, but also with how these resonate with universally human experiences of hearth and home and a sense of place (Buttimer 1976; Relph 1970; Tuan 1975). Out of our common immersion in the lifeworld, for example, we achieve a form of shared understanding, or intersubjectivity; our sense of place is formed by the human need to give meaning to the bonds between the cultural and natural worlds. Methodologically, humanism involves interpretations of signs (symbols, gestures, and utterances), of the meanings humans invest in nature (including animals), and of the creative activities of people, especially those practices that shed light on place-meanings, such as are found in the arts, literature, and architecture.

## Critical realism

**Critical realism** in geography emerged as a critique of spatial science and its assumption that causality could be "read off" from spatial variation (Sayer 1992). In contrast to spatial science's realist (or "naively given") model of representation, critical realism adopts a depth model, seeing both images and meanings as reflective of wider, deeper structures. It also offers a critique of what is said to be humanistic geography's overwhelming focus on the individual as the agent for change. Moreover, though critical realists recognize the need for hermeneutic understanding (seeing individuals as part of wider worlds), they maintain nonetheless that there is an objective reality of both natural and social relations that is analytically reachable. Objects in critical realism are not spatial variations but events – thing that happen in the world that cause it to change. This, they maintain, is the only way to gain insight into causal forces. Importantly, the "depth" or layered ontology of critical realism suggests that these events can only be understood as having been produced by deeper structural forces and their causal mechanisms. Thus we cannot understand the causes for a factory strike – an event that requires explanation – by simply referring to the anger of the workers. We have to situate that anger within wider forces such as capitalism, which would help explain the workers' reactions *as well as* the exploitative practices of the factory owners. Under critical realism, such explanations tend to be case specific: the particular conjuncture of social identities (workers with similar politics for example) and exploitative practices (lack of bathroom breaks for example) cannot be known without in depth case study, and these concrete conditions have themselves to be situated within the mechanisms of worker solidarity and capitalist exploitation. Importantly, human beings are thought to be capable of considered reflection upon these wider structures; that is, they can ponder and thus change the conditions of their own existence.

## Poststructuralist geographies

Poststructuralists critique widespread assumptions that natural and social objects exist as pre-given, singular entities with fixed identities (Willems-Braun 1997). They offer instead an elaboration of the social processes (e.g., power) that have led to their production and dissemination *as* entities. Taking very little for granted in terms of "object-ness," poststructuralists offer a fluid conception of social power that, in its broadest sense, can be defined as the potential to construct difference, that is, to draw the boundaries between the epistemological categories of the *Certain-Real-Material-True* and the *Uncertain-Ghostly-Imaginary-False*. This power is also extensive throughout our common ontological designations, such as *Culture-Nature* and *Individual-Society*, as well as through the objects these binaries seem to designate. The critical aspect of **poststructuralism** is derived from the fact that the power to designate difference lies with those in positions of social dominance, whether by virtue of their class, race, and gender, or by virtue of their social *position* in sites of knowledge production, such as universities and the media. Through deconstructive analyses, the constructed (contingent, tentative, and uncertain) character of these pillars of domination – including those lodged in Western thought – can be exposed (Dixon and Jones 1998).

In Table 2.4 we offer a summative table of these paradigms. First, we highlight their key ontological and epistemological assumptions. Second, we flag some of the possible data sources and analytic techniques researchers tend to adopt. Finally, we sketch what relationship you, as the researcher, shares with the object of analysis investigated. Taken together, Table 2.4 provides a succinct overview of some of the main theoretical statements made in this chapter, and serves as a useful reference to return to as you read the rest of this book.

## Conclusion

This chapter has not been prescriptive in terms of telling you whether or not to use remote sensing for your study, or whether a focus group would be better than an in-depth interview. That's what the rest of the book is for. Instead, we hope to have provoked you into thinking through how the research process is always "in the middle" of your own experiences and endeavors as well as a paradigmatic body of literature, and how in turn these are underlain by ontological and epistemological assumptions about the world. Paradigms, ontologies, and epistemologies should not therefore be thought of as existing apart, but

**Table 2.4** Understanding paradigms in geography

| Geographical paradigm | Theoretical assumptions | Data and analytic approaches | Researcher–object relations |
|---|---|---|---|
| Spatial Science: "mapping spatial relations" | Space is viewed as a two-dimensional container within which places exist as discrete phenomena with distinct attributes. Variations across and interactions between places are measurable, usually quantifiable. Each object of analysis has a series of distinct characteristics, including its location within this two-dimensional container. A descriptive mapping of the spatial locations and characteristics of objects is the first step in assessing causal relations. If theory suggests a causal relation and characteristics are found to spatially vary in systematic ways, then the theory has empirical support. | *Data:* Census (socio-demographic, economic); field methods (tree ring width, snow-pack depth); remote sensing (vegetation cover; slope); social surveys (consumer behavior, political opinion). *Analytic approaches:* Descriptive and multivariate statistical analysis; mathematical analysis; simulation; GIS overlay; survey research methods; behavioral modeling; carbon dating. | Subjectivity controlled in favor of objective forms of research analysis and presentation. Researcher maintains a distance between her/himself and the object under study. |
| Humanism: "interpreting places of meaning" | Space as a two-dimensional container is but one way of looking at the world, commonly used by scientists but also governments. In the everyday sense, place is a more useful concept because it describes the attachment people have to particular parts of the world such as their home, workplace, or car, but also their neighborhood and country. A "sense of place" differs for each individual and changes through time. Specific objects are associated with a particular "sense of place," or produced as a means of expressing this sense. Specific objects are placed within the landscape in order to represent particular emotions and feelings. | *Data:* Archival material (diaries, local histories); oral and other folk traditions; landscape reconstruction; place histories; individual life stories and experiences. *Analytic Approaches:* Ethnographic immersion; participant observation; textual analysis; structured and unstructured interviews; culturally-specific interpretation; "thick description." | Researcher rejects the notion of objectivity and focuses on developing an empathic understanding of individual's experiences as they relate to space. Focus can also be on what meanings imply "to me," the researcher who is charged with the task of interpretation. |

| | | | |
|---|---|---|---|
| Critical Realism: "uncovering the structures and mechanisms behind events" | Events in the world are embedded within wider structures (e.g., capitalism and patriarchy) and their causal mechanisms (e.g., worker exploitation, gender division of labor).<br><br>For many physical processes, explanation is of a "necessary" type: the causal powers operated the same from one place to another.<br><br>For social relations, the mechanisms for causing change are usually contingent, or context specific.<br><br>Paying attention to contingent contexts, including the different spaces in which things happen, requires in-depth case study.<br><br>Social structures and mechanisms are differentiated across local and global contexts. | *Data:* Participatory observation and ethnography; interviews; field study of context-specific causal operations.<br><br>*Analytic Approaches:* Reflection on and interpretation of the nature of events and the wider mechanisms that embed them, as well as the causal forces (structures) from which they emerge; in the physical sciences, researcher remains open to the possibility that causal mechanisms are complexly, contingently, and non-linearly related. | In social analysis, researchers assume that individuals have a situated capacity to reflect upon structures and mechanisms, even though they might not have a well-developed theoretical language for articulating them.<br><br>Even though the world exists objectively, our capacity for knowing it in an objective way is limited by our own contingently situated knowledge. |
| Post-structuralism: "interrogating the production and import of spatial discourses" | Discourses – the socially constructed ways we have to describe the world – name or "fix" the character of phenomena as well as the boundaries between the insane and rational, the possible and the impossible.<br><br>Discourses have material effects; it "matters" how people and objects are characterized and brought into relation with one another.<br><br>Though they are the vehicles through which power-laden relations are formed and operate, discourses are variable and therefore open to transformation.<br><br>Space is socially constructed from the unequal relations of power that seek to fix the nature of people and things.<br><br>Not just discourses, but matter too is a force in the world. | *Data:* Texts and visual data; ethnographic immersion and participant observation; descriptive accounts of micro-scale actions (practices) and events.<br><br>*Analytic Approaches:* Textual and visual interpretation; genealogy of the emergence and deployment of discourses; deconstruction (analysis of the work of binaries). | Researcher and researched, human and non-human, are thoroughly embedded within wider fields of relation and meaning.<br><br>Researcher eschews the notion of total knowledge, understanding that his or her views are from "somewhere" and thus are just as partial as their subjects.<br><br>Reflexivity includes a monitoring of the researcher's positions with respect to the settings of which they are a part. Researcher is her/himself an object of analysis. |

*Note:* Adapted from Del Casino et al. (2000) and Del Casino (2009) rather, as intermingling forces that saturate, guide, and limit the research process. Are you intent on explanation, prediction, empathetic understanding of others, or social transformation? If prediction is your goal, then you must be concerned with matters of causality; your objects of analysis, for example, must be described in terms that emphasize their discreteness, while you as researcher must consider how best to remain detached from them so that objectivity can be retained. If, however, understanding the experiences of others is your objective, then you will find yourself operating within an embedded notion of phenomena, as well as yourself as researcher, such that the picture that emerges is a series of relationships that connect you and your subjects in various ways, as well as with the world in general. Perhaps the question, "how do you see the world?" should be the first step in any research process. Whether you are aware of it or not, the moment you start the research process, you're already answering this important question.

# References

Abler, R., Adams, J., and Gould, P. (1971) *Spatial Organization: The Geographer's View of the World*. Englewood Cliffs, NJ: Prentice-Hall.

Buttimer, A. (1976) Grasping the dynamism of the lifeworld. *Annals of the Association of American Geographers* 66: 277–84.

Del Casino, V. J. (2009) *Social Geography: A Critical Introduction*. Chichester: Wiley-Blackwell.

Del Casino, V. J., Hanna, S., Grimes, A., and Jones III, J. P. (2000) Methodological frameworks for the geography of organizations. *Geoforum* 31: 523–38.

Deleuze, G. (1993) *The Fold: Leibniz and the Baroque*. Minneapolis: University of Minnesota Press.

Descartes, R. (1996) *Meditations II*. http://www.wright.edu/cola/descartes/intro.html [accessed August 27, 2009]. Copyright David B. Manley and Charles S. Taylor.

Dixon, D., and Jones III, J. P. (1998) My dinner with Derrida, or spatial analysis and poststructuralism do lunch. *Environment and Planning A* 30: 247–60.

Foucault, M. (1970) *The Order of Things: An Archaeology of the Human Sciences*. New York: Pantheon.

Hartshorne, R. (1939) *The Nature of Geography*. Lancaster, *PA: Association of American Geographers*.

Marx, K. (1976) *Capital, Vol. 1*, Fowkes, B., trans. London: Penguin.

Massey, D. B. (1994) *Space, Place and Gender*. Minneapolis: University of Minnesota Press.

Massey, D. B. (2005) *For Space*. London; Thousand Oaks, CA: Sage.

Relph, E. (1970) An inquiry into the relations between phenomenology and geography. *Canadian Geographer* 14(3): 193–201.

Sayer, A. (1992) *Method in Social Science: A Realist Approach* (2nd edn). London: Routledge.

Tuan, Y.-F. (1975) Place: an experiential perspective. *The Geographical Review* 65(2): 151–65.

Willems-Braun, B. (1997) Buried epistemologies: the politics of nature in (post)colonial British Columbia. *Annals of the Association of American Geographers* 87(3): 3–31.

# Additional Resources

Castree, N. (2005) *Nature*. New York: Routledge. Whether your interests are in critical theory, human-environmental geography, or physical geography, this book is an excellent place to begin situating your own stance with respect to the nature–culture binary.

Gregory, D., Johnston, R. J., Pratt, G., Watts, M., and Whatmore, S., eds. (2009) *The Dictionary of Human Geography* (5th edn). Oxford: Wiley-Blackwell. Now in its fifth edition, the *Dictionary* is a valuable resource for anyone in geography, but especially for human geographers.

Harvey, D. (1996) *Justice, Nature, and the Geography of Difference*. Oxford: Blackwell. Probably the best place to go if you want to know how geographers understand the embedded relations that are **dialectics**. Harvey's work has been more important than any other in showing how space is works in dialectical thought. See chapter 2 of the book.

Livingstone, D. (1992) *The Geographical Tradition: Episodes in the History of a Contested Enterprise*. Oxford: Blackwell. An excellent book on the history of modern (especially pre-World War II) geography. Very strong in showing how geographical knowledge is part and parcel of an era. Complements nicely Peet (1998) and Rose (1993) as a trilogy.

Rose, G. (1993) *Feminism and Geography: The Limits of Geographical Knowledge*. Minneapolis: University of Minnesota Press. An important book in the trajectory of feminist theory in geography. It offers an historical assessment of the discipline from the standpoint of feminist theory, with an emphasis on the epistemological coordinates underwriting most of twentieth-century geographical knowledge.

Peet, R. (1998) *Modern Geographical Thought*. Oxford: Blackwell. A jam-packed and highly personalized assessment of the history of geographical thought, with an emphasis on the post-World War II era. Dated for some of the ways that the poststructuralist turn is interpreted, Peet's book is nonetheless a good starting point for students interested in contemporary geographical theory.

Thomas, D. S. G., and Goudie, A. (2000) *The Dictionary of Physical Geography* (3rd edn). Oxford: Blackwell. Like its human geography counterpart, the *Dictionary* offers expert summaries of key concepts, techniques, and approaches across a full range of topics in physical geography.

## Exercise 2.1   Theorizing the 2008 Global Economic Crisis

In order to animate the paradigms outlined in Table 2.4, we have taken one particular topic of interest, the "global economic crisis" that exploded in the autumn of 2008, to tease out the diverse approaches and research questions that geographers face in conducting research. In particular, we highlight the key ontological and epistemological assumptions that underpin paradigmatic approaches in spatial science, humanism, critical realism, and poststructuralism. The exercise provides a case study for investigating the four paradigms; it also asks *you* to write down a few ideas about how you would analyze the economic collapse in their terms. The effects of the crisis are only beginning to unfold with major job losses and unemployment, homelessness, shrinking economies, social unrest, calls for protectionism and regulation, and unprecedented levels of government borrowing and bailouts. So how exactly is the researcher able to translate what seems like a gigantic problem into manageable objects of analysis? The answer, of course, depends upon the paradigm that you opt for.

### Spatial science

As a spatial scientist, you might be interested in collecting country-wide data on the changes in gross domestic product (GDP) at points before and after the onset of the crisis. Comparable data of this sort for different countries (from, for example, the World Bank) can enable descriptive comparisons. To these data you raise additional questions: do changes in GDP (per capita) seem to relate to the sectoral concentration of the countries' economies (e.g., percent employment in primary, secondary, or tertiary activities); to levels of human capital (e.g., educational attainment levels); or to their integration in the global economy (e.g., international trade levels, membership in organizations such as the European Union)? Similar analyses can be undertaken at the intra-national level, with county or state unemployment levels obtained from national government agency websites. Task: (a) Develop a set of research questions or hypotheses that you want to raise from within this paradigm; (b) How would you collect the data for this study?

## Humanistic geography

A humanistic approach can be used to investigate affects of the crisis on the lived experiences of people in different places. The geographer needs to interpret people and place (including landscape) not only as a site of economic restructuring, but also of symbolism and meaning. For example, questions may form around the rapid increase in foreclosures and the rise in homeless families throughout the US. We could also add everyday geographies and think of the experiential accounts of people throughout the country, including their fears, hopes and perceptions. What happens to notions of community in towns and neighborhoods wrecked by a wave of foreclosures and plant shutdowns? Here, in-depth interviews are important for constructing key meanings and themes associated with everyday (and often banal) details of people's lives. Task: (a) Pick a place for in depth study of the impacts of economic restructuring. (b) Write about how you would approach this study: what questions would you ask, to whom would you speak, how would you approach the collected data?

## Critical realism

Under critical realism, the 2008 economic crisis is an "epiphenomenal" (surface level) event undercut by a much wider and deeper set of causal structures: the inherently uneven, contradictory, and geographically differentiated process by which capitalism grows and contracts. In this view, crises are inevitable; the only question is when and where they will arise, and how they can be faced. Accordingly, a critical realist might ask of the crisis: What are the context-specific or contingent mechanisms that led to this event, at this time and in the countries most affected? Is it best theorized as a series of local events that "went global," or as the collapse of a global process with many local impacts? Are there other factors – such as government regulatory mechanisms – that need to be taken into account in explaining the crisis? Task: (a) What are some of the objects of analysis and research entry points that a critical realist paradigm might lead you to? (b) How would go about identifying events of interest, and what would be some of the underlying causal structures and mechanisms that you might examine?

## Postructuralist geography

Discourses about the impacts of the economic crises often coalesce around dominant themes – such as the greed of Wall Street bankers, the lack of regulation in the financial industry, or the state of the over-unionized US car industry. Here, the geographer could do a genealogical analysis into the source and dispersions of the multiple discourses and representations surrounding the crisis. All of these debates and forms of arguments can be narrowed down by focusing on a particular newspaper, a news show, or an online blog.

Also viable sources of data are documentaries and other visual representations (e.g., photographs) about the people and places of decline. Task: (a) Select some "texts" – written or visual – about the collapse. How has it been represented? What kinds of people are shown most often, how are they depicted, and in whose interest are these stories and images produced? (b) What underlying binary relations (*Orderly–Chaotic, Predictable–Random, Individual–Society, Self–Other*) seem to propagate within these documents?

# Chapter 3

# Observing Our World

*Bruce L. Rhoads and David Wilson*

- Introduction
- The Philosophy of Observation
- Observation, Theory and Geography
- Methodological Aspects of Observation in Contemporary Geography
- Conclusion

Keywords

Abduction
Cycle of erosion
Deduction
Empirical
Environmental determinism
Humanism
Life worlds
Logical positivism
Marxist theory
Observation

Paradigm
Phenomenology
Poststructuralism
Realism
Regionalism
Relativism
Spatial analysis
Structuralism
Theory-laden

## Introduction

As a scholarly enterprise, geography is a discipline based on inquiry that is driven, above all else, by the impulse to comprehend the human and physical attributes of the world in their interacting complexity. Inquiry is often conducted within a framework of information that is relevant to the issues driving curiosity and can be used to formulate answers to research questions. Information about the world is gathered through **observation**, which is essential to the production of knowledge. The era of "classical" geography, which extended from the ancient Greeks, and "geographers" such as Strabo and Pausanias, to the early part of the nineteenth century and the great works by Alexander von Humboldt and Carl Ritter, had a rich, but largely uncritical, tradition of observation. Within this tradition, observers collected information using their sensory apparatus (eye, ears, etc.) and were widely held to be neutral, impartial decipherers of the world. As we will learn, however, and as the nineteenth century German philosopher Friedrich Nietzsche contended, knowledge depends on the perspective from which it is viewed (also see Chapter 5). Thus, observation can be considered a fluid, dynamic process, a specific practice embedded in a communal sense of appropriateness and utility. To this end, a variety of different perspectives on observation have been utilized since modern geography emerged with the establishment of university departments of geography in the late nineteenth and early twentieth centuries.

The evolving practice of observation within modern geography is examined in this chapter. We will see how perspectives on observation have changed over the past century, as well as how viewpoints regarding philosophical perspectives have changed. Similarities and differences on observation in human and physical geography are also emphasized. Before proceeding, however, a word of warning is in order. Since about 1970, perspectives on observation in human geography and physical geography have unfolded along different and typically unconnected paths. The perception that physical and human geographers deal with different subject-matters (the "physical" and "human" worlds) has helped to minimize intellectual interaction between the two parts of the discipline. For this reason, recent debates on the nature of observation in human and physical geography largely have occurred independently of one another. Some, including one of the authors (Rhoads 2004), have argued that the potential for connectivity between the two parts of our discipline remains, but in this chapter the roles of observation in contemporary human and physical geography are discussed separately.

## The Philosophy of Observation

Inquiry within geography can fruitfully be viewed as the interplay between theory and observation. Theory consists of ideas about how the world is structured and how it works. Observation, in the most general sense, involves human interaction with the world. Its objective is to obtain information about the structure and dynamics of the world so that this information can be compared with human ideas, or theory. The comparison of observational information with theoretical ideas constitutes the core of scientific inquiry and represents the process by which scientific knowledge is generated. Regardless of the differing opinions that have emerged about the interplay between theory and observation, most geographers accept that this interplay is central to geographic inquiry.

Specific conceptions about the role of observation in geographical inquiry are influenced by presuppositions about the world, the truth of which are taken for granted. These presuppositions are embodied in the different philosophical perspectives on scientific inquiry. Perhaps the most well-known of these perspectives is **logical positivism**

**Table 3.1**   Tenets of logical positivism

---
- Observation provides the foundation for all knowledge
- Observation is based on human sensory experience (empiricism)
- Observation is untainted by theoretical presuppositions (theory-neutral observation)
- Observation yields basic facts (atomistic bits of objective knowledge) about the world
- Theory is developed inductively from the bedrock of veridical (truthful) facts
---

(sometimes known as logical empiricism), which was promoted by members of the Berlin and Vienna "Circles." Scholars in these two groups shaped and later dominated the philosophy of science in the first half of the twentieth century (Table 3.1).

In many ways, logical positivism formalized the inductive, **empirical** approach to inquiry that marked the beginnings of modern science in the sixteenth century. It conforms to the classical view of the scientific method (Chapter 5) in which scientific inquiry begins with observation, and theories develop through attempts to organize facts into systematic explanatory relationships. According to positivism, observation, because it is both objective and veridical (that is, phenomenon are viewed without distortion and conform to fact, both now and in the future), provides the "bedrock" for the development of scientific knowledge (Rhoads and Thorn 1996). In particular, the theoretically neutral information derived from observation can, through the process of hypothesis testing, be used to adjudicate among competing theories.

By the mid-twentieth century logical positivism was subject to increasing criticism (by philosophers such as Karl Popper), and the recognition that observation is not necessarily theory neutral, and **paradigms** or prevailing theoretical frameworks often strongly influence what scientists choose to observe reduced its appeal as a means of verifying hypotheses about the way the world operates. By contrast, post-positivist perspectives on science emphasize that observation is **theory-laden**; that is, the information content of all observations is at least partly influenced by "a priori" theoretical commitments (Rhoads and Thorn 1996). A priori knowledge is independent of experience, whereas "a posteriori" knowledge is dependent on experience, and these two forms of knowledge differentiate the epistemological notions of "inductive" and "deductive" reasoning (and, as we shall see in Chapter 5, have a bearing on the way we formulate research questions). Differences between these perspectives depend on the degree to which observation is viewed as "theory-dependent." If observation is highly dependent on theory, especially a theory under test, information derived from the observation may be biased and unhelpful in evaluating the theory. Thus, whenever theory-dependent observations are used to evaluate theory, the research process can be said to be somewhat circular. This view laid the ground for relativistic conceptions of science which maintain that the relative value of theoretical ideas must be judged on the basis of social criteria since observation, which is biased by theory, cannot be used to adjudicate among different ideas. A corollary of **relativism** is that objectivity no longer is preserved in any absolute sense; all ideas fall or stand based on their relative merits as defined within a social context.

**Realism** lies at the opposite end of the post-positivist spectrum. Realists embrace the theory-laden nature of observation, but maintain that this neither jeopardizes objectivity nor their capacity to identify the theories that mostly closely reflect the true nature of reality

(Rhoads and Thorn 1993). Realism, in fact, argues that the theory-ladenness of observation is *necessary* for objectivity: only by viewing the world through the lens of theory can we hope to discern what the world is really like. In other words, science without concepts is blind and the notion of building knowledge from theory-neutral observations, even if such observations could be obtained, is fundamentally misguided. This necessitates that both background knowledge and observational information serve as forms of evidence in the testing of a theory; empirical data are not ignored, but neither are they viewed as the absolute arbiter when a theory is evaluated. Viewed in this light, theory-dependence does not threaten objectivity, but instead provides the basis for collection of appropriate observations for theory testing (Rhoads and Thorn 1996).

Currently in the natural sciences, including physical geography and geographic information science, observation is viewed not as information obtained directly through the senses (vision, hearing), as the logical positivists maintained, but as data gathered using elaborate theory-dependent processes that often involve instrumentation (consider the images gathered by recording or real-time sensing devices mounted on aircraft or satellites, that have no intimate contact with the objects they are observing). To this extent, most scientific observations now consist of a causal chain that links a human observer to natural phenomena via the technology they employ for data collection, management and analysis (Rhoads and Thorn 1996), and that draws on background knowledge about how specific technologies can generate data on the phenomenon. This view emphasizes that there is distinction between "data" and "phenomena" that complements the difference between "observation" and "theory" (Rhoads and Thorn 1996). Science then can be construed as a search for phenomena through the acquisition of data that relies on theory-dependent technology and techniques.

## Observation, Theory and Geography

Geography in Europe and the United States emerged as a scholarly discipline at the turn of the twentieth century. Since that time several major intellectual frameworks, or paradigms, have guided research and inquiry within the discipline. Initially, two theories, the **cycle of erosion** and **environmental determinism**, provided the primary bases for explanation in physical and human geography, respectively (Rhoads 2005). Both frameworks were championed by William Morris Davis, who, through his position as a professor of physical geography at Harvard University, his extensive writings on geographic research and education, and his status as a founding member and first president of the Association of American Geographers, had a major influence on the early development of geography (and geology; Davis was also Sturgis Hooper professor of geology at Harvard and a president of the Geological Society of America). According to the cycle of erosion, all landscapes and landforms evolve predictably through time (passing through youth, maturity, and old age) and their current characteristics are a function of structure, process and stage. Environmental influences, especially climate, were also thought by environmental determinists, such as Ellen Semple and Ellsworth Huntington, to explain patterns of human behavior. For example, according to the "equatorial paradox" a country's degree of economic development is dependent on its distance from the equator (though this obviously ignores historical patterns as well as the role in under-development played by European colonizers). Observations obtained through inquiry guided by these theories consisted mainly of highly descriptive visual illustrations of landforms and human activities that conformed to the tenets of logical positivism. Little or no effort, however, was made to systematically use the information derived from these observations to rigorously evaluate the basic propositions embedded in the underlying theories. Instead, the theories were assumed to be valid and, for the most part (often despite their authors' protestations to the contrary), the observations were interpreted in a manner that invariably lent support to the theories. In other words, the extremely theory-laden nature of the observations did not permit the validity of these theories to be objectively evaluated. Thus, for example, in her book, *Influences of geographic environment*, which was published in 1911, Semple used observations made by Strabo and more contemporary writers as a basis for suggesting that "among mountain as among desert peoples, robbery tends to become a virtue; environment dictates their ethical code" (p. 588). Similarly, Rhoads and Thorn (1996) suggest that the cycle of erosion, as applied by Davis, was not a theory to be tested, but a set of regulative principles for governing what is and is not an acceptable geomorphological explanation, and they provide a detailed discussion of the bias in geomorphic inquiry this introduced (Davis never identified the types of evidence that would require him to acknowledge that the cycle of erosion was flawed).

During the 1920s environmental determinism came to be viewed as problematic, at best, and human geography began to seek new frameworks for inquiry; although European geographers, such as Paul Vidal de la Blache rephrased the argument and suggested that the environment constrains, but does not determine, human activity. Two alternatives subsequently emerged: the historical-cultural perspective championed by Carl Sauer (in which the cultural landscape was considered to have been created from the physical landscape by a cultural group, such that change in the landscape reflected cultural changes through time); and the **regional** perspective advocated by Richard Hartshorne (who sought to delineate the unique physical and human characteristics of a particular area). Both perspectives were overtly concerned with separating geographical inquiry from the interests of other fields by focusing on a unique subject matter: cultural landscapes or regions, respectively. Neither was explicitly theoretical, which was a welcome characteristic in the light of environmental determinism's perceived theoretical bias, nor did either openly embrace logical positivism. Observation essentially was viewed as the unproblematic, mainly visual, identification of geographical "facts" that could be used to objectively identify: (a) the sequence of human occupation of the landscape over time, or (b) the interactions of physical and human elements in their regional association. To the extent that knowledge about geographical phenomena was derived from spatial and temporal patterns in observed facts, rather than from interpretations guided by a priori theory (Rhoads 2005), both the historical-cultural and regional perspectives on geography were highly empirical (that is, they relied on information derived from sensory experience). During this period, physical geography languished and, in some circles, came to be viewed as necessary only to provide the environmental context for the study of landscapes or regions.

The implicit influence of logical positivism on geography became apparent between 1950 and 1970, as both the human and physical sides of the discipline adopted the quantitative analysis of data as the preferred modus operandi of geographical inquiry. This era, referred to as the "quantitative revolution," shifted the emphasis from simple visual observation and qualitative accounts of observational facts toward the production of quantitative observational "data" (measured and recorded on a numerical scale). The analysis of relationships among quantifiable variables relied heavily on statistical methods (most notably chi-square and t-tests, regression and principal components analysis, see Chapters 17 and 18), which permitted causal relations to be established through inductive inferences made on the basis of the statistical results.

More theoretically, quantitative data and the use of statistical analysis were viewed as tools for preserving the objectivity of geographical inquiry and for establishing geography as a legitimate science that sought generalities over descriptions. The explicit connection of geography to logical positivism did not, however, occur until the late 1960s, by which time Popper's concept of defining scientific statements in terms of "falsifiability" (the notion that knowledge can be disproved by observation and experimentation) and other ideas had begun to challenge its viability as a philosophy of science (Box 3.1; Rhoads 2005).

---

### Box 3.1   Falsification

Falsification is a methodology of science expounded by the well-known philosopher, Karl Popper. To appreciate his contributions, consider that a basic tenet of logical positivism is the verification theory of meaning, whereby only statements verifiable through direct veridical experience, i.e. observations, can be said to be scientifically meaningful. In this sense, logical positivists subscribed to the notion that scientific theories must be "proven" through repeated testing that yields evidence in support of the theory. Popper challenged this methodological approach to science by claiming that verification is an inductive process in which generalizations are derived from and supported by repeated observations of regularities. The problem with induction is that there is no logical principle that can justify the truth of a universal generalization, or law, derived from repeated observations. As a simple example, consider the statement "All swans are white." No matter how many white swans we observe, no number of sightings can establish the truth of this statement. In other words, no matter how much evidence we have in favor of this statement, there is no logical principle that justifies the conclusion that indeed we have proven or verified it. On the other hand, as Popper argued, all we need is one counterexample, for example an observation of a black swan, to show that the statement is false. Thus, according to Popper, scientific statements are those that are capable of being refuted, and the goal of science should be to falsify our theories, rather than verify them. Our attitude toward theories should be a critical one and we should constantly be devising tests that could yield evidence refuting these theories. If after exhaustive testing no such evidence emerges, the best we can say is that the available evidence *corroborates* the theory, rather than verifying it. In geography, both the cycle of erosion and environmental determinism were guilty of emphasizing verification over falsification; the geomorphologists looked to landscapes to confirm Davis's theory, while the human geographers sought evidence for the conformance of culture to environmental patterns.

# Methodological Aspects of Observation in
# Contemporary Geography

Since the early 1970s, human geography and physical geography have taken divergent perspectives on observation. Very visibly, human geography has experienced a major upheaval in its use of philosophies and theoretical perspectives; the era of single, dominant "paradigms" has been replaced with an assortment of views on the process of advancing geographic understanding, and the positivist conception of observation has been challenged and supplanted. In particular, alternative notions have followed the rise of new perspectives (see Chapter 2), such as **humanism**, **structuralism**, and **poststructuralism**. As a result, no dominant, long-term unity has been maintained about notions of what reality is or how we can best reveal its true nature and causes. Meanwhile, physical geography has continued to rely on a data-driven approach to observation. In the sense that the choice of what and how to observe is often guided by theoretical considerations, it is widely recognized that observation is theory-laden, but physical geographers generally subscribe to the notion that these considerations do not compromise the objectivity of hypothesis testing. Sophisticated instruments and equipment also are increasingly being used to generate quantitative data and, to this extent, physical geographers are moving further from the practice of relying on (qualitative) visual observation as a data-gathering method. The trend reflects the post-positivist stance that scientific observation is much more complex and sophisticated than simple sensory interaction between humans and their surroundings. According to this view, visual observation uses nothing more than crude sensors (eyes) to detect a very narrow band of electromagnetic energy (Rhoads and Thorn 1996). Whereas science, through the theory-guided development of technology, has produced more reliable sensors with a greater range of detection capabilities than the human eye possesses. The following discussion serves to illustrate the differences in contemporary perspectives on observation, first in human geography and then in physical geography.

## Observation in human geography

The influence of spatial analysis – which focuses on measurements of properties and relationships that incorporate or are expressed in space – and its vision of observation have persisted as popular, viable ways to practice human geography. Nowhere is this influence more evident than in the application of geographical information systems to research problems in human geography. However, despite the use of sophisticated statistical procedures and "high-tech" data gathering devices, the spatial-analytical perspective is still inclined to view observation in the traditional way. That is, observation is considered to be the ideal and necessary way to verify notions about the world's character, forces, and mechanisms, and supposedly "reflects" back to geographers the true content of the world which can, in turn, be used to facilitate counting, classification, and model-building. Moreover, even though the task of observation may substantially rely on secondary-data sources (for example, census data; see Chapter 11), empirical observations are necessary to anchor and validate these data sets.

This sense of observation is underpinned by two key notions. The first is that the world consists of an ordered arrangement of "landscapes" that can be easily deciphered and understood. That is, the world contains discrete and distinguishable elements (such as people, cultural attributes, regions, neighborhoods, and houses) that stand out unambiguously for easy comprehension, counting, and cataloging. In this perspective, the world objectively and neutrally "speaks" to those whose aim is to record reality. Nothing veils their true essence. Knowing the world, then, is less an interpretive enterprise than an exercise in diligent observation ("eyeing," "measuring") and reporting. The world, in the final analysis, is a real (not abstract or ideal) space filled with many separate, often disparate elements; any false information about the nature and composition of reality (ontological deception) emerges only because a ready-to-apprehend reality is misread.

The second notion, of a potentially unobtrusive and untainted observational being, is an equally crucial provision; observers who produce the secondary data that "feed" quantitative projects have to dispassionately and objectively apply a sensory apparatus and capture the essence of a decipherable reality. These "field-workers" are servants to the empirical, "blank slate beings" who are involved in the neutral process of reporting and do not intervene or interfere with the world; they impose nothing (no a priori theories or categories), and nothing (be it politics, a priori expectation, or privilege) taints their gaze. In effect they spurn subjectivity, discard self-evident and intuitive theories, purge bias, and simply "take in" the world by accurately observing it.

This view was strongly challenged in human geography with the rise of humanistic studies (in **phenomenology**, existentialism, and hermeneutics). As a result, a long-term debate revolving around the "crisis of positivist observation" was initiated. A central question in the minds of many geographers was what, if any, essential truths have quantification and spatial analysis revealed about humans and the human environment? Humanistic studies were offered as an alternative to advancing geographic understanding. According to Yi-Fu Tuan, for example, we inhabit a world of meaning, not a framework of geometric relationships. Thus, proponents of humanistic studies, such as Jonathan Smith (1996) and John Eyles (1989), reject the reduction of lived worlds to the spatial analysts' simplifications of numbers, equations, points, and lines (that is, they reject objectivity and veridical facts as foundations for knowledge). Moreover, to some humanists **spatial analysis** is a dehumanizing project. The alternative is to focus on people's worlds of meaning, perception, attitudes and images which, it is asserted, capture both the sense of an elusive and variegated set of realities and the notion of the world as an arena for the lived and the humanly experienced.

Questioning the value of positivist observation is part and parcel of this new focus, in which the world is no longer seen as an objective reality that is inert and external to people's actions, lives, and beliefs. Instead, reality is something groups and individuals construct and live through. Reality, in other words, is not an experience that is common to all, but a subjective, meaning-infused set of worlds that is sculpted by human initiative, perception, and values. For this reason, humanists offer, it is necessary to move beyond reliance upon empirical appearances in favor of methods that uncover the in-depth constellation of meanings, perceptions, and values that give rise to the variegated worlds that individuals inhabit. In other words, geographers must be suspicious of observed appearances and delve into people's **life worlds** (which are composed of the conscious projects that shape human existence and the passive realities that impinge upon it). Viewed from this perspective, knowing the world involves excavating beneath appearances to unearth the repositories of human meaning and interpretation that underpin lived realities.

The rise of structuralist studies in human geography (undertaken by **Marxists**, realists, structurationists and critical theorists) also helped deepen the crisis of positivist observation that flows from the sense of the world's complexity; the leading proponents of which, such as David Harvey (2000) and Richard Walker (2004), continue to attempt to unearth the deep, empirically elusive forces that guide human actions and the transformation of human and physical landscapes. Here, however, the emphasis is on the economic and political structures that underpin places and societies (Marxism, for example, provides a theory and methodology for understanding the political economy and the imperatives of capital accumulation, social reproduction, and political legitimization). Once again, observation, when taken to be a task that meaningfully records a transparent reality, is called into question. In particular, it is held that empirical observations cannot capture the essence of processes and causes because they ignore the power relations in which visible spatial relationships are embedded.

Instead, "raw appearances" have to be comprehended as complex outgrowths of deeper, empirically elusive mechanisms, because rather than embodying causation, the domain of the empirical reflects it (also see Chapter 5). Beneath the exterior of appearances is the domain of real structures that cannot be directly discerned, but which, nevertheless, embed the mechanisms that guide human actions and ultimately transform the world. In his critique of observation, for example, Sayer (1984) addresses the problem of relying upon the empirical to explain the world (a position that the positivist notion of observation embraces). To Sayer, "what causes something to happen has nothing to do with the number of times it has happened or has been observed to happen and hence whether it causes a regularity." The essential point is that "seeing the world" constitutes a relatively simple first step in the difficult task of coming to understand causes and effects.

In the 1980s poststructuralist studies (postmodernism being the movement's most publicized branch) came to the forefront in human geography. Like humanists and structuralists, poststructuralists consider the positivist vision of observation to be a naïve exercise in data capture that is not capable of portraying the complexities of the world. Many poststructuralist studies identified a new, blended reality of complexity, heterogeneity, particularity, and uniqueness, and this implied that the world was not susceptible to simple and efficient seizure by empirical observation. To postmodernists, in particular, the dilemma is intractable, for the world purportedly reflects a messy, fragmented and disjunctive disorder that positivist science systematically disregards or denies. Viewed from this perspective, observation is an academic intervention that imposes the vantage point of "academic positionality" on the data-gathering task; a person's class, race and gender inflects their simplified representational schemes, and reductionism inevitably follows. To ensure that the essence of the complex world is not reduced to the sums of simpler phenomena, a different kind of observation is needed, one that recognizes the role that categorization and social power impose on the world. Specifically, a heightened sensitivity to complexity, disorder, and academic positionality is required. To proponents like Dear (2001) and Strohmayer and Hannah (1992), this theoretical sensitivity connects the observational process to an awareness of time-space specificity (that ranges from the unique to the ubiquitous) and of poly-causality and causative contingency, acuity to the imposition of language, and the realness of disordered spaces. Observations, then, are dependent on the observer's relative position and, rather than being straight-forward perspectives of a world that timelessly communicates its essence through easily created categories, are a complex product of human experiences (see Chapter 2).

## Observation in physical geography

Contemporary approaches to observation in physical geography can be separated into three major categories: field observations, physical experiments, and remote sensing techniques. Field observations include measurements of forms, fluxes, flows and ages made with human-operated equipment or sensors, which in turn yield quantitative data on these attributes. Observation typically occurs through the medium of systematic field measurement campaigns that are guided by a research design and address a specific set of research objectives. The data collected in these campaigns typically focus on the form and function of Earth's surface (geomorphology), biological organisms (ecology, biogeography), the atmosphere (climatology), or hydrosphere (hydrology).

Temporal scale often plays an important role in determining the relevance of particular types of field data for specific investigations. In geomorphology, for example, geo-historical studies may be directed towards the analysis of changes in landform characteristics over time spans ranging from thousands to millions of years. It is not possible to directly measure the processes (fluxes, flows) that produced these changes. Instead, attempts may be made to reconstruct/retrodict past forms in order to establish how and at what rate change occurred. A common way of doing this is to substitute space for time – that is, to infer the historic operations of physical processes from "snapshots" of spatial varying phenomena (Box 3.2). In contrast, field investigations that highlight process dynamics or

### Box 3.2    Ergodic Reasoning

The principle of ergodicity was developed by physicists to deal with problems in thermodynamics. It holds that sampling across an ensemble of systems, i.e. sampling many systems – each of which is characterized by a particular spatial configuration at the time it is sampled – is equivalent to sampling the configuration of a single system through time. In this sense, ergodicity has been referred to as "space for time substitution" – that sampling systems over space yields information on how they change over time. An example in geomorphology would be to use information on drainage density from many different drainage networks to infer how drainage density in a single drainage network might change through time. Whether or not spatial information can be used to represent temporal dynamics in systems of interest to the physical geographer has proven controversial (see Paine 1985, in Additional Resources).

relations between process and form typically allow direct measurements of fluxes and flows to be made over time scales of minutes to years. The observations made in these types of studies are often organized into the framework of field experiments, or measurements conducted under controlled field conditions. Field experiments can have different levels of "control," ranging from those that attempt to manipulate natural conditions and isolate processes of interest, to those that rely on informed judgment and experience to guide data collection and generate the evidence required to test a hypothesis (Slaymaker 1991: 7–16). It may also be possible to accommodate natural variability by using statistics to guide a data collection program, and this approach may be especially relevant to field studies that involve a large number of measurements or, conversely, where there are limitations on the number of samples that can be collected. Note that a sample is an element, or group of elements drawn from a population, that may be analyzed to ascertain or, in a statistical sense, estimate the characteristics of a population as a whole, whereas measurements assign numbers to the population or elements of it (see Chapter 6).

Laboratory experiments are also used in physical geography. However, the approach has been adopted cautiously because geography, by its very nature, deals with interrelations among phenomena in the natural world, not the artificial and often highly constrained conditions that exist in the laboratory (see Chapter 21). A key consideration is the extent to which measured data can be related to conditions in the real world that the laboratory experiment is meant to represent (Peakall et al. 1996). Experiments may be conducted using scale models, unscaled models, or physical analogues. Physical analogues attempt to reproduce significant aspects of the form and function of natural phenomena, whereas scaled and unscaled models provide a physical representation of a designated feature and may involve the use of a prototype (that is, an object, such as a laboratory flume, or a small segment of the real world, such as a New Jersey landfill). The properties of scale models are defined according to the principle of similarity (that is, there is some degree of symmetry in the appearance and/or behavior of an object).

Remote sensing constitutes the third major category of observational technique used in contemporary physical geography. Twenty years ago, aerial photography served as the primary means of obtaining information about the Earth from above, but today, a plethora of air- and space-borne remote sensing technologies, including multispectral, infrared, microwave, and LIDAR systems, yield observational data that are of interest to both physical and human geographers (see Chapter 10). Remote sensing has emerged as a new research specialization within physical geography, and the large amount of observational data (on, for example, atmospheric characteristics, topography, surface water, and land cover) has also fueled the development of geographic information systems to manage and analyze large spatial data sets, and make them accessible to a non-technical audience (see Chapter 22).

Although it is acknowledged that observations are "theory-laden," in the sense that the choice of what to observe and the methods of observation themselves depend to some extent on theory, observational data continue to be viewed as the metric against which theories and hypotheses must be evaluated. In general, two types of scientific reasoning are prevalent in physical geography: **deductive** and **abductive** arguments (Rhoads and Thorn 1993). Deductive arguments are based on general relations (deterministic or probabilistic laws) which generally have a cause-effect structure and which, for the purpose of testing, are accepted as valid:

If A, Then B
If A is the case
Then B should occur

Although many deductive relations are far more complex, consider the following simple law-like statement: "if water is subjected to a gradient of gravitational potential energy, such as that associated with a sloping surface, then it will flow down that gradient." Observations can be made of water on sloping surfaces, either in nature or in the laboratory, to see if it flows downhill. Any observation of stationary water or water flowing uphill would cast doubt on the validity of the initial premise. However, repeated confirmation of the law-like "if – then" relation lends support to the idea that the gravitational potential energy gradient associated with a sloping surface causes water to flow downhill. Physical models based on physical laws and their interrelations expressed in mathematical form may also serve as vehicles for deductions (see Chapters 19 and 21), because application of such models yields predictions of outcomes. Consider, for example, a model of a meandering river which, if the mathematical relations accurately characterize the processes involved, might reveal how the channel planform (outline or shape when viewed from above) changes over time. The deductive validity of the model may be tested using field measurements or remote sensing to determine if the observed change in channel pattern conforms to the predicted pattern.

Abductive reasoning, in which empirical observations take center stage, are a common component of many geo-historical studies and also play a role in contemporary process studies (Rhoads and Thorn 1993). In this case an effect is observed, and "cause–effect" relations are considered to infer a cause that could have produced the effect.

B is observed
If A, then B
A is the cause of B (?)

In other words, reasoning proceeds from effect to cause, rather than from cause to effect as in deductive reasoning. Abductive inferences are inherently less certain than deductive inferences (hence the question mark) because reasoning from an effect to a cause, based on a cause–effect relation, does not guarantee the cause; another cause could have produced the same effect (see Chapter 5). Moreover, in the case of geohistorical studies, causes usually cannot be observed because they occurred in the distant past. Thus, for example, Grove Karl Gilbert suggested the large crater, now known as Meteor Crater, in Arizona, could have been formed in two ways: by a meteorite impact or volcanic action (Rhoads and Thorn 1996). He concluded, incorrectly, that Meteor Crater was the result of volcanic action. Abductive inferences, however, are commonly revised when, for example, subsequent process-based studies reveal new cause–effect relations.

## Conclusion

Geography has a rich tradition of observation aimed at gathering information about the complex world we inhabit. In both human and physical geography, the notion of observation as the unbiased visual perception of phenomena dominated until the 1950s. Ironically, this highly empiricist perspective on observation, which preceded and was coincident with the development of logical positivism as a philosophy of science, developed independent of this influence. During the 1950s and 1960s, in the era of spatial analysis, geography embraced the notion that observations consist of quantitative facts, obtained primarily through various forms of measurement. Since that time, observation in physical geography has focused on the measurements of forms, flows, fluxes and ages using increasingly sophisticated types of instruments in the field and laboratory, and by remote sensing. The spatial-analytic perspective persists in human geography, and many human geographers now claim that objective, unbiased observations of human systems cannot be made by human observers. They gain a multiplicity of insights on the complexity of geographic phenomena by employing a variety of theoretical constructs to guide the act of observing and to interpret observations. Both human and physical geographers recognize that observation is theory-laden but, whereas physical geographers implicitly subscribe to the notion that this does not jeopardize the objectivity of observational outcomes, many human geographers view objectivity rather more circumspectly.

What does the future hold for the evolving concept of observation in geography? In physical geography, the trend toward the use of instruments to collect quantitative data on natural phenomena is likely to continue. Indeed in the natural and earth sciences, in general, much scientific observation is undertaken with the aid of instruments, and the capacity to develop increasingly sophisticated instrumentation to obtain quantitative data is one of the major factors driving "progress." New technological developments can also influence the development of a scientific field, as has occurred in physical geography over the past 50 years with the development of computer and remote-sensing technologies, and field and laboratory instrumentation. Moreover, unless there is a dramatic shift in the perceived relationship between data and theory within the natural sciences, physical geographers will continue to view data derived from scientific observations as the metric against which theories and hypotheses must be evaluated. Indeed, they, like other scientists, favor instrumental observation because it is considered to be free of personal idiosyncrasies and, therefore, more capable of producing empirical "facts," which serve as the basis for evaluating how deductive or abductive arguments fair in relation to information about the real world.

In much of human geography there is an emerging consensus that views observation as an important, but theory-circumscribed operation. In short, the sense of an untainted, discourse-free eye that can fruitfully "scan" the world will likely continue to dissipate. In its place, an alternative perspective is beginning to emerge, one that views theoretical orientations as kinds of discourse (be it approached from the perspective of spatial analysis, Marxism, postmodernism, existentialism, etc.) that observes a world made whole, intertwined with processes that are illuminated by the orientations themselves. Through these orientations, researchers employ distinctive orders, mechanisms, and ways of seeing that make observations relevant to a particular orientation. That is, observational data, their use in the knowledge-building enterprise, and the phenomena of interest are linked, and observations reflect the presuppositions and values placed in the research environment. To an extent, then, observationally-informed research may become as much a contest of ideas as a quest to further clarify reality. This is not a retreat into relativism (whereby no observations count because all capacity to make decisions is lost). Rather, human geographers recognize that the knowledge they advance through informed observation is hopelessly bound up with the assumptions, theories, and predilections involved in the research act. Observation, in this frame, becomes a complex kind of human intervention in a world whose content inexorably reflects the desires, views, and subjectivities of researchers as much as a "pure" essence of reality.

Finally, this chapter has treated the topic of observation in human and physical geography since the 1970s separately. There are some similarities between the two parts of the discipline, for example, each recognizes that observation is theory-laden. But intellectual segregation nonetheless continues based on different perspectives regarding the implications of this "theory-ladenness." What opportunities exist for bringing them together within an observational integrated framework? Geography's long-standing concern with interrelations between nature and society, the human–environment tradition, provides an obvious domain for interaction between human and physical geographers but, thus far, such interaction has been limited. Some middle ground has been occupied by using coordinated spatial-analytic methodologies to examine the effects of human activities and environmental change on land use and land cover (such as in the Amazon Basin, Aral Sea and lowlands of southern Iraq). However, much more collaboration between human and physical geographers is required if the challenges to producing meaningful connections between such disparate approaches as the social-theoretic and physically-based styles of inquiry in human and physical geography ultimately are to be overcome.

# References

Dear, M. (2001) *The Condition of Postmodern Urbanism*. Oxford: Blackwell.

Eyles, J. (1989) The geography of everyday life. In *Horizons in Human Geography*, Gregory, D. and Walford, R., eds. Basingstoke: Macmillan, 102–17.

Harvey, D. (2000) *Spaces of Hope*. Edinburgh: University of Edinburgh Press.

Peakall, J., Ashworth, P., and Best, J. (1996) Physical modeling in fluvial geomorphology: principles, applications and unresolved issues. In *The Scientific Nature of Geomorphology*, Rhoads, B. L. and Thorn, C. E., eds. Chichester: Wiley, 221–53.

Rhoads, B. L. (2004) Whither physical geography? *Annals of the Association of American Geographers* 94: 748–55.

Rhoads, B. L. (2005) Process and form. In *Questioning Geography*, Castree, N., Rogers, A., and Sherman, D., eds. Malden, MA: Blackwell, 131–50.

Rhoads, B. L., and Thorn, C. E. (1993) Geomorphology as science: the role of theory. *Geomorphology* 6: 287–307.

Rhoads, B. L., and Thorn, C. E. (1996) Observation in geomorphology. In *The Scientific Nature of Geomorphology*, Rhoads, B. L. and Thorn, C. E., eds. Chichester: Wiley, 21–56.

Sayer, A. (1984) *Method in Social Science: A Realist Approach*. London: Hutchinson.

Semple, E. C. (1911) *Influences of Geographic Environment*. New York: Holt.

Slaymaker, O. (1991) *Field Experiments and Measurement Programs in Geomorphology*. Rotterdam: Balkema.

Smith, J. M. (1996) Ramification of region and senses of place. In *Concepts in Human Geography*, Earle, C., Mathewson, K., and Kenzer, M. S., eds. Lanham, MD: Rowman & Littlefield, 189–213.

Strohmayer, U., and Hannah, M. (1992) Domesticating postmodernism. *Antipode* 24: 29–55.

Walker, R. (2004) *The Conquest of Bread*. New York: New Press.

## Additional Resources

Goodchild, M. J. (2004) GIScience, geography, form and process. *Annals of the Association of American Geographers* 94: 709–14. Provides a good overview of contemporary GIScience and its relationship to geography as a whole.

Gregory, K. J. (2000) *The Changing Nature of Physical Geography*. London: Arnold. An excellent overview of the development and current status of physical geography.

Haines-Young, R. H., and Petch, J. R. (1986) *Physical Geography: Its Nature and Methods*. New York: Harper and Row. Reviews the nature and methods of scientific approaches to physical geography, emphasizing the critical rationalist method of Sir Karl Popper.

Johnston, R. J., and Sidaway, J. D. (2004) *Geography & Geographers: Anglo-American Geography since 1945* (6th edn). London: Arnold. Excellent historical summary of the paradigms and people of post-World War II geography as it developed in the United States and United Kingdom.

Kuhn, T. S. (1962) *The Structure of Scientific Revolutions*. Chicago: University of Chicago Press. Proposes that the historical development of science is punctuated by abrupt changes, or revolutions, interspersed with periods of stability.

Paine, A. D. M. (1985) "Ergodic" reasoning in geomorphology: time for a review of the term? *Progress in Physical Geography* 9: 1–15. Describes the origin of ergodic reasoning in science and provides an overview of strengths and limitations of the concept as applied in geomorphology.

Shapere, D. (1982) The concept of observation in science and philosophy. *Philosophy of Science* 49: 485–525. A classic treatise on the nature of observation in contemporary science and the role that observation plays in the scientific process.

Turner, B. L. (2002) Contested identities: human-environment geography and disciplinary implications in a restructuring academy. *Annals of the Association of American Geographers* 92: 52–74. Argues that studies of human interaction with the biophysical environment lie at the core of geography's identity as an academic discipline.

## Exercise 3.1   Observation in Physical and Human Geography

The objective of this exercise is to enable you to critically evaluate the ways in which observation can be viewed from different perspectives. Choose two research articles from the recent (post-1970) literature in geography (represented by, for example, *Annals of the Association of American Geographers, Transactions of the Institute of British Geographers*), one with an emphasis on physical geography and one with an emphasis on human geography. How is observation used in the two papers (that is, what role does it play in the investigations) and how are the ways in which observation is used similar or different? Are these similarities or differences related to subject matter? To what extent do you feel the observations, including the collection and interpretation of information, are guided by preconceived ideas or theories, or to what extent are they independent?

# Chapter 4

# Measurement and Interpretation

*Sent Visser and John Paul Jones III*

Keywords

| | |
|---|---|
| Communication model | Positionality |
| Data matrices | Reliability |
| Dependent/independent variable | Validity |
| Levels of measurement | Variability |
| Mixed methods | |

## Introduction

As we saw in Chapters 2 and 3, there are differences of opinion among geographers – and in fact among members of the research community more broadly – regarding the nature of the world, the objects in it, and how we ask questions of it. In its broadest sense, however, science is about acquiring knowledge, and within that larger charge we usually find two major objectives in what we call research.

First, there is explanation: the task of identifying the causes behind "events" (as in a series of plant closures) or the "states of the world" that surround them (like high levels of worker unionization or similar types of spatial variation). Most of what we traditionally know as science proceeds within this view, but its proponents nevertheless remain divided. On the one hand, there are some who believe that science cannot explain a single event. For them, events such as the attack on Pearl Harbor in 1941 cannot be an object of science because the event did not occur repeatedly, such that a relationship between multiple attacks and their numerous potential causes could be effectively analyzed. In this view, it is **variability** – that is, research objects with high levels, medium levels, and low levels in the values corresponding to events or states of the world – that becomes the entry point to explanation. A good deal of research in spatial analysis in geography conforms to this view, and geographers working within it often collect data across a large number of spatial units (e.g., census tracts, weather stations) in their effort to "explain" spatial variation (e.g., of poverty levels, or temperature anomalies; see, for example, Chapter 18).

As we saw in Chapter 2, however, critical realists reject this view, believing that singular events are worthy of explanation. In fact, they offer the rejoinder that the number of times something happens, or the magnitude at which it happens, actually tells us little about causality (Sayer 1985). For these theorists, causality can only be determined by rigorous, in depth attention to the presence of the causal powers that produce particular outcomes. The search for causes must therefore proceed on a case-by-case basis. This does not mean that we might not find repetition in causal forces, but we cannot assume them in advance (in practice, regularity is more likely to be encountered when studying the causal forces in nature). Thus under critical realism a detailed description of the confluence of spatial, historical, and cultural contexts, together with an understanding of the motivations of the individuals involved, could indeed produce an "explanation" of the event of December 7, 1941.

A second objective within the broad category of science is that of understanding. In this perspective, the goal of knowledge is meaning and its social and geographic differences, as in: "What does this old place mean to me?"; or "Why is this culture so obsessed with fountains?" Within the rubric of understanding, meaning is an object of interpretation (rather than measurement), with the feelings and experiences described ranging from the highly personal, such as those generated through certain styles of biography, to the socio-cultural and historical, as when we situate individuals within those wider contexts. When meaning is the object of inquiry, the methods and practices of *hermeneutics* are often invoked. This is a venerable area of scholarly inquiry developed in classical ages for the interpretation of texts, such as the Bible. Hermeneutics requires that researchers abandon notions of detached objectivity in the acquisition of knowledge, and instead rely upon the generation of empathetic and situated, or grounded, understandings. In addition to focusing on the production of meaning as an object of inquiry, this approach also requires that the contexts within which meanings are *evaluated* also come under scrutiny. Hence, researchers interested in interpreting meanings are encouraged to engage in a form of self-reflexive (i.e., self-evaluative) analysis that puts their own **positionality** as interpreters under as much critical study as the meaning they are examining. For only in this way will they be able to evaluate that meaning's resonance in terms of its culturally specific origins and its differently situated audiences (including the researcher).

This chapter takes up explanation and understanding under the twin banners of measurement and interpretation. As we go along, it will seem that measurement and interpretation are cleaved according to a strictly quantitative versus qualitative binary opposition. And indeed, many years ago that seemed to be a significant distinction in geography. As time has gone on, however, a few changes have been made. One of the most important of these has been the rise of **mixed method** approaches in geography (Mattingly and Falconer-Al-Hindi 1995; Rocheleau 1995). After some heated debates in the literature, researchers began to see the value of both quantitative and qualitative approaches in geography, and now many researchers, especially those trained in the past decade, have some experience in both. Second, at the theoretical level, there have been challenges to the older model of science that sanctioned only quantitative approaches in explanation (for examples, see Cloke et al. 1991; Dixon and Jones 1998; Sayer 1985). These include areas such as humanistic geography, Marxist geography, feminist geography, political ecology, and poststructuralism (Peet 1998). All of these broke the theoretical hold that quantitative spatial analysis had on the discipline, such that it is now accepted practice to conduct research that relies partially or even solely on qualitative approaches. Finally, it has come to be accepted among many geographers that even quantitative approaches – which rely on discrete classifications and careful measurements and counts for developing empirical data – are themselves a "language." Yes, it is one that accepts numbers as its currency, but it is nonetheless a particular way of describing the world, much like qualitative approaches. So while there may be differences in precision and accuracy between quantitative and qualitative analysis, many scholars now recognize that simply because a piece of "data" is a number rather than a narrative description does not prima facie make it better for doing "science," at least in terms of the broader view proposed here. With this in mind, let's get started with measurement.

## Issues in Measurement

### Variables and relationships

Most explanatory science approaches causation from the viewpoint of what causes the values of an attribute or characteristic to vary from observation to observation. The values of an attribute measured for a class or type of observation are called a *variable*. Theory attempts to explain why variables vary the way they do. If, in a theory, an **independent** (or causal) force or factor (X) causes a **dependent** (or caused, or response) effect (Y), then in the real world, a measurable empirical relationship should exist between the variables that are X and Y. In particular, X and Y should co-vary with one another if a causal relationship exists. If the relationship does not exist then the theory is false, although there may have been problems in the attempt to study its existence. If the relationship is shown to exist in a certain situation, then the theory is not false. Note that this is not the same as saying that the theory is proven true, because there may be other situations in which the relationship does not exist, or there may be a reason for the existence of the relationship other than that postulated in the theory.

What is a *relationship*? If observations with a certain value of an attribute X also tend to have a certain value of attribute Y, then that is a relationship. If land parcels with a forest cover tend to have deeper soils than grass-covered land parcels, or vice-versa, those are relationships. If you randomly select urban intersections, and find that when there is a supermarket there, there is no other supermarket within a half-mile, then that is a relationship. There is a relationship if shopping centers with a shoe store tend to have more than one; that is, they have several or none. If every time you see John you see Mary, there is a relationship, although you do not know its nature; if you never see them together, that is a relationship too.

It is called a positive relationship if observations that have high values of an attribute X tend to have high values of an attribute Y, and observations that have low values of X tend to have low values of Y. It is called an inverse or negative relationship when observations that have high values of attribute X tend to have low values of attribute Y, and observations with relatively low values of X tend to have high values of Y. People who have more years of schooling tend to have higher incomes than those with less education. That is a positive relationship. The greater the distance from the downtown of US cities, in all directions on average, the lower the percentage of residents who are non-white. This is an example of an inverse relationship.

Relationships are average tendencies found in a large number of observations. Statistics that describe a relationship, either its closeness, nature, or both, are called *correlations*. The selection of the observations to ascertain whether there is a relationship is a tightly controlled process necessary to avoid bias in their selection, and so to avoid discovery of a relationship that does not actually exist or vice-versa (see Box 4.1).

## Units of observation

The entity being studied is the observation, or unit of observation. It is also called an "element" or "case." If you are analyzing some characteristic of individual human beings, then the individual person is the observation unit. People are common units of observation. For example, you might have been one if you participated in an experiment in an introductory psychology class.

More often than not geographers study the characteristics of areas. They are interested in why characteristics vary from place to place (i.e., explaining spatial variation). Common spatial units of observation in human geography are those for which governments collect data – countries, states (or provinces), counties (or parishes), MSAs (metropolitan statistical areas), urbanized areas, cities, census tracts, and blocks. In physical geography, common units of observation include river basins, weather stations, pixels from a remotely sensed satellite, and field-based samples. In research, however, anything may be a unit of observation. Newspaper articles, books, book publishers, dogs, buildings, cars, gas stations, beer distributors, even scientific experiments themselves can be units of observation. If you are looking at variations in attributes between "things," those "things" are the units of observation, the entity for which you measure some attribute or characteristic.

## Attributes, characteristics, and variables

We have been using the word "attribute" or "characteristic" interchangeably. The attributes of individuals that you wish to study may include their age, hair color, GPA, class rank, number of siblings, gender, ethnicity, marital status, etc. In the case of areas we may be interested in the characteristics of the areas themselves, of individuals in those areas, or of other things in those areas. Examples for county units of observation may include the average rainfall during July each year, the area cultivated, the number or percent of farm operators over 65, the divorce rate last year, the percent of the population under 15 years of age, the dollar volume of retail sales last year, etc. For spatial units of observation the

**Box 4.1**   Causality and Correlation

This example of causality and correlation was developed by the sociologist Charles Bonney; it is reported here in adapted form from Earl R. Babbie, *The Practice of Social Research* (1998: 74–5).

If you have hypothesized a cause and found the relationship to support it, have you proven the cause? Assume that in a study of college students you have found an inverse relationship between smoking pot (MJ) and their grade point average (GPA). Presumably the causative reasoning is that since marijuana adversely affects memory, pot smoking lowers grades. If, however, you cannot determine which occurred first, it can also be argued that low grades causes pot smoking as an escapist behavior.

Alternatively, the inverse relationship may be a spurious correlation, the result of emotional problems – a "missing variable" (i.e., a variable not yet accounted for) that causes *both* low grades and pot smoking. If this missing variable (e.g., withdrawal, negative emotions, or EMO-) can be controlled for, the logic says, the inverse correlation between MJ and GPA may disappear. Or, perhaps the students who smoke pot do so because they are not very bright; in this case intelligence is causing MJ and GPA simultaneously. This is another spurious correlation, with intelligence the real cause. Or even still, the relationship between MJ and GPA could be coincidental.

These explanations have all appeared in the press at some time to explain the inverse relationship between MJ and GPA. Unfortunately (or fortunately), none are correct, because the correlation between GPA and MJ is actually positive. Knowing this, however, does not necessarily lead us to a quick causal conclusion, for it can be argued that: (a) marijuana relaxes people, clears away other stress, and allows more effective study (MJ causes GPA); (b) marijuana is used as a reward for getting good grades (GPA causes MJ); (c) a high level of curiosity (e.g., outgoing, positive emotions, or EMO+) helps learning, and leads to investigation of taboos (EMO+ causes both MJ and GPA, and neither MJ nor GPA are causally linked); (d) people who smoke pot know they are in danger of being called out for bad grades, so they compensate by studying more; or (e) the sample of students contains a lot of bright, industrious students who happen to smoke pot.

In short, if X causes Y, a relationship between them must exist, but if a relationship does exist, it doesn't necessarily mean that X causes Y.

data are usually made up of averages of some other unit of observation such as individuals (e.g., percent poor) or of the total number of some other unit of observation (e.g., retail square footage). The time of observation almost always matters. Most attributes – except those such as areal extent, elevation, etc. – change over time for any spatial observation unit. To describe the characteristics of an area it should be clear when it had these characteristics. Unlike physics and chemistry where the nature of matter and energy are invariant with respect to human time frames, the characteristics of areas do vary over time.

If the measured value of an attribute or characteristic varies from observation to observation, then that measure constitutes a *variable*. If the observation unit is a location or area and the attribute varies geographically, the measurement is *spatially variable*. So often do we work only with things that vary that we often find ourselves talking only about variables and not attributes and characteristics. In fact, if something does not vary it is usually not clear why we would be interested in it! Here we will continue to use them interchangeably.

To say anything "objective" about the attributes of some group of observations, the attributes must be measured. We can measure anything that is "real." By "real" in this context we mean that the distinctions between observations in terms of the degree to which they possess some attribute is apparent to everyone. Philosophers debate the possibility that nothing is real, that the existence of all things (ontology) is dependent on human perception or upon the languages humans use to describe their existence (epistemology) (see Chapter 2). We are not talking about "real" in these senses, however. We are simply talking "real" in the sense that if something is apparent to one of the five senses of the vast majority of people, then it is real. Thus, age and height and marital status of a person are real, the distance you commute to school is real, and tons of cotton grown in a county last year is real (even if there was none).

What is "unreal?" Political attitudes measured as conservative, liberal etc. are not "real" in this sense, though they can be described and we can attempt to devise measurements for them. The same is true of prejudice, religiosity, and sense of place. These are attributes that are "unreal" in the sense we cannot define them in a way that everyone will agree upon. So some people may consider an individual conservative while others do not. These terms are human percepts; they are not real to the senses. They can be, however, objectively measured in the sense that you can devise an index for them such that some other person can duplicate your measurement and classify the observations the same way that you do. For instance, to measure conservatism, you may devise a survey containing five attitudinal questions, and categorize an individual as conservative if they answer strongly agree on five out of five questions. Another researcher may use the same questionnaire, administer it the same way (e.g. personal interview), and they should end up categorizing the same people the same way. This questionnaire is objective, in the sense that it was constructed without willful bias, but whether or not you have actually measured conservatism is subjective, and open to debate.

**Validity** raises the question of whether or not your questionnaire actually measures the concept or construct that you think it is measures. For example, in the case of conservatism, consider this diagram:

Political Positionality

Conservative-Moderate-Liberal

Five Question Scale

The underlying theory is based on the idea that there is a "thing" called political positionality, which we know fundamentally is multi-dimensional. We want to understand how it works, in theory. The best way we understand politics, in this case, is through a conservative-moderate-liberal continuum, and this is our construct: it is still a concept (on a continuum), but it is more concrete than simply "politics" or political positionality in the abstract. Finally, our variable is the concrete measure we get for each person, after having asked them to answer our five questions. So in general, we have a cascade of increasingly more concrete notions:

Theory

↓

Construct

↓

Attribute or Characteristic or Variable

Validity, then, is how well our attributes match our constructs, which themselves are a bit nebulous, emerging as they do out of even more abstract theory. Social scientists spend a lot of time thinking about the constructs embedded in theories, and about the extent to which we can measure these constructs with variables. Take human capital theory, for example. It states that individuals have very different capacities and that these can ultimately influence not only their life courses but the trajectory of the regions they live in. The theory invokes several different constructs, including: wealth, income, health, educational level, native intelligence, creativity, entrepreneurial spirit, communication skills, ability and propensity to migrate, aversion to risk, willingness to innovate, etc. With all these in play, how would you go about measuring, on a concrete level, the human capital for a single region? What variables would *you* collect for your region? How well do you think a variable like "Percent over 25 with a college degree" measures knowledge, or even intelligence? How well does it measure a multidimensional notion such as human capital?

**Reliability** refers to whether the measure or measurement device measures the attribute the same way for each observation, or the same way each time or place it is used. If everyone understands perfectly the five questions in the same way, the measurement is reliable across the observations. If you give the same questionnaire to the same people at a later date, and assuming they cannot remember what they answered the first time and that their attitudes have not changed, were they categorized the same way as before? If so, the measure is reliable, irrespective of whether it is valid (i.e., really measures conservatism). A measure can be reliable without being accurate. If a device measures a 2 inch rainfall as 1.9 inches every time, the device is reliable, but not accurate.

The *unit of measurement* is the category or metric used to distinguish one observation as different from another with respect to some attribute. For rainfall at some location over some period of time, the unit of measurement is inches or centimeters. For gender, the units of measurement are the categories, male and female. These categories may be given number labels such as male = 0, female = 1. For a number or metric measurement such as inches, notice that we are still categorizing observations. Some may fall into the category 3.6 inches, some into the category 3.7 inches, none into the category 3.8 inches. However, 3.8 inches is still a category; so we can say that the unit of measurement is denoted by the categories into which we divide our measure. All measurement is in essence the categorization of observations with respect to some attribute(s) of those observations, and *classification*, therefore, constitutes measurement.

*Precision* is the degree of fineness used to define the categories of measurement. Rainfall may be measured as 3 inches, 3.3 inches or 3.317986 inches. The last is the most precise. Precision has nothing to do with accuracy, however. If rain is measured as 3.3179 inches, but was actually 3.1762 inches, the measurement is precise but inaccurate.

Scientists often employ a variable language. They describe causes in terms of variables: if one variable causes another, then the variables must have a relationship; they must co-vary. Co-variation can only be established via repeated measurements of an attribute. Repeated observations are required to create variables. This is why, in some versions of explanatory science, we need multiple observations to assess whether or not causal forces are operating. We cannot find causality if: (a) a variable does not vary (i.e., if we have an attribute that is a constant); or (b) if we have only one observation (in which case, there are not enough cases to discover co-variation). As mentioned in the introduction, these are not fully agreed upon assumptions, but many in the scientific community adhere to them.

Tables 4.1 to 4.3 below are portions of **data matrices**. They illustrate how data are organized for analysis, and they also serve to clarify the nature of variables and observations. A matrix is any tabular array of numbers, but in a data matrix the observations go down the side of the table such that a row of values constitutes the values of several variables measured for that observation; that is, the observations are the rows. A column in

**Table 4.1**  Simple data matrix, individual level

| Individual | Age (years) | Gender | Class rank | Annual parental income |
|---|---|---|---|---|
| John | 37 | 0 | 4 | 117,000 |
| Mary | 19 | 1 | 2 | 38,000 |
| Peter | 21 | 0 | 3 | 27,000 |
| Libby | 23 | 1 | 2 | 54,000 |

*Notes*: Gender coded as 0 = male, 1 = female; Class rank coded as: 1 = freshman, 2 = sophomore, etc.

**Table 4.2**  Simple data matrix, county level

| County | Dist KC | Soil | Rain | Farm area | Cap ex ($) |
|---|---|---|---|---|---|
| Aurora | 4.8 | Us | 21.82 | 396,987 | 5,792,596 |
| Beadle | 5.4 | Us | 19.44 | 785,949 | 14,860,173 |
| Brown | 6.3 | B | 19.10 | 1,068,332 | 16,394,836 |
| Douglas | 4.6 | Us | 22.66 | 272,949 | 5,493,126 |
| Haakon | 6.3 | Us | 15.12 | 1,189,330 | 3,501,982 |

*Notes*: Dist KC = distance of county midpoint to Kansas City, inches (map scale = 1:5,000,000); Soil = predominant soil type in county: E = Entisol, B = Boroll, Us = Ustoll, Ud = Udoll; Rain = average annual rainfall, inches; Farm area = total farmland area per county, in acres, 1969; Cap ex = total farm capital expenditures per county, 1969

*Source*: Visser (1979; 1980)

**Table 4.3** Simple data matrix, country level

| Country | Pop80 | CBR | IMR | GDPpc ($) |
|---------|-------|-----|-----|-----------|
| Argentina | 27,064 | 25.2 | 40.8 | 1,935 |
| Barbados | 253 | 17.0 | 25.1 | 2,686 |
| Belize | 162 | 38.7 | 33.7 | 750 |
| Bolivia | 5,600 | 46.6 | 77.3 | 567 |
| Brazil | 123,032 | 36.0 | 46.6 | 1,664 |

*Notes*: Pop80 = population estimate, 1980, in 000s; CBR = crude birth rate, year varies per country (1973–80); IMR = infant mortality rate, year varies per country (1973–80); GDPpc = gross domestic product per capita, 1980, US dollars
*Source*: Boehm and Visser (1984)

the data matrix, then, is a variable. By convention, the first column in a data matrix is used to identify the observations, and in these instances it is not technically a variable, but a marker so that you know the observation to which you are referring (i.e., state two letter code, census tract number, etc.).

A data matrix may be entered in a spreadsheet, but we hesitate to call the data matrix a spreadsheet, because of the rules that row entries must only refer to the observation unit in that row, and that the column values must all be for the variable in that column. By convention, we tend to enter a zero or an unlikely figure (9999) in the column to indicate that the data is missing or unavailable.

In Table 4.1 the observation units are college students. In Tables 4.2 and 4.3 they are respectively South Dakota counties and Latin American countries. Reading down any column, it can be seen that values vary from observation to observation – these are, there-fore, variables. The college student data may be considered raw data. Table 4.3 includes average values (an average is a statistic) for the people in the countries, but the variation between the countries in the values of these averages (like crude birth rate) can be analyzed with the same techniques that we use on raw data, although interpretation of the results is hazardous because of a problem called ecological fallacy (see Chapter 11).

## Measurement quality

As mentioned above, measurement means creating categories or classes for an attribute and assigning observations to those categories. Thus the gender (or sex) of individuals is typically categorized as male or female; religion can be measured in several additional number of categories; and these measurements are altogether different from variables like crude birth rates or rainfall, which vary continuously. In these ways, measures differ in terms of how much information they provide. Higher quality measures permit more pow-erful analyses of the information that the measures contain.

There are four recognized **levels of measurement**. They are hierarchical. All measures have the properties of the lowest level, but some measures have extra properties. The lowest level of measurement is *nominal*. This is the classification of an attribute into relatively homogeneous groups. The intent is simply to differentiate observations with respect to that attribute; for instance, Gender in Table 4.1 and Soil in Table 4.2. The gender variable is a dichotomous nominal variable in that an observation is female or not-female. Nominal variables have no explicit or implicit ordering (e.g., from better to worse, larger or smaller). The Soil variable is a polytomous nominal variable, but it can be made dichotomous by classifying counties 1 = Ustoll, or 0 = not Ustoll. We might want to do this if we are interested in the specific effects of Ustoll soils on agricultural production, but when we make such a conversion we lose information, which technically speaking lowers the quality of the measurement.

Two rules apply to the categories of a nominal measurement. The categories must be *exhaustive* and *mutually exclusive*. Exhaustive means there must be a category that an observation can be assigned to. Table 4.2 was part of a study where all the counties fell into one of the four soil types. If a county did not then another category would have to be added. That category could be "other." If the exhaustive criterion is not met, some observations will be wrongly assigned to categories that they do not belong in, or their attribute will not be recorded.

Mutually exclusive means that an observation may only be assigned to one category and one category only. If people are being categorized with respect to religion, and two of the categories are Shinto and Buddhist, then we will have a problem if someone can be both Shinto and Buddhist. If they are categorized in both, then the total number of observations in the variable will be greater than the number of actual observations. The solution is to either create a separate "both Shintoist and Buddhist" category, or a separate "multiple religions" category (see Box 4.2).

The categories of a nominal variable may be identified with number labels but the number labels are not metric. Since there is no implied ranking among nominal variables, arithmetic operations on them have no meaning. Subtracting a 2 = Boroll from a 3 = Ustoll does not create a 1 = Entisol. It seems absurd that we need mention this, but this mistake has been made in statistical tests assessing the degree of difference between observations with regard to some attribute.

*Ordinal* measurement is the second level. An ordinal variable is one which is nominal with the added property that the categories of the variable have an inherent order. There is no inherent order to the categories of Soil, so it is not ordinal. Class Rank in Table 4.1 is ordinal, however.

Another example of an ordinal variable is military rank; individual military personnel can be categorized as private = 1, lance-corporal = 2, corporal = 3, etc. Similarly, houses may be assessed in the following way: uninhabitable = 1, inhabitable but in need of repair = 2, fair condition = 3, good condition = 4, excellent condition = 5. If any attribute is ranked it is an ordinal variable. There may be many observations at a given rank in the order, or the observations may be ranked such that there are only a few observations, or one, at each rank of the variable.

A frequently used ordinal variable is the Likert scale. Usually a question is asked of a respondent in a survey to which the answer is Strongly Agree, Agree, No Opinion, Disagree or Strongly Disagree. These categories may then be assigned number values, such as Strongly Agree = 1, Agree = 2 etc. The numbering may also be reversed. Which category is highest or lowest does not matter as long as there is an inherent order to the categories. The number labels are the ranks of the categories. They provide the position of the category in the order. This example is that of a five point Likert scale. It can be made a seven point

**Box 4.2** Measurement Affects the Value of the Variable

In November of 2005 an ABC News/*Washington Post* poll asked respondents, "In making its case for war with Iraq, do you think the Bush administration told the American public what it believed to be true, or intentionally misled the American public?" Fifty-five percent went for misled. Would a majority of American adults think the Bush administration lied if the measure had been exhaustive; that is, if it had permitted "don't know" or "no opinion" as a response? Did this majority think that they had been misled prior to being asked if they had been?

Survey questions often generate invalid and unreliable data because of bad questions, which do not permit mutually exclusive or exhaustive answers. If asked, "do you think someone lied because…", how do you answer if you think that they lied, but not for the reason given? Or what if you do not think they lied? And, did they have an opinion before the respondents were asked the question? In many cases they did not, and so the measurement or observation actually created the attribute for the observation unit.

Seemingly objective surveys can give meaningless results and measure things that do not exist. If manufacturers are asked to rank the importance of location factors in locating their factory, they will rank factors highly that they were only made aware of via the list in the questionnaire. Factors that are really important to them but not listed will not be revealed.

Attitudinal and values data are most susceptible to creation or bias as a result of observation. Where there was no opinion before the question was asked, and "no opinion" is not a choice, an attribute is being created. When people are asked about things they know nothing about, and "don't know" is not a response category, an attribute is being created. Even with such categories there are problems: answering "no opinion" or "don't know" may be embarrassing to the respondent. Respondents want to give answers that please the researcher. A recent attempt to redo the famous Kinsey report on sexual behaviors had only 1% of adult respondents admitting to being primarily gay. Apparently something we know to be prevalent goes out of existence when it is measured.

These issues are not confined to the social or behavioral sciences. Physicists have known for a century that some things do not exist until humans observed them, and that the nature of an object is changed by the act of measurement. To observe short-lived sub-atomic particles, protons are smashed into each other at high velocities, and the recorded electromagnetic traces of what emerges are identified as this or that particle. Do these entities exist when we do not observe them? Would their nature be different if they were created some other way?

Likert scale by adding the categories, Slightly Agree and Slightly Disagree. This can increase variation in the responses because the number of ranks is now 7. It also increases precision but not necessarily accuracy.

If a variable has values that not only place observations in a certain order, but also indicate the magnitude of difference between two observations, then it is at the *interval* level of measurement. As an example, the difference between 30 degrees F and 60 degrees F is 30 degrees, and that temperature difference is exactly the same as the temperature difference between −30 degrees and −60 degrees, or between 180 and 210 degrees. Likewise, the difference between 1940 AD and 1990 AD is 50 years, as is the difference between 250 BC and 200 BC − 50 years in each case is the same period of time. In contrast, using the 5-point Likert scale, the difference between Strongly Agree and Agree is a difference in rank of 1 and the difference between Agree and No Opinion is also a difference in rank of 1. But the two differences in opinion are probably not the same. If we used a 7 point scale, is the difference between Strongly Agree and Agree (= 1), the same as the difference between Agree and Slightly Agree (= 1), and are these differences twice that of the difference between Agree and No Opinion (= 2)? That is unknown. Obviously, if you add Strongly Agree = 1 to Agree = 2, you do not get 3 (= No Opinion). So for ordinal data we know how to rank, but we do not have a consistent metric, as we do for interval variables like temperature or time.

If the order of the magnitude of difference between two values of a variable can also be stated, and it has real meaning, then the variable has a *ratio* level of measurement. Ratio variables shown in Tables 4.1 to 4.3 include Age, Annual parental income, Dist KC, Rain, Farm area, Cap ex, Pop80, CBR, IMR, and GDPpc. A county with an annual average rainfall of 30 inches has three times as much rainfall as one with 10 inches a year. This is what is meant by order of magnitude. The variable has a true zero. The value zero means no rainfall. A true zero means there is no quantity of that attribute, even if no observation could conceivably have this value of a variable (e.g. 0 ft height of an individual). Unless temperature is measured as degrees Kelvin, zero does not mean that a substance contains no heat energy. A temperature of 100 degrees C is not twice as hot as 50 degrees C, nor is 2000 AD twice as late as 1000 AD. Temperatures, measured as degrees Celsius or Fahrenheit, and historic time are interval variables, but they are not ratio variables. The order of magnitude of difference cannot be stated. The Kelvin temperature scale, however, does have a true zero, and on that scale 400 degrees is twice as hot as 200 degrees. In general it means that a substance has twice as much heat energy as another, although temperature does not actually measure energy content.

All measures can be lowered in level. A ratio variable can be made ordinal one, and an ordinal variable can be made nominal. Information is lost doing this, however, sometimes with consequences (see Box 4.3). Consider for example the Köppen climate classification system. If we use this polytomous nominal variable instead of the continuous data that goes into its construction (namely temperature, precipitation, and seasonality), we have lost some of the fine grained variability that is included in its climate categories. Finally, you can only change a lower level of measurement into a higher one by changing the unit of observation. An individual may be measured as male or female (nominal), but percent of the population of counties that is male is a ratio variable.

If the values of a metric variable can be stated to an infinite degree of precision then the variable is continuous. Thus rainfall can be measured as 3.7 inches or 3.7120485985388958 inches. The number of people in a country only comes in whole units; so this variable is discrete. Only a finite number of values are possible. The distinction between discrete and continuous is generally of no significance. Discrete variables measured at a ratio level of measurement are treated statistically as if they were continuous.

## Box 4.3   Bad Classifications

In the first half of the nineteenth century geographers copied the natural sciences and engaged in a great deal of classification to gain some understanding of the phenomena they were studying. Carl Linnaeus's famous eighteenth-century classification of plants based on their morphology was influential in this regard. He discovered that his plant groupings reflected real differences in other scientific criteria; that is, differences in form reflected real species differentiation. In geography, some famous classifications include Whittlesey's world agricultural types, and Thornthwaite's climate classification. Regionalization across many different variables proceeded apace. The idea was to maximize the homogeneity of the multiple attributes within a spatial boundary and to maximize the heterogeneity between the attributes of other areas.

In the rush to regionalize, however, geographers started to overlook the fact that their classifications of space are measurements, and that they were arbitrarily making continuous phenomena discrete (converting ratio data into the Corn Belt or the Central Business District). The term "natural boundary" came to be used a lot, and the fact that attributes are transitional or continuous across those boundaries was simply ignored. The idea of "natural regions" also came into existence. Regions had unique assemblages of attributes, and a major task was to ascertain why they had these characteristics, that is, what made them different from other regions. Pretty soon the regions became "reified" (made real), even though they were an artifice of measurement and the result of taking continuous phenomena and converting them to discrete categories. Only later were geographers to see that the data they collected, or that governments collected for them, could not be argued to exist as "natural" in any way.

A social parallel to geography's bad classifications is found in Linnaeus's own categorizations of humans. He came up with the concept of race based on skin color and region of origin, and assigned other attributes to the races – European, Asiatic, African, and American. A later classification by Linneaus's protégé, Johann Friedrich Blumenbach, added a fifth race – Malay (which was said to include Pacific Islanders and Australian Aboriginals). As Stephen Jay Gould (1994) reports:

> By moving from the Linnaean four-race system to his own five-race scheme, Blumenbach radically changed the geometry of human order from a geographically based model without explicit ranking to a hierarchy of worth, oddly based upon perceived beauty, and fanning out in two directions from a Caucasian ideal. The addition of a Malay category was crucial to this geometric reformulation….With this one stroke, he produced the geometric transformation from Linnaeus's unranked geographic model to the conventional hierarchy of implied worth that has fostered so much social grief ever since.

The popular view of race defined according to skin color does not reflect a real or significant biological distinction between humans (Diamond 1994). This is why we often see the term written in quotation marks (i.e., "race"). Dividing a continuous measure into discrete groups has reified a minor difference between individuals into an enormous difference with harmful social consequences. Socially speaking, "race" is still a meaningful measure insofar as many humans think of themselves and act toward others as if it exists. But would people today opt to have a racial identity if science had not created the measure in the first place?

## Issues in Interpretation

### Language matters

Why would we elect to describe the world in textual form, and offer an interpretation of the meaning it presents to us, when we have the option of measuring its variations and "getting on" with some science? An answer lies in the idea that the languages we use to describe the world are not neutral in how that world works. This reply points to a host of issues surrounding the question of representation, that is, how we *re-present* the world as a concrete reality. For many geographers working since the poststructuralist and cultural turns of the 1990s, processes of representation establish the parameters within which elements of the world (see ontology, Chapter 2) are named, classified, and related to one another. Without attention to the social powers behind such processes, and to the inherent openness and indeterminacy of language, we would be forced to conduct our trade with a realist (or "mirror") model of realist representation, unaware of the role that politics, ideology, and hegemony (e.g., the status quo) have in framing reality. We would also overlook the slippages and elisions in our always situated and partial descriptions of the world.

If, on the other hand, we open ourselves up to the gaps between reality and its representation, and indeed to the impossibility of ever fully and impartially grasping the richness and complexity of reality within the limited domain of language, then we begin to see through the "naturalizations" that give space the appearance of transparency, or that convey upon it the status of truth and objectivity. Importantly, in de-naturalizing space by pointing to its representational character, we do not at the same time mean to deny material geographies. Rather, we recognize that space both embeds and conveys social meaning (Natter and Jones 1997).

If, finally, we act in the world on the basis of these understandings, then surely the world is at least partially "socially constructed" through them. So to ignore the role of representations (visual, textual, or whatever) in the production of space and in the construction of meanings about space, is to miss a large part of the geographic enterprise. But how do we go about making methodological choices in the face of the interpretative challenge that representation invites?

### The communication model

In the broadest sense, interpretative research in geography begins with a three part model composed of the addressor, the message, and the addressee (Adams 2009). We can represent these components in a connected way as follows:

Addressor → Message → Addressee

The **communication model** states that there is someone who, or something that, produces a message for an addressee, and that the person who receives the message somehow has to make sense of it in order for understanding to take place. This last component is the interpretative moment, but it is only one aspect of the model. In fact, there are a host of theoretical, methodological, and analytic implications that ensue from it. These variously affect how we select objects of analysis (or units of observation) in geography, and how we develop questions surrounding them. Geographers concerned with interpretation tend to find that their "messages" are qualitative in character, usually textual or visual, but this does not mean that they are only interested in description or understanding. Many geographers (e.g., critical realists, see Chapter 2) doing interpretative research are just as interested in explaining how the world works as their fellow scientists whose research relies on large data matrices filled with numbers.

Let's begin with some concrete examples of this model in action, as depicted in Table 4.4. It illustrates a wide range of "messages" available for geographic research; for each one, it also includes a listing of potential addressors and addressees as examples. Note the variety of textual and visual communications possible, from political speeches and print advertising to videos posted on websites. Also see the diversity of actions taken as messages, such as street protests (these too are filled with symbolic content: they operate as "sign systems"). With some thought you can doubtless come up with many more examples.

A question you might ask yourself, however, is this: How is research into messages of these sorts geographical? Doesn't something have to vary in order for it to be geographical? How could these examples really be part of the discipline of geography? The answer, as alluded to above, is that we live in a world of interpretation, and how we engage meanings has an impact on how we organize the world, live and move through it, and alter it, for better or worse. We are literally born into a world of signs, words, pictures, and symbolic actions – most of our life is about negotiating these signs, making sense of them, and using them in our daily geographic activities.

As Table 4.4 shows, various places, parts of nature, built environments, regions, nations, even the Earth itself (Cosgrove 2001) are constantly represented, both textually and visually, in commercial and other forms of maps, photography, video, film, prose, poetry, journalism, and oral descriptions. A whole host of questions can be developed to analyze the underlying geographies of these representations, including:

- How are the spaces in the message(s) portrayed? Negatively or positively, developed or under-developed, modern or traditional, civilized or barbaric, safe or dangerous, orderly or unruly, exciting or boring, natural or artificial, wasteful or sustainable, egalitarian or exploitative?
- What wider social and political connotations are linked to these portrayals? Are people with certain social characteristics (age, gender, "race," non-western) positively or negatively associated with these portrayals?
- Given that any representation relies equally on decisions about what is *not* to be included, what was left out of the narrative (story) or the photographic frame? Not only what is, but what is not represented?

**Table 4.4**  Assorted examples of the communication model

| Addressor(s) | Message(s) | Addressee(s) |
|---|---|---|
| Newspaper conglomerates, business journalists | Nightly business reports, stock tables, economic editorial | Businesspersons, government officials |
| Retail industry firms, advertising firms | Radio and television commercials, print advertising | Shopping public, divided on cultural and economic criteria |
| Architects, developers, planners | Buildings, suburban developments, streetscapes | Motorized, bicycling, and ambulatory publics |
| Entertainment firms, software industry | Video games | Game playing public, divided on socio-demographic criteria |
| Political parties, politicians, military-industrial complex | Political speeches, signs at pro-war rallies | Voting public, divided on political criteria |
| Screenwriters, directors, entertainment firms | Television episodes, films | TV and movie watching public |
| Musicians, entertainment firms | CDs, music videos, downloadable songs | Music listening publics, divided on "taste" and "genre" |
| Unions, labor organizers | Plant shutdowns | Concerned public, business owners, government officials |
| Fashion designers, dress companies, retail firms | Hemlines, necklines, body types | Fashion conscious member of the same or opposite sex |
| Property owners | "No Trespassing" signs | Potential intruders |
| Anarchists, revolutionaries | Rallies against globalization, slogans against "the man" | Corporations, capitalists, other potential anarchists |
| Refugees, victims of atrocities and natural disasters | Accounts of atrocities, descriptions of events | Global public opinion, newsmakers, politicians |
| Professors, teaching assistants | Course content | Enrolled students |
| Non-governmental organizations, non-profits | Philanthropic materials, donation websites | Concerned public willing to give to various causes |
| Bloggers | Testimonies, opinions, deliberate falsehoods | "Blogosphere," public opinion makers |
| Cartographers | Maps and cartographic like products (GPSs) | Publics in need of wayfaring assistance, government agencies, businesses, spies |
| Government census agencies, census takers | Census data | Publics in need of socio-demographic and other data |
| Nature photographers | Nature photographs | Publics ready to purchase images of pristine nature |
| Pollsters | Political opinion data | Politicians, general public, news personnel |

- How do the elements of the representation fit together to make a whole? How does the represented space *work*, as a space? Does the portrayal come off as a coherent and tidy whole, fitting together like a puzzle, or is something intentionally juxtaposed to jar the addressee into rethinking her or his relationship to place, society, or environment?
- How does the message fit or not fit within other representations of its genre? Does it explicitly or implicitly draw from, repudiate, make ironic gestures toward, or otherwise comment on previous representations of its sort?

Another set of questions that you might ask about representational geographies revolve around the addressors, or "authors": Who made the representation, and why did they do so? Why did they include X but not Y, and why did they inflect the image or language in this or that way? If, moreover, we place the addressor within larger social, cultural, economic, political, and even geographic contexts, then what does that tell us about them? Whose larger interests (institutional, class-based, nation-centered) are being served, and with what purpose? In this way, and with the knowledge that no representation is ever neutral, that no photograph or newspaper article (much less a video game, a "No Trespassing" sign, or a political speech) is ever "just a representation," we come to a series of questions about addressors that go directly to their wider purpose. This includes their class, gender, "race," and national origin backgrounds, as well as their situatedness as geographic actors, including the spaces they hail from and purport to represent.

Finally, there emerges from this communications model a set of questions related to the addressee. These "audience" type questions revolve around issues of reception: What kind of audience did the addressor imagine they would reach with the message? How is that audience part of a specialized community of interpreters (based on location, scientific knowledge, or social characteristics, for example)? What is the impact of these larger contexts, including the everyday geographies of reception, upon the audience? How do some audiences resist, reframe, or misunderstand messages? While all these questions apply to intended addressees, they also apply to unintended audiences, such as those who encounter the message unexpectedly, secretly, or even purposively as part of a research project. Thus, even the researcher is part of the audience, and within that role he or she needs to put not only the addressor and the message under investigation (which is normally part of the research enterprise), but also the relationships these have with herself/himself. Of such relationships, ethical researchers must ask: What is the social-cultural-political-economic-geographic "distance" between me and the addressor? How do these distances influence my interpretation of the intentions of the addressor and the message? Am I reading too little, too much, or too differently given these different contexts? In short, researchers cannot afford to stop analysis at simply the addressor or the message itself; they must interrogate the situational contexts within which their interrogations emerge.

## Conclusion

The processes of gathering measurements and interpretative materials on the world follow directly from the process of observing that world (see Chapter 3). In this chapter we have provided the student with a set of meso-level concepts and directions for gathering "data," in the broadest sense. Some of this data is quantitative in character, and relies on measurements that vary in their degree of precision, accuracy, reliability, and validity. Some of them are interpretative in nature, and prompt questions about meaning – its origins, character, and deployments – and the work it does in the world. Thought broadly, these are both part of the wider scientific enterprise: explaining and understanding the world.

# References

Adams, P. (2009) *Geographies of Media and Communication*. Oxford: Wiley-Blackwell.

Babbie, E. R. (1998) *The Practice of Social Research* (8th edn). Belmont, CA: Wadsworth Publishing.

Boehm, R., and Visser, S. (1984) *Latin America: Case Studies*. Dubuque, IA: Kendall Hunt.

Cloke, P. J., Philo, C., and Sadler D. (1991) *Approaching Human Geography: An Introduction to Contemporary Theoretical Debates*. New York: Guilford Press.

Cosgrove, D. (2001) *Apollo's Eye: A Cartographic Genealogy of the Earth in the Western Imagination*. Baltimore: Johns Hopkins University Press.

Diamond, J. (1994) Race without color. *Discover Magazine* http://discovermagazine.com/1994/nov/racewithoutcolor444 [accessed August 25, 2009].

Dixon, D., and Jones III, J. P. (1998) My dinner with Derrida, *or* spatial analysis and poststructuralism do lunch. *Environment and Planning A* 30: 247–60.

Gould, S. J. (1994) The geometer of race. *Discover Magazine* http://discovermagazine.com/1994/nov/thegeometerofrac441 [accessed August 25, 2009].

Marx, K. (1845/2000) Theses on Feuerbach. In *Karl Marx: Selected Writings*, McLellan, D., ed. Oxford: Oxford University Press.

Mattingly, D. J., and Falconer-Al-Hindi, K. (1995) Should women count? A context for the debate. *The Professional Geographer* 47: 427–35.

Natter, W., and Jones III, J. P. (1997) Identity, space and other uncertainties. In *Space and Social Theory: Interpreting Modernity and Postmodernity*, Benko, G. and Strohmayer, U., eds. Oxford: Blackwell, 141–61.

Peet, R. (1998) *Modern Geographical Thought*. Oxford: Blackwell.

Rocheleau, D. (1995) Maps, numbers, text and context: mixing methods in feminist political ecology. *The Professional Geographer* 47: 458–66.

Sayer, A. (1985) Realism and geography. In *The Future of Geography*, Johnston, R. J., ed. London: Methuen, 159–73.

Visser, S. (1979) *The Spatial Dynamics of Agricultural Intensity*. Unpublished PhD dissertation, The Ohio State University, Columbus.

Visser, S. (1980) Technological change and the spatial structure of agriculture. *Economic Geography* 56: 311–19.

# Additional Resources

Berry, B. J. B. (1964) Approaches to regional analysis: a synthesis. *Annals of the Association of American Geographers* 54: 2–11. Is there a clearer scalar framework for the measurement of all manner of human and physical data, or for organizing a discipline around their incessant variations? We don't think so.

Buttimer, A. (1976) Grasping the dynamism of the lifeworld. *Annals of the Association of American Geographers* 66: 277–92. A classic theoretical call for the interpretation of people's everyday lives in the discipline of geography.

Gould, P., and White, R. (1974) *Mental Maps*. London: Penguin. Full of examples of mental maps, one of geography's earliest and most significant forays into interpretation.

Gregory, D. (1996) *The Geographical Imagination*. Oxford: Blackwell. A classic book in the representational-epistemological armory of the 1990s. Beautifully written, argued, and illustrated, this is a must read for anyone interested in the cultural turn in geography.

Haggett, P. (1995) *The Geographer's Art*. Oxford: Blackwell. A gem full of discoveries by a multi-dimensional lover of geography.

Hubbard, P., Kitchin, R., Bartley, B., and Fuller, D. (2002). *Thinking Geographically: Space, Theory and Contemporary Human Geography*. New York: Continuum. Here's a good introduction to contemporary theories in geography. Chapter 5, "Geographies of Text," provides additional support for some of the materials developed in this chapter.

Natter, W., and Jones III, J. P. (1993) Pets or meat? Class, ideology and space in *Roger and Me*. *Antipode* 25: 140–58. Michael Moore's first celebrated documentary – on the social and economic effects of automobile plant closures in Michigan during the 1980s – remains surprisingly fresh after all these years. This early piece in filmic analysis in geography focuses on the spatial and ideological cleavages exposed in the documentary.

## Exercise 4.1   Measure and Interpret Your Friends!

This is a "mixed method" exercise. Collect together a half dozen of your friends, roommates or classmates. Create a data table for them in the style of Table 4.1. Include these variables: age, gender, number of years in current place of residence (i.e., in current town or city), number of years with car in current place of residence, current distance traveled to school, current distance traveled to work (if not applicable, enter 0). For all variables except gender, calculate the averages of these measures. Describe their variability in rough terms: which variables tend to have the most widely dispersed values, and which variables tend to be more clustered?

Now ask each of your respondents to draw a "mental map" of their town or city. The map must be drawn on a regular sheet of paper, and they have five minutes to complete it. They should give as detailed and as accurate a map as they can in the allotted time. Once they are done, do a content analysis of the map. First, score each map in terms of the number and types of structures shown (entertainment, retail, other businesses; institutions and schools; homes and apartments); count the number of streets, parks, intersections, and major landmarks, as well as the number of external orientations provided (compass points, directions to other cities, surrounding mountains, etc.). Second, on the basis of this analysis, produce an ordinal scale value for the quantity and quality of information provided on the map (e.g., rank order as "poor," "fair," "good," "excellent").

Now compare your quality and quantity values to the other measurements you have collected. Do you see any associations between the quantity and quality of the mental maps with age, gender, years in residence, length of time with automobile access, length of journey to school and to work? How else might you account for the varying level of information on the maps? Are any of the respondents geographers? Did they do better on the exercise than those without geographic training?

# Chapter 5

# Operational Decisions

*Andrew Herod and Kathleen C. Parker*

Keywords

Abduction
Deduction
Epistemology
Extensive research
Hermeneutics
Induction
Intensive research
Kantianism
Leibnizianism
Newtonianism

Ontology
Phenomenology
Positivism
Realism
The scientific method
Theory-determined
Theory-laden
Theory-neutral
Verstehen

# Introduction

Putting a plan of research into action is not a straightforward task, for this is the point at which theory (conceptions of how the world is; see the section on "Ontology," Chapter 2) meets practice (efforts to validate or challenge those conceptions). Bearing this in mind, in this chapter we address six crucial matters that all researchers (whether human or physical geographers) must consider as they begin the process of empirical investigation:

- What is the purpose of the research?
- How do we understand our objects of analysis?
- How do we operationalize the research process (for example, do we use a case study approach)?
- What is the relationship between the questions we pose and the answers we generate through research?
- What implications are there of using different types of data collection?
- What value might there be in comparative approaches to research?

These questions do not stand alone and cannot be posed or answered in a conceptual or philosophical vacuum. Indeed, although we might try to avoid issues of philosophy and simply "get on with the research," good research cannot be conducted in a philosophical or epistemological void. In fact, given that research is a process by which we make claims about how the world is, any such claims can only make sense when interpreted through particular epistemological or philosophical lenses; that is, what we understand to be a "causal relationship" or a "valid statement" is shaped as much by the epistemological framework within which we work as it is by any empirical evidence we may discover "in the field." (The term **epistemology** refers to the frameworks of knowledge through which we come to know the world; for example, "positivism," "realism," "idealism," etc. Thus, epistemology addresses the nature of knowledge, including its presuppositions and what leads us to believe certain statements are "valid" or "truthful." This contrasts with, but is related to, the term **ontology**, which is concerned with the nature of "being" and what objects actually exist (are real).)

Before considering the aforementioned matters further, two key issues need to be outlined. First, in broad terms there are two quite different approaches to understanding the world: one that argues that research is about "revealing" its underlying nature, and one that maintains that research "constructs" the world through the narratives it produces. Researchers adopting the former approach assume that there is an objective world "out there" and that the job of research is to "disclose" this world to us. To do so, they might use quantitative and/or qualitative methodologies and may draw upon different epistemologies, such as **positivism** or **realism** (see Chapters 2 and 3). Through engaging in research, they claim to be getting closer to the ultimate "truth" of how the world is structured. For their part, researchers adopting the latter, "constructivist" approach equally may employ quantitative and/or qualitative methods and draw upon different epistemologies. However, where they differ is that they argue that one can never know for certain how close one has come to "the truth" of how the world works, since to do so would require knowing what the truth is before starting the research – a case of putting the cart before the horse. Consequently, researchers using this second approach tend not to make claims of having discovered "the truth" or that they have revealed the "real" nature of the world. Rather, they contend that they are constructing "narratives" about the world. Such narratives can be evaluated as being more or less coherent and/or more or less consistent with empirical observations of the world, but they make no claim to having taken us closer to some pre-existing (capital T) "Truth" about the world.

Second, different epistemologies draw upon different knowledge bases when making claims about the world. For example, positivism is based upon the premise that the only valid statements a researcher can make about the world derive from direct observation. Conversely, realism is based upon the premise that not everything that shapes how the world works is necessarily observable. Thus, while the effects of exercising political power may be observable, power itself is not directly observable. Likewise, humanistic approaches contend that, rather than relying upon observation, "experience" and "understanding" (***verstehen*** – understanding the intention and context of human action) can serve as valid bases for creating knowledge. Hence, we may know something intuitively and create interpretations of the world based upon that knowledge, even if we cannot necessarily observe and test it (as when we "know" what it is like to be "happy" or "sad," without being able to observe empirically the variables that would determine which of these states of feeling we are experiencing). These issues are important because they cut to the heart of debates in the physical and social sciences about the bases for making valid claims about the world, and whether physical and social sciences require different methodologies for understanding the world or, conversely, whether there is a unifying methodology to knowledge production, such as **the scientific method** (see Box 5.1).

## What is the Purpose of Research?

At first glance, this question might seem a strange one, for most of us probably assume that the purpose of research is to tell us about the world around us. Certainly, that is an important function of research, but a little more thought reveals that there are other aspects to this question that deserve consideration and that can dramatically affect how research questions are chosen and operationalized. For example, some traditions within geography argue that the purpose of research is only to interpret the world to discover its inherent truths, although this instantly raises the question of whether the way we go about interpreting the world impacts what we think we can say about it and whether we can, in fact, provide interpretations that are simply "naïve reflections" of the way the world is. Other traditions have argued for the practice of what is called "emancipatory science"; that is, they argue that the goal of research should not simply be to produce "value-free" knowledge that passes no moral or political judgments but, rather, to produce knowledge that is designed to transform the extant conditions of the world. This latter perspective is perhaps most famously expressed in Karl Marx's (1845) *Theses on Feuerbach*, wherein he stated that "The philosophers have only interpreted the world in various ways; the point, however, is to change it" (Ludwig Feuerbach (1804–72) was a German philosopher). Critics of this position have argued that it simply reduces science to the level of ideology, and that there needs to be a separation between "facts" and "values." Its defenders, however, contend that failing to take an advocacy stand does not imply that one remains unbiased

## Box 5.1   The Scientific Method

The scientific method is a process by which we acquire new knowledge and refine our existing understanding about a **phenomenon** we observe. It involves observation, use of existing knowledge to develop hypotheses to explain the observed phenomenon, experimentation to test those hypotheses, and careful analysis of the data gathered during experimentation to reach valid conclusions. Because the scientific method is considered a means for developing knowledge, an integral component of the procedure is the communication of results to other researchers so that they can perform independent tests of the hypotheses through further experimentation. Therefore, the scientific method is an iterative approach in which our understanding of phenomena and the underlying causative relationships are continuously tested and refined. Thus, for example, a biogeographer studying cactus population dynamics might observe the absence of seedlings in a locale. Noticing an abundance of rodent burrows and knowing the tendency for rodents to browse on young cacti, the researcher may hypothesize that the absence of cactus recruitment is caused by rodents, whose browsing prevents seedlings from becoming established. The researcher may test this hypothesis by conducting a census of the rodent population to determine the population pressure (hence potential browsing pressure) and by establishing both control and experimental rodent-exclosure plots that are, in all other respects, identical to each other. If rodent populations are high and cactus seedlings are able to become established in the exclosure plots, but not in the control plots, the researcher could logically conclude that browsing by rodents is preventing cactus regeneration.

and neutral but, rather, that one is committed, whether consciously or not, to the status quo. For the defenders of emancipatory science, then, one can be both objective, in the sense that research (whether in human or physical geography) is designed to reveal the world as it really is, *and* partisan, in the sense that knowledge is produced with the intent of acting effectively to change the world.

A second concern revolves around the issue of whether research is designed to look for spatial patterns in phenomena, or whether it is designed to understand matters of causality. This distinction harkens back to the philosophical question of whether or not it is possible to claim that a particular action *caused* another. Thus, for most positivists, "action A" is said to cause "event B" if "event B" occurs with some frequency after "action A"; this is termed the "regularity" approach to causality. However, for realists, even if it occurs with some regularity, just because "event B" occurs after "action A" does not mean one can claim "action A" *caused* "event B." Likewise, the absence of a causal relation between "action A" and "event B" cannot be assumed just because "event B" does not occur after "action A," for there may be countervailing forces preventing "event B" from occurring (for example, while under normal circumstances striking a match will cause it to burn, lack of oxygen [a countervailing force] means it will fail to light). In the geographical realm, this means that whereas geographers influenced by positivist epistemology might contend

**Box 5.2**   Equifinality

Equifinality occurs within systems when a certain outcome can result from several different sets of initial conditions, or through the action of different processes. Hence, in geomorphology, similar landforms may develop under contrasting environmental parameters whereas in human systems diverse suites of historical influences may lead to the emergence of similar agricultural practices among traditional farmers. In cases where equifinality arises, teasing apart the subtle distinctions among scenarios becomes a major challenge to understanding causative factors generating the observed phenomenon. This is exemplified by the debate that surrounds global warming, which arises, at least in part, because a number of different processes, both natural and human-induced, can lead to increased planetary temperatures.

that research to find causal relationships involves looking for spatial regularities in particular phenomena, those influenced by realism argue that just because two phenomena are coincident in space does not imply that they are causally related, while their lack of spatial coincidence does not mean they are not causally related. In other words, depending upon the contexts within which they occur, similar processes may produce different outcomes/ geographic patterns, and similar outcomes/geographic patterns may be produced by different processes (see Box 5.2). Consequently, for realists the starting point has to be the search for causal mechanisms; that is, their focus is on processes and structures rather than on outcomes and patterns.

## Objects of Analysis

How we conceive of the objects we intend to study (landforms, cities, ecosystems, the location of corporate headquarters, etc.) can have a significant impact upon the research process and its outcome. For instance, Hacking (2002) has shown that how researchers classify individual humans can significantly shape their behavior and that, in turn, this behavior can challenge such classifications. Hence, classifying someone as a criminal may lead them to act in a criminal manner, while their behavior pattern may sometimes cause us to reconsider and redefine the category "criminal." Such considerations in-and-of themselves relate to fundamental ontological questions about the nature of objects of study, and **hermeneutical** questions about the meaning we ascribe to particular objects. For example, are the objects of study discrete entities that exist independently of each other, such as alpine glaciers and mountains? Or do they exist only in relation to other objects, in which case separating them from these other objects or their surroundings can transform their very nature? (Slaves, for example, can only be slaves as a result of their relationship with another object, namely a slave-owner – the one cannot exist without the other.) Furthermore, at the level of meaning (rather than existence), our understanding of particular objects, such as slaves and alpine glaciers, may only have significance because of our ability to contrast and connect them with other objects, such as slave-owners and mountains. It is important to note, however, that sometimes these two levels, that of existence (the "ontological" level) and that of meaning (the "hermeneutic" level), coincide. Thus, slaves can only exist if slave-owners do (the ontological level), while our understanding of what a slave is only makes sense if we have a concept of what a slave-owner is with which it can be compared (the hermeneutical level). On other occasions, however, the ontological and the hermeneutical levels operate separately; thus, alpine glaciers and mountains may exist independently of each other, although our conception of the one may help us make sense of the other.

Such questions may appear esoteric, but they have great significance for how research is conducted and especially for the practice of abstraction, which refers to the manner in which we conceptually isolate particular objects of study so that we may better understand the causal relationships between them by filtering out what are considered to be "contextual" factors. Thus, abstracting should produce what Sayer (1984) has called "rational abstractions," that is abstractions in which the causal linkages between objects are retained, rather than "chaotic conceptions" (i.e., abstractions in which causal linkages are broken and/or unrelated objects/phenomena are associated with one another). The only way to know whether one has identified rational abstractions or chaotic conceptions is through empirical investigation. For instance, "the service sector," which is often an object of analysis, is really a chaotic conception, because it places in a single category ("the service sector") many different types of activities (serving hamburgers, managing financial information for a large bank, being a personal assistant to a union boss) that have little in common. In this way, explanations of decreased union activity that point to the growth of the "service sector" may miss entirely the intricacies of causal chains that would be maintained through rational abstractions (e.g., whereas fast food employers are generally antiunion, and hence their employees are less able and likely to engage in traditional union politics, this is less likely the case for the personal assistant, who is also in the service sector). Likewise, seeking to explain the cause(s) of population decline in neotropical migratory birds by looking only at factors that affect their breeding grounds in North America, and ignoring processes (such as deforestation) that impact their wintering grounds south of the Tropic of Cancer, conceptually isolates the wintering and breeding environments. Neither of the latter two approaches provides rational abstractions because the service sector lumps together different objects and activities into a single category, and the connection that migration creates between the different environments is ignored. Such considerations are important for two fundamental reasons.

First, they influence whether we focus our analyses upon objects in-and-of-themselves, or whether we seek to understand such objects in relation to other objects (and, if so, which ones). For instance, if we wished to understand why particular countries are "underdeveloped," we might decide to focus our attention upon the characteristics of the countries themselves, by looking at how their economies or political systems are organized, or their rates of population growth. However, such an approach fails to recognize that the ways in which such countries fit into the broader global economy can have a significant impact upon their mode of economic development, a shortcoming that may lead to different sets of conclusions. Focusing attention upon the characteristics of the countries, then, locates the cause of their lack of development within the countries themselves (e.g., a high birth rate), whereas focusing on the connections these countries have with the broader global economy shifts the explanation to other factors, such as how global commodity markets operate or how colonial practices underdeveloped their economies. Equally, to understand the cause of gully incision at a given location, we might decide to examine the physical characteristics of the underlying hillslopes and the processes that act on them. But while this approach may provide insight into the site-specific aspects of gully formation, it does not account for large-scale phenomena, such as land use or climate change, that may have also influenced gully development. Significantly, then, in both human and physical geography the cause(s) of locally manifested conditions may lie outside the local area, and we should be open to this possibility in our research design.

Second, these considerations have implications for how we conceptualize the spatial integrity of human and physical landscapes. Three primary ways of conceptualizing space have influenced geographic thought and shaped ideas about how landscapes are structured:

1   The Newtonian view, in which space (and time) are absolutes and objects move around in them (space is merely a container for social/natural objects).
2   The Leibnizian view, in which space and time are constituted through the interactions between objects (space does not exist prior to the social/natural relationships that bring it into being).
3   The Kantian view, in which space and time are simply mental frameworks for imposing order on the world (spatial structures are mental constructions) (see Curry 1996).

Once again, these may seem esoteric distinctions, but they too have significant implications for research. Hence, if we take a common geographical unit of analysis, such as the region, we can see that each approach conceives of regions in quite different ways. For those who follow a **Kantian** approach, regions are simply convenient mental devices for dividing up the world but do not exist in any ontological sense. For those following a **Newtonian** approach, regions are real but they serve as little more than spatial boxes within which social and natural processes unfold; that is, the process of region formation is not connected to the natural and social processes that occur within the region. Finally, for those who adopt a **Leibnizian** approach, regions are real things, created by social/natural processes; consequently their spatial resolution changes (regions may grow or contract in size) as the various social and natural processes constituting them unfold.

These distinctions also have methodological implications, for whereas both the Kantian and Newtonian conceptualizations allow for landscapes to be divided up analytically into smaller and smaller parts (space is seen as more or less infinitely divisible) or amalgamated together into larger and larger parts, the Leibnizian approach implies that a certain analytical coherence must be maintained and that arbitrarily dividing or amalgamating various parts of the landscape into "regions" does not result in sound abstractions. The result is that simply delineating an area on a map as one's "region of study" in accordance with the Kantian or Newtonian approach has the potential to create spatially chaotic conceptions, either by placing objects together that happen to have propinquity but which are unconnected by processes, or by separating places connected by processes that are located far from each other. Thus, an analysis of the "New York region's" financial geography that focuses solely on the northeastern United States will exclude places that are intimately connected to New York's financial institutions yet are physically distant (e.g., London), but it will include places that are physically close to New York yet which have little connection to its financial institutions. Likewise, characterizing the southwestern United States as forming a vegetative region because the pre-Columbian cultures there all used fire to manipulate botanical resources ignores both important contrasts among vegetation types that are caused by natural fires and spatial and temporal variations in the cultural practices of the people that inhabited the region before Europeans arrived. Of course, in some ways mapping such rationally abstracted regions necessitates that the cart be placed before the horse (how can we know what a rationally abstracted region is before we have done the research that shows us how places are connected?), but this is a question that speaks to the methodological problem of drawing boundaries rather than to the ontological question of whether regions really "exist."

## Case Studies?

When thinking about how to begin one's research geographers typically face the question of whether to use an **extensive** or **intensive** approach to data collection. This decision has implications for making broader claims about the world. For its part, extensive research requires a large sample of observations and seeks to identify patterns in the phenomenon of interest, often through the use of inferential statistics or numerical analysis (see Chapters 18 and 19). As such, repeated data sampling may be used to confirm or refine any patterns that are detected. While this approach may identify general trends, it has relatively weak explanatory power because neither correlation nor consistent association demonstrates causation. In contrast, the objective of intensive research is to determine how processes operate to produce an observed pattern. This approach typically relies on smaller samples, such as case studies, and may involve qualitative methods to characterize social processes or detailed measurements to uncover specific physical processes (Box 5.3).

Generally speaking, how "extensive" and "intensive" approaches are regarded is dependent upon the research model (epistemology) to which a researcher adheres. For those who contend that the goal of research is to make generalized statements about how the world works, cases studies serve as specific examples of the processes that operate at a broader scale, although these processes may not lend themselves to generalization because of (local) idiosyncrasies in the way they operate. Such a viewpoint is common among researchers who adopt a positivist epistemology, where the goal of research is to develop laws. However, for other researchers who, for example, follow a realist epistemology, case studies are not meant to be empirical generalizations. Rather, they are used to throw light upon the processes that operate at broader scales; that is, case studies are used to determine issues of causality and process, and are not intended to be representative of broader patterns.

This distinction, then, between whether or not using a case study approach serves as a valid basis for creating knowledge, has its origins in epistemological differences of opinion. Specifically, given that positivism understands causality to be demonstrated through the regular, temporal occurrence of events ("event B" always occurs after "action A"), case studies are seen to be too particular and not sufficiently replicable and/or generalizable to provide much insight into causal relationships between actions/processes and outcomes; that is, they simply do not represent a large enough sample of observations. For realists, on the other hand, who view causality not in terms of the regularity with which "event B"

**Box 5.3**    More on Correlation and Causation

In physical geography, Hack's law refers to the empirical relationship between the length of the main channel of a stream (L) and the area of its basin (A), and can be expressed as: $L = 1.4 A^{0.6}$. Although data collected over a broad area that encompasses substantial variation in L and A may fit the relationship, data points plotted on a graph often show appreciable scatter about the line representing Hack's law, particularly if streams have been sampled over a small area. Such empirically derived morphometric laws have been criticized because they are statistically based and merely demonstrate correlational relationships – as opposed to being based on an intensive analysis of physical processes that shape drainage networks, such as fluvial dynamics, sediment transport, and channel development. Furthermore, the relationship between stream length and basin area is often scale-dependent.

In human geography, an intensive approach may involve conducting interviews with corporate executives to understand the reasons for closing a particular manufacturing facility. Thus, whereas a researcher might conduct a correlation analysis for metropolitan areas that relates plant closure rates to wage and unionization levels, such an extensive approach will tell us little about the causal forces for specific plants. Consequently, we may have to conduct interviews with various corporate decision-makers to determine what factors led them to close their plant, factors which may be quite different from those affecting others (for instance, whereas most firms' respondents may have indicated in the survey that it was a downturn in the economy that led them to shutter their plant, the specific firm executives interviewed may indicate that their local plants were closed as a result of their company being purchased by a competitor).

occurs after "action A" but, rather, in terms of how "action A" and "event B" are connected through particular mechanisms, case studies are the only way to get beyond the statistical association between two events and into the nitty-gritty of causality.

The methodological starting points of these two approaches are also quite different. Thus, "[e]xtensive research focuses on *taxonomic* groups, that is groups whose members share similar (formal) attributes but which need not actually connect or interact with one another" (Sayer, 1984: 221). Accordingly, individual members "are only of interest in so far as they represent the population as a whole." For example, a researcher may examine the characteristics of a subset of university students, carefully replicating a university's student body in terms of sex, race, disability, age, etc., with the intent of having the subset say something about the broader population of students of which it is presumed representative. On the other hand, "[i]ntensive research focuses mainly ... on groups whose members may be either similar or different but which actually relate to each other structurally or causally." For example, the researcher may focus upon certain individuals, who may or may not be "representative" of the broader student population, because their experiences might inform us about particular aspects of student life, such as the challenges faced by disabled students. Thus, in the latter approach, "[s]pecific, identifiable individuals are of interest in terms of their properties and their mode of connection to others" rather than their representativeness, and "causality is analyzed by examining actual connections" rather than by "relying upon the ambiguous evidence of aggregate formal relations among taxonomic classes." These different objectives mean that in extensive studies the procedure for selecting the sample must be decided upon ahead of time and adhered to rigorously to ensure that the sample population is representative. By contrast, in intensive studies the individual members of research populations need not be "representative" and may be selected as the researcher proceeds, based upon what the unfolding research reveals about the nature of the causal relationship(s) between the objects of analysis. Neither approach is necessarily superior to the other; rather, the different approaches reveal different things about the world to us. Nor is the adoption of an "extensive" and "intensive" approach mutually exclusive; the different approaches can be complementary when used, for example, in an iterative manner at different stages in a research project to address related questions. So, for example, the discovery of statistically significant differences of opinion between white males and non-white females on a hot button political topic may be interesting to us, but it does not necessarily mean that those gender and "race" characteristics are directly causal in nature. The real causes might lie in the socio-economic contexts within which the survey respondents were raised, the patterns of socialization that took place at the family dinner table, or the different personal experiences they had while negotiating the world while being coded as "white male" or "African American female." Likewise, a biogeographer studying how genetic variation and structure in pines vary with different disturbance regimes might initially conduct a broad survey of pine populations across the species range to determine the population age, disturbance agent, and genetic variability of each. In addition to informing the research question, this extensive sampling may reveal a subset of populations that could then become the focus of more intensive data collection to determine within each population spatial patterns of genetic relatedness, genetic contrasts among different cohorts, the population age structure, and spatial patterns of recruitment relative to windthrow gaps and fire, thereby providing a more complete picture of the stand-scale processes that underlie the observed patterns of genetic variation.

## On the Relationship between Questions and Answers

A further consideration when seeking to implement a program of research is the matter of how questions shape the answers we get. This issue relates to how the theoretical constructs we use to interpret data affect the process of interpretation and whether, in order even to ask questions, we need to have some conception of a likely outcome. There are three different positions on this issue. First, some approaches to research (such as positivism) argue that research can and should be **theory-neutral**; that is, how we ask questions and understand what we find should be unaffected by the theoretical framework we adopt. Such approaches assume that the process of interpreting "the facts" is unproblematic and that everyone, regardless of their theoretical perspective, will interpret things unambiguously and in the same manner. However, although some scientists argue that it is possible to engage in unproblematic interpretations, they often fail to recognize how their claims to objectivity are tied to their own perceptions of "the facts." Thus, as Stephen J. Gould noted (1996: 106): "Science is rooted in creative interpretation. Numbers suggest, constrain, and refute; they do not, by themselves, specify the content of scientific theories. Theories are built upon the interpretation of numbers, and interpreters are often trapped by their own rhetoric. They believe in their own objectivity, and fail to discern the prejudice that leads them to one interpretation among many consistent with their numbers."

A second approach to this issue suggests that questioning and answering is **theory-determined**; that is, how we ask questions and understand what we find are determined by the theoretical framework we adopt. At the heart of this approach is the belief that mutually exclusive epistemologies structure the way we observe and interpret the world. It is important, though, to distinguish between the theory-determinance of meaning and of observation. Thus, the philosopher of science Paul Feyerabend explored how the study of meaning was theory-determined; the term "mass," for example, means different things in classical and relativistic mechanics. In contrast, Thomas Kuhn, who directed attention to the theory-determined nature of observation, argued that our theoretical framework actually determines how we see the world. If we accept the idea that observation is theory-determined, however, then we must also accept that it may not be possible to compare directly two or more opposing theories that seek to explain the same phenomena, nor for someone immersed in one particular theoretical approach to interpret the data generated by another (Longino 1990: 26–7).

With regard to the process of observation, the art critic John Berger (1972: 8) has argued that there are culturally-specific "ways of seeing":

> The way we see things is affected by what we know or what we believe. In the Middle Ages when people believed in the physical existence of Hell the sight of fire must have meant something quite different from what it means today. Nevertheless their idea of Hell owed a lot to the sight of fire consuming and the ashes remaining – as well as to their experience of the pain of burns.

Equally, how researchers have been trained to observe the world shapes how they see it. Hence, Sayer (1984: 51–2) argues that a biologist looking down a microscope may see quite different things than a layperson does; bacteria as opposed to blobs, for example. That is:

> The distinction between the observable and the unobservable is therefore not simply a function of the physical receptivity of our sense organs: it is also strongly influenced by the extent to which we take for granted and hence forget the concepts involved in perception.

The final approach views questioning and answering as **theory-laden**, and suggests that even though how we ask questions and understand what we find are shaped by the theoretical framework we adopt, they are not determined by it. To put this another way, such an approach argues that the theory we use to interpret a particular situation or set of "facts" certainly shapes how we view them and what we understand them to be, but it does not determine what we find. This must still be established empirically. Hence, as Sayer (1984: 68) states, "whereas a scientist's theoretical framework may shape what s/he understands money to be, in order to find out how much of it s/he has in their pockets, s/he must still actually look."

These issues might appear abstruse but they have significant implications for conducting research, because they relate to the matter of whether or not research conducted under one epistemological approach can "talk" to research conducted under another and what this means for the ability of researchers steeped in one tradition to evaluate the results of workers who adhere to another. For instance, if we believe that data interpretation can be conducted in a "theory-neutral" manner, then we are more likely to adopt a rhetorical position which claims that all data can be interpreted and made sense of using a single epistemological approach; that is, "the facts can speak for themselves" and, through these "facts," we can objectively evaluate competing theoretical positions. If, however, we subscribe to an epistemological position which contends that data can only be interpreted within the context of a particular theoretical framework (the theory-determined position), then it is more likely that we will adopt a rhetorical stance that dismisses objections with the retort that critics "simply do not understand" (or are incapable of understanding) the data in the same way the researcher does because the critics have adopted a different or "the wrong" theoretical framework. By way of contrast, adopting a position that accepts that the process of questioning and interpreting data is theory-driven will likely yield a rhetorical position in which different perspectives are seen as capable of talking to each other (at least to a degree), and in which different theories may be evaluated according to some common set of criteria (for example, how well they live up to their own predictions or how well they follow their own rules for validity).

## Approaches to Data Collection

To a significant degree the approach to data collection taken in the research process will itself depend on whether the researcher uses an inductive or deductive approach to knowledge production. These two approaches differ primarily in the degree of confidence one has about the premises that reinforce the conclusion(s) of the research. For their part, **deductive** approaches typically start with a theoretical framework in which premises are derived from definitions, laws, formal logic, or mathematical expressions. In a valid deductive argument, if the premises are true, the conclusion cannot be false. This is not the case with **inductive** approaches, in which premises involve a greater degree of uncertainty because although observations of previous events may allow us to identify regular patterns, which may serve as the basis for predicting future events, we cannot be certain that the next observation will be consistent with the previous observations. Hence, based upon past experience we can fairly confidently infer that the Sun will rise tomorrow, although one day in the future (when our Sun finally dies) we will be wrong. For this reason inductive research often relies on inferential statistics and probability to quantify the degree of certainty involved in a conclusion. One approach that seeks to move beyond such limitations is termed **abduction**. It involves searching for patterns in phenomena in order to suggest hypotheses that might reasonably explain new or unpredicted data. Developed by the nineteenth-century philosopher Charles Peirce, abduction is the process whereby a person develops a rational conjecture about what may be happening in a particular context, using a set of logical operations to determine a hypothesis, with the approach being continuous and cumulative. For example, from a Peircean perspective the shift from a view that perceived the Earth to be flat to one that saw it as round did not involve throwing out "old knowledge" but, rather, adding new observations until enough inconsistencies were found in the explanatory power of the flat-Earth perspective. These inconsistencies could lead to modifications to the existing perspective so that it can accommodate and explain the new facts, or a new perspective entirely (round-Earth) could be developed. Thus, as Senglaub et al. (2001: 11) have put it: "Abduction assumes the role of idea generation while deduction evaluates the hypotheses and induction provides the mechanisms for justifying a hypothesis with data."

As with the case of extensive and intensive research, many research questions can be addressed using either a deductive or an inductive approach, and neither is necessarily superior to the other. Rather, the approach used will reflect the existing theoretical framework and the researcher's perspective as much as it does the focus of the research topic. Such matters, however, do have important implications for how data are gathered. For example, researchers adopting an inductive approach may begin by "mining data" to determine which, if any, variables have associations with each other. This may sometimes provide interesting insights, but naïve data mining by researchers who regress every variable against every other one to see if any are correlated in statistically significant ways is generally a poor model for research, in part because it tends to substitute the "law of large numbers" for clarity of thought about causality. (This "law" suggests that if we look at enough variables the chances are that some will show a degree of correlation; in fact, it is quite likely that statistically significant ones will be revealed even when there is not a real correlation, much less a causal relation!) Such an approach may also reveal spurious correlations, which may have predictive power but little explanatory power. A classic in this regard is the infamous "Hemline Index" discovered by University of Pennsylvania economist George Taylor in the 1920s. He purported to show that the length of women's skirts is related to stock prices, a correlation that may allow one to predict (perhaps even with some degree of accuracy) market trends by observing women's fashion but that tells us little about how markets actually operate. Similarly, there are correlations between arm length and shirtsleeve length, and ones between ice cream sales and summertime shark attacks. If, however, we were to believe that such correlations constituted causal relationships, we would expect that wearing long-sleeved shirts would make our arms grow, or that sharks prefer feeding on people who eat ice cream.

Conversely, deductive approaches typically begin with a set of premises, with the goal being for the researcher to seek out specific data that either support or contradict a particular conceptual/theoretical position. These data may be drawn from the results of experiments, or from archival documents or interviews with individuals who might be expected to have particular insights on certain matters. This raises several important issues, namely: how are the data interpreted?; do researchers see only what they want to see so as to advance support for a particular position?; are the initial premises, in fact, valid?; and how does a researcher account for the unexpected (although in the context of this latter issue, the deductive and inductive approaches to knowledge production are both susceptible to similar limitations, though for somewhat different reasons)?

## Comparative Analysis

At some point in the process of preparing to conduct research, the issue of whether or not to engage in comparative analysis may present itself. Inevitably, opinions diverge about the value of comparative analysis and whether we should engage in it. For some, comparative analysis's goal is simply to illuminate the broader spatial patterns that are thought to reflect particular natural or social processes. In this sense, comparative research is designed to elucidate whether or not similar processes play out in geographically (spatially) different contexts, with the intent of determining whether or not specific processes or phenomena can be generalized so that some law-like statement can be made about them. Hence, after examining the controls on vegetation patterns on alluvial fans in semiarid environments, one might investigate fans in humid climates to determine whether the relationships that are apparent in deserts extend to other environments. Conversely, for others the goal of comparative research is not to determine how generalized particular processes or phenomena are but, rather, to use different sets of surroundings to provide insights on particular processes in the search for causality. Comparative research conducted in this tradition recognizes that similar patterns in different places may result from dissimilar processes and, likewise, that just because different places lack similar phenomena does not mean they are experiencing dissimilar processes (contextual factors may cause similar processes to play out differently in different places). Viewed from this standpoint, the purpose of comparative research, then, is to ascertain how the existence of similar and/or different phenomena/processes in different places serves to illuminate deeper phenomena/processes, such as the nature of capitalist accumulation or biological evolution. Thus, comparative research in this vein may reveal how two places at either end of a multinational corporation's supply-chain can experience quite different economic conditions, even though they are tied together through the same corporate structure, or how the process of evolution is shaped by different physical environments.

## Conclusion

This chapter has addressed several "epistemological," "philosophical," and "methodological" questions that are important to consider as one begins to think about operationalizing a research project. These philosophical issues cannot be separated from their methodological counterparts, for how we investigate and make sense of the world is deeply shaped by our understanding of "causality," "validity," "objectivity," and a host of other factors. Such questions are not just important for determining how a research project might be implemented, but also for evaluating and interpreting its results.

## References

Berger, J. (1972) *Ways of Seeing*. London: Penguin.

Curry, M. R. (1996) On space and spatial practice in contemporary geography. In *Concepts in Human Geography*, Earle, C., Mathewson, K., and Kenzer, M. S., eds. Lanham, MD: Rowman & Littlefield, 3–32.

Gould, S. J. (1996) *The Mismeasure of Man*. New York: Norton.

Hacking, I. (2002) *Mad Travelers: Reflections on the Reality of Transient Mental Illnesses*. Cambridge, MA: Harvard University Press.

Longino, H. E. (1990) *Science as Social Knowledge: Values and Objectivity in Scientific Inquiry*. Princeton: Princeton University Press.

Sayer, A. (1984) *Method in Social Science: A Realist Approach*. London: Hutchinson.

Senglaub, M., Harris, D., and Raybourn, E. M. (2001) *Foundations for Reasoning in Cognition-Based Computational Representations of Human Decision Making*. Albuquerque, NM: Sandia National Laboratories.

## Additional Resources

Baker, V. R. (1996) Hypotheses and geomorphological reasoning. In *The Scientific Nature of Geomorphology*, Rhoads, B. L., and Thorn, C. E., eds. New York: Wiley, 57–85. Examines the applicability of a philosophy of science based on physics to field-based disciplines, like geology, and discusses the development and testing of hypotheses in geomorphological inquiry.

Chalmers, A. F. (1999) *What Is This Thing Called Science?* Indianapolis: Hackett Publications. A clearly written introduction to the nature of science that has been widely read since it was first published in 1975.

Chorley, R. J. (1962) Geomorphology and general systems theory. *US Geological Survey, Professional Paper* 500B. This paper discusses systems theory in the context of geomorphology and emphasizes analysis of relationships of landforms to the broader systems of which they are a part.

Chorley, R. J. (1978) Bases for theory in geomorphology. In *Geomorphology: Present Problems and Future Prospect*, Bembleton, C., Brunsden, D., and Jones, D. K. C., eds. Oxford: Oxford University Press, 1–13. Discusses uses of systems theory in the physical sciences and summarizes several bases for the development of theory.

Frodeman, R. (1995) Geological reasoning: geology as an interpretive and historical science. *Geological Society of America Bulletin* 107: 960–8. Frodeman reviews the philosophy of science as it relates to geology and argues that geology has developed a set of logical properties (geological reasoning) that have enhanced its nature as an interpretative and historical science.

Giere, R. N. (1988) *Explaining Science: A Cognitive Approach*. Chicago: University of Chicago Press (Chapter 8, "Explaining the revolution in geology"). This chapter shows how the process of research and making scientific discoveries can be shaped dramatically by the social, geographical, and historical context within which researchers work.

Haines-Young, R., and Petch, J. (1983) Multiple working hypotheses: equifinality and the study of landforms. *Transactions of the Institute of British Geographers* 8: 458–66. The authors explore the concept of "equifinality" – the notion that similar landforms may result from different processes – and address the issue of the relationship between process and form, specifically with regard to matters of causality.

Keat, R., and Urry, J. (1982) *Social Theory as Science*. London: Routledge & Kegan Paul (Chapters 1–4, "Conceptions of science," "Positivist philosophy of science," "Realist philosophy of science," "Forms of conventionalism.") Four chapters that provide a comprehensive overview of these different epistemologies (and how they view the world) and a good introduction to the tenets of positivism and realism.

Massey, D. (1999) Space-time, "science" and the relationship between physical geography and human geography. *Transactions of the Institute of British Geographers* 24: 261–76. Massey argues that because geographers adopting a positivist epistemology have long viewed the field of physics as the appropriate model of science for research, they have incorrectly theorized time and space. Thus she advocates a different model of science and a different way to think about space, time, and what she calls "space-time." Doing so, she maintains, may reveal a commonality of research approach in physical and human geography.

O'Hear, A. (1989) *An Introduction to the Philosophy of Science*. Oxford: Clarendon Press. A comparatively jargon-free introductory treatment of the philosophy of science that explores the major debates about the nature of science.

Peet, R. (1998) *Modern Geographical Thought*. Oxford: Wiley-Blackwell (Chapter 2, "Existentialism, phenomenology, and humanistic geography"). This chapter explores some tenets of humanistic approaches to geography.

Phillips, J. D. (1999) Methodology, scale, and the field of dreams. *Annals of the Association of American Geographers* 89: 754–60. Phillips examines the relationship between method and results in physical geography, and specifically argues for a more sophisticated understanding of how scalar issues impact research.

Rhoads, B. L., and Thorn, C. (1994) Contemporary philosophical perspectives on physical geography with emphasis on geomorphology. *Geographical Review* 84: 90–101. The authors argue for philosophical awareness among physical geographers, and stress the value of adopting methodological diversity. They suggest that knowledge of the tenets of social constructivism, postpositivist empiricism, and scientific realism (philosophical approaches typically thought of as being associated with human geography) can be beneficial to physical geographers because they challenge traditional assumptions about how we interpret the world.

Sayer, A. (1982) Explanation in economic geography. *Progress in Human Geography* 6: 68–88. This article discusses the implications of adopting a realist approach in social science inquiry and addresses the notions of chaotic conceptions, rational abstractions, and models of causality.

## Exercise 5.1   Making Operational Decisions

A good way to become familiar with the issues this chapter addresses is to design a research project. This exercise provides an opportunity for you to work in groups (of two to four) to design a research project that explores the geography of a particular process or phenomenon in the community where your college or university is located. This may involve analyzing its physical or its human geography (or both). Doing this will help you learn to formulate clear research questions; understand the role that an investigator's epistemological framework plays in shaping his/her methodology; and learn to develop a specific methodological plan for data collection and analysis that directly addresses the research question and permits valid conclusions to be drawn about it. Specifically, you should develop a plan for investigating your community using at least two different epistemological frameworks (perhaps a positivist and realist approach; see Chapters 2 and 3).

The first step is to articulate clearly the question you wish to answer. The question should be sufficiently specific to offer guidance as you develop your methodology and make decisions about the type of data you seek to acquire. For example, if you were conducting research into your community's environmental problems, what specific sets of questions would you want to ask? "We want to look at environmental problems in our community" is *not* a research question. "Does our community experience poorer environmental quality than that of the nation as a whole?" *is* a research question, though it leads to a number of subsidiary questions: How might we measure environmental quality locally and nationally? Are different measures of environmental quality equivalent and, if not, how should we weigh them? How might we determine the factors that lead to poorer environmental quality?

The second step is to formulate a plan for collecting information that addresses your research question. As you begin to think about your research project, consider some of the advantages and disadvantages of the different epistemologies you have chosen, how you might need different types of data to answer different types of questions, and what sources of data may be available to you. Depending upon your research question, you may have to use: archival sources (such as maps, historical documents, or perhaps personal letters that speak of historical impacts on the environment, the locations of particular industrial facilities, or the former positions of rivers and streams); data you collect yourself through experimentation and/or observation (such as taking samples to measure water or air pollution); or data from government sources (such as the census or the national environmental protection agency: see Chapter 11). If, on the other hand, you are conducting research into, say, whether people feel safe in different parts of the community, you might need to use quite different data sources. These might include interviews with residents and local community leaders (politicians, community activists, assault center volunteers, etc.) to determine their perception of different neighborhoods (which would require you to think about how these people would be chosen and what you would ask them; see Chapter 12) and police statistics on crime (perhaps to compare people's perceptions with the actual geography of crime).

Finally, consider whether your data will truly address your research questions and provide clear answers. As simple as this might sound, if you do not pay attention to designing your research methodology, you may spend considerable time collecting data, only to find, upon analysis, that the data do not directly inform your research questions. If, for example, you are particularly interested in assessing whether the environmental quality of surrounding neighborhoods was physically changed by the construction of a landfill, you would ideally need to acquire information about the environmental variables of interest to you over a time sequence, and place your data within the broader context of other potentially relevant land-use changes that might also have occurred during the same time period. Limiting your data collection to the current time period and the immediate vicinity of the landfill might permit you to *describe* the current quality of the area around the site, but you would be unable to address any of the past changes in environmental quality (or land use) that have occurred. In short, if it is designed well your research project should answer clear, unambiguous questions posed by an investigator who understands the influence that the epistemological framework will have on his/her research. You should have a carefully developed methodology for data collection that permits conclusive research results to be obtained.

# Chapter 6

# Sampling Our World

*Ryan R. Jensen and J. Matthew Shumway*

- Introduction
- The Target Population
- Sampling Methods
- Determining the Sample Size
- Representing Geographic Data
- Secondary Data Sources
- Data Standards
- Errors and Accuracy
- Conclusion

Keywords

| | |
|---|---|
| Census | Random sample |
| Degree of variability | Sample |
| Inherent error | Sample size |
| Metadata | Sampling unit |
| Operational error | Stratified random sample |
| Population | Systematic sample |
| Purposive sampling | |

## Introduction

Geographers are motivated by a curiosity about the Earth's human and physical aspects. They have and will continue to contribute important theoretical ideas about how the world works, but to the extent that this work ultimately relies on information derived from observations, geography as a discipline has a strong empirical focus. As Gersmehl and Brown (1992: 78) suggest, the purpose of geographic observation is to "record the traits and locations of features with enough precision for valid pattern visualization and analysis." One of the most intellectually challenging facets of geographic research concerns the issue of how best to represent different aspects of the world. Representation commonly involves measurement, the process of assigning a number to different objects in a population (both human and non-human objects, including spatial partitions). Measurement functions as a conduit for information and helps link the empirical and the theoretical (see Chapter 4). Often, however, the **population** as a whole is much too large for us to obtain information about in its entirety and must necessarily be sampled. Even **censuses** – catalogs of entire populations – can contain **samples**. During the year 2000, for example, the US Census Bureau attempted a count of every person in every housing unit; each person was asked basic information such as age and gender. Information on educational attainment and length of residence, however, was derived from a sample of one-in-six persons and housing units. Sampling thus involves selecting a small piece of reality or a small number of objects from a population for specific investigation. This chapter summarizes the methods geographers use to sample the world, and discusses some of the inherent limitations in sampled data. The chapter also describes secondary data and some of its limitations as well as data standards and data errors.

## The Target Population

In keeping with the objectives of a research project (see Chapter 5), the first step in sampling is to identify the "target population" that is of interest to us and about which we wish to draw conclusions. For example, if we are interested in the narrative explanations that homeless people give in accounting for their predicament, then the target population might be homeless people. The target population should be sufficiently specific to permit each individual component to exhibit the same defining characteristics, and for this reason the target population may also be identified on the basis of location (whether, for example, we are interested in the cause of homelessness on a national, regional or local scale, or in an urban or rural setting), the time period involved, and other significant characteristics (such as age group or gender if we are interested in the causes offered by a specific population). Similarly, rather than attempting to calculate the sinuosity of all rivers in the United States, for example, a researcher might choose to focus on "alluvial rivers of the Great Plains" or "small mountainous rivers." There are no firm guidelines or rules for selecting the target population about which one wishes to draw conclusions. In most cases logic and experience dictate that it will be prescribed by the purpose of the research, existing literature, and how the researcher understands the objects of analysis (see Chapter 5), but constraints imposed by logistics, cost, and time also often place practical limits on how the target population is defined.

Once the target population has been defined it is necessary to identify the method by which access can be gained to it by delimiting a "sampling frame," or physical representation of the target population, which may take the form of a list that provides access to the target population. The list should enumerate all the members of the target population from which the sample is to be selected, but there are clearly problems associated with the completeness of coverage afforded by any list. Such a list simply may not exist, be difficult to compile, or exclude or under-represent some members of the population the researcher is interested in. For example, by definition, homeless people have no home, so it is not possible to locate them through postal codes, the electoral register or any other list based on street addresses. Also, if the telephone directory is used as a basis for identifying a target population, households without a traditional telephone or who have chosen not to have their phone number listed will be excluded. Normally lists should not be outdated, duplicate, misidentify, or over-represent components of the target population. Area sampling frames – those based on geographic partitions – provide indirect access to a target population because the individuals who reside within them (such as a census tract or block group) are identified in advance of the method used to access them. Thus a researcher who has a map of showing all the rivers crossing the Great Plains must still identify the alluvial ones, and from that select a sample. For this reason area and list sampling frames are sometimes utilized in conjunction with one another.

Having defined the target population and identified a sampling frame a researcher may additionally have to identify the **sampling unit** about which data will be collected and around which the sampling process will revolve. Sampling units are typically individual components of the sampling frame, such as households in a city or pebbles on a beach that have the potential to be clearly identified and unambiguously selected. In some cases the sampling unit will also contain a "respondent" who provides the necessary information, and a "unit of analysis" on which information is collected. Thus, a researcher evaluating the health of preschool children in an inner city housing development might find that a list of addresses is the sampling frame, the physical structure the sampling unit, a parent or guardian the respondent, and the child the unit of analysis.

## Sampling Methods

The next step is to select the method that will be used to obtain a representative sample of the target population from the sampling frame. In this context, the term "representative" implies that, except for random error, any information derived from sampling the target population will be the same as the information derived from a census of the target population. Random errors are unpredictable variations in magnitude and sign (that is they may be either positive or negative) that affect the reproducibility of results by imparting variability to individual measurements (but not to the average measurement for the group as a whole), and are often a product of inherent limitations in the equipment or technique(s) the researcher is using. In practice, because they are usually small and positive or negative in equal proportion, if it is assumed that the values in a large sample are dispersed according to the normal distribution (see Chapter 17), repeat measurements can be used to investigate their effect. In that case, since the mean,

$$\bar{x} = \frac{\sum_{i=1}^{n} x_i}{n} \tag{6.1}$$

is the most probable value of a repeat measurement (where $n$ is the number of measurements) and the standard deviation,

$$\sigma = \sqrt{\frac{\sum_{i=1}^{n}(x_i - \bar{x})^2}{n-1}} \tag{6.2}$$

is a measure of the dispersion of the repeat measurements about the mean, the standard deviation (or error) of the mean, $\bar{\sigma} = \frac{\sigma}{\sqrt{n}}$, which serves as a measure of the overall amount of uncertainty in the repeat measurements, decreases as the number of measurements increases (in proportion to the square root of the number of measurements). Simply put, even though it may not always be practicable, random errors can be reduced by repeated sampling and averaging over a large number of observations. Systematic errors, by contrast, cause an observation to consistently deviate from its true value by a fixed amount. They are commonly introduced by imperfections in the equipment or technique(s) a researcher is using. Because systematic errors often have a consistent sign and magnitude, they are often referred to as "sampling bias."

Bias introduced through the operational actions of a researcher who, for example, only selects people with blonde hair to interview or shiny black pebbles on a beach to collect (note that researchers are presumed to be competent *humans*, and mistakes, muddles, and lapses do not constitute bias). These can usually be minimized by adopting a standard procedure and training the researcher to adhere to it. Bias that is introduced by a device used to measure a property can be avoided by carefully calibrating, setting up and operating measurement devices and, in a more general sense (since questionnaires are a form of measurement device; see Chapter 12), by ensuring that, for example, respondents are not embarrassed, asked to hypothesize, given the opportunity to paint themselves in a better light, or prompted to express certain opinions. Most often, however, bias is introduced at the point when the target population is defined and/or the sampling method is selected. In the former case, as we have seen, this can be the result of over- or under-representing the target population in the sampling frame. For example, since they are part of the population of homeless people to which our research findings are to be applied, we could sample the homeless population of a city by interviewing the residents of shelters; but this approach will exclude any population subgroups that the shelters do not cater for as well as any homeless people who choose not to go to a shelter. Populations of interest to geographers also often vary in both space and time so that, for example, the circumstances that led to a person being a resident of a shelter for the homeless might be expected to vary from season to season and from place to place. Spatial and temporal variations also often occur in a systematic manner. For example, the pattern of scour and fill in a river channel as water discharge increases during a flood event is different in pools, which scour at high flows, and on riffles, which fill. Thus, a researcher sampling downstream of a pool, where the highest bed load transport rates occur prior to the flood peak, will observe a different relation between the bed load transport rate and water discharge during a flood event to that documented by a researcher sampling downstream of a riffle, where the highest bed load transport rates occur during the falling stage of the flood hydrograph (bed load is sediment that moves in almost continuous contact with the river bed).

The goal of all sampling is to obtain an unbiased (representative) sample of the target population, and the methods used to accomplish this fall into two categories: probability and non-probability sampling. The essential difference between the two methods is that in probability sampling every component has an equal (or known) opportunity of being selected, whereas in non-probability sampling every item does not. In the former case, because components are randomly selected from the target population it is also possible, for example, to accurately estimate the chance (probability) that a component will be included and how different the sample is from the target population. For this reason non-probability sampling methods typically are used to obtain descriptive comments about the sample itself rather than to make general comments about the target population as a whole; for example, to generate hypotheses in exploratory studies.

A basic assumption of non-probability sampling is that the characteristics about which a researcher wishes to enquire are evenly distributed throughout the target population; any sample of the target population can be considered representative if this is the case, leaving a researcher free to arbitrarily select components from the target population. A common method of non-probability sampling is convenience (haphazard or accidental) sampling in which components are drawn from the target population simply on the basis of their accessibility; such as the people at the front of a queue waiting to purchase tickets to, or the spectators seated in the front row at a sporting event. In most cases the advantage of convenience will be offset by the lack of information a researcher has about the representativeness of the sample, but this is not always the case if the target population is known to be homogeneous (in which case any of its components should yield similar information). For example, any water sample obtained from a reservoir in which the water is well mixed can be used to determine whether or not the water supply is contaminated. Other non-probability sampling methods are: volunteer sampling; snowball sampling; **purposive** or judgment sampling; and quota sampling. In the case of volunteer sampling no sampling frame is used and, as the name implies, the participants make all the decisions by volunteering to be components of the sample population. Snowball sampling is typically used when the target population consists of a hard to reach group, such as an ethnic minority, when the only way a researcher has of contacting its members is to ask those people who have been located (by whatever means) if they know of anyone else with the same characteristics. In the case of purposive sampling researchers use their judgment to select a sample they believe is representative of the target population, so that objectivity (or the lack of it) becomes a critical issue since a large bias can be introduced if a researcher's preconceptions about the target population are inaccurate and/or are at the heart of the selection process. Maximum variation (diversity or heterogeneity) sampling is a variant of purposive sampling that attempts to represent all the extremes of the target population. The assumption is that, if the components of the sample population are deliberately selected because they are as different from one another as possible, the aggregate information obtained will be typical of that gained from the target population as a whole. Quota sampling attempts to be even more representative of the target population by replicating its structure and sampling a predefined number (quota) of components in mutually exclusive sub-groups that have been selected on the basis, in the case of people, for example, of age or gender, or lithology, in the case of beach pebbles. It remains that, since the characteristics of the target population about which a researcher wishes to be informed are assumed to be evenly distributed, the components of the sample population are selected arbitrarily (that is they do not have an equal or known chance of being selected) and so it is not possible to make statistical inferences about the target population from which they are drawn.

By contrast, probability sampling uses a random selection method to set up a process or procedure that assures the different units in the target population have equal probabilities of being selected. This eliminates bias and in so doing allows us to calculate each unit's probability of inclusion, determine errors and make inferences or general conclusions about the target population as a whole. Ideally the sample should measure and represent all of the variation present in the population. Random sampling with replacement occurs when the researcher places a measured observation back into the pool (or population) to be measured. This is done so that the probability of an observation being sampled remains constant. For example, assume a population of 100 individuals from which a researcher is to make 50 observations. If the researcher was conducting a simple random sample without replacement, the probability that any individual would be selected for measurement would initially be .01 (or 1 out of 100). However, by the 25th measurement, the probability of any individual being measured would be .013 (or 1 out of 75). By the 50th measurement, the probability increases to .02 (or 1 out of 50). Therefore, as you increase the number of observations – without replacing the measured individuals – the probability increases that a single individual will be selected for measurement. This problem is solved if you simply replace each measured individual in the population. **Random samples** can be generated in a large variety of ways. Spreadsheets, such as Microsoft Excel, and other software programs can generate random numbers given a set of parameters. For example, random numbers from a phone book can be generated using the number of pages, columns per page, and lines per column. GIS and other software can also generate random numbers or points in a user-defined area. One drawback to random sampling is that small pockets or areas of differences may not be sampled. In other words, there is a possibility that observations selected may be unrepresentative of the entire population (Burt and Barber 1996). This is especially a concern when the random sample is small, and there are very small pockets of extreme variation found within the population.

Other sampling designs help to minimize this effect by not allowing the composition of the sample to be left entirely to chance. **Systematic samples** are chosen according to a rule (every 5 km, for example). This kind of sampling involves systematically sampling an area or individuals. For example, assume you were using soil point sampling in a large field to determine average soil pH. If you generate simple random points to calculate your average you might not account for pH in all portions of the field. Conversely, a systematic collection of points (one point every 10 meters, for example) would help to ensure that all parts of the field are sampled. A stratified sample is implemented when researchers know that the population contains different sub-populations and he/she samples within each of these. Stratified sampling helps to ensure that all of the variation present in a group is measured in the sample. This is done first by determining any specific groups or sub-groups that may exhibit different characteristics than the rest of the population. For example, when conducting a political poll, it might be necessary to ensure that all socio-demographic groups (income, race, gender, etc.) are sampled to reflect the real opinion of the people. In a **stratified sample** the samples can be done either randomly or systematically within each sub-population.

Sampling with probability proportional to size is done to ensure that the probability of being selected is not dependent on the size of a group or sub-group being sampled. For example, assume you are collecting surveys from several cities and towns within a state. This scheme helps maintain the probability that if you are one of 500,000 people within a large city that you have the same probability of being selected if you were one of 500 people in small town. When there are natural groupings within a population, cluster sampling can be employed. Clusters are first defined within a population, and a sample of these clusters is collected. Cluster sampling is especially useful for marketing research when specific groups are of interest. In terms of efficiency, stratified samples are best and cluster samples are worst, with random samples somewhere in the middle of the two (Burt and Barber 1996).

Sometimes the sampling scheme can be broken into several stages. Multi-stage and multi-phase sampling usually take members of an original sample and sample these individuals more extensively. This enables researchers to more completely understand what they are studying. For example, assume you collected surveys from 100 individuals that described their feelings about global climate change. Of the 100 people initially surveyed, you randomly selected 25 of them for more in depth questioning. These additional 25 people surveyed may shed additional light on global warming.

When every individual in a sample carries the same weight, the sampling design is "self-weighted." Most sampling schemes use this kind of weighting. However, there may be times when different weights must be given to different individuals or observations within a sample. For example, assume that you are examining socio-demographic characteristics in different neighborhoods throughout a city. Depending on the number of individuals you are able to measure in each neighborhood, you may weigh some higher than others. Assume that you are not able to sample many households within a "wealthy" neighborhood, while you were able to sample many more households within a "less-than-wealthy" neighborhood. In this case, you might decide to weigh the "wealthy" neighborhood measurements more than those from the other neighborhood. Whenever a decision like this is made, you should always describe why you did what you did and hopefully ground your decisions in the literature.

## Determining the Sample Size

Another issue with samples is the number of observations necessary for a given study. The determination of the appropriate **sample size** is important because often the most costly (time and money) part of a project is sampling. Sample size is contingent on the amount of variation that exists in the population being studied, the actual size of the population, and the types of questions being asked. Other factors that determine the sample size for a given project are the level of precision required and the needed confidence level. Level of precision is the range where the true population parameter is thought to be, and it is sometimes referred to as "sampling error." Level of precision is often referred to as a percentage (as in samples of likely voters reported with "plus or minus 4 percentage points"). Confidence level is based on the central limit theorem that states that when a population is repeatedly sampled, the average of the samples is the true population parameter (see above). Also, these individual values (of each sample) are normally distributed about the mean of the population. Therefore 95% of the sample values will fall within two standard deviations of the population parameter. So, if a 95% confidence level is selected, it is assumed that 95 times out of 100 that the sample value will have the true population value within its precision. **Degree of variability** refers to the amount of variation present in the population. Generally speaking, the greater the variability in the population, the larger the sample size needs to be so that all of the variability is measured (Agresti and Finlay 2009; IFAS 2008).

After considering each of the aforementioned criteria, you can then begin to determine the sample size required for a study. This can be done through a census, a published table, or a formula. A census involves measuring all possible objects in a population. Censuses are usually not possible when the number of possible objects or individuals is large. However, for smaller populations – e.g., a college class or a church congregation, it is usually feasible to measure all individuals within the population. Published tables can also provide guidance for sample size. These tables are often calculated using common equations. For example, equation 6.3 can be used to guide the number of samples where the confidence level is 95% and the proportion in the population is 0.5:

$$n = \frac{N}{1 + Ne^2} \qquad (6.3)$$

where $n$ is the number of samples required to sample the population $N$ with precision $e$ (such as 3%, 5%, etc.; IFAS 2008.). Other equations have been developed to determine sample size for specific studies. For example, to estimate the number of sample points necessary to determine the accuracy of a classified map one can use equation 6.4 based on binomial probability theory:

$$n = \frac{z^2 pq}{E^2} \qquad (6.4)$$

where $p$ is the expected percent accuracy of the map, $q = 100\text{-}p$, $Z$ is the Z-score 95% confidence, and $E$ is the allowable error (Jensen 2005).

After deciding what your sample size should be, you may now begin sampling your population. While sampling sounds like a simple process (especially when you are conducting a "simple random sample") care needs to be taken to ensure that your sampling methods are consistent throughout the process. For example, after collecting the first 50 of 100 surveys you decide that you would like to omit one of the questions in a 20 question survey and modify two others. This is not good because the last 50 surveys would not be consistent with the first 50. This would make it very, very difficult to compare the respondents. If there is some methodological flaw in the survey then all 100 should be given the opportunity to complete the same survey.

## Representing Geographic Data

Geographic data typically comprise measurements of features on or near Earth's surface. A fundamental challenge is to decide what features need to be represented, in what form, and in what detail. For example, consider an urban landscape consisting of a myriad of heterogeneous objects, such as houses, roads, bridges, pavements, shops, offices, industrial buildings, etc., positioned in different physical settings (floodplains, hillsides, etc.). All of these human and natural features may be surveyed on the ground or documented using remote sensing technology (e.g., NSGIC 2006; see Chapter 10), but each typically will have to be characterized or classified before it can be represented and/or portrayed on a map (see Chapter 16). Real features such as buildings and city parks, for example, can be represented as polygons, but invented features that one cannot perceive when looking at the urban landscape, such as the lines that show the elevation of land, piezometeric (groundwater) and other "surfaces," may also need to be represented. Using the terminology applied in a GIS (see Chapter 22), spatial data consist of points, lines, and polygons (vector data) or a matrix of cells or pixels with identical dimensions (raster data), the base units being the geographic $(x, y)$ coordinates that define the object(s) of interest: coordinates are points; connected points are lines; and lines with the same beginning and end points are polygons or areas (similarly, one can think of raster data as regularly spaced points). Some components of the mosaic of features are omnipresent and vary continuously through space (and time), and our ability to measure and represent these variations enables us to understand their respective role(s) and relation(s) to one another and what happens as they change. To do this we may make measurements at discrete points, that may be selected randomly or dispersed at regular intervals along a line or transect, or within a prescribed area. Such measurements inevitably capture only a part of the real variation and often must be interpolated to create a continuous surface or delimit a feature, and can be made in the context of a spatial, temporal, or thematic frame of reference; to characterize variations that occur from place to place (such as the different soil types that occur within a given area), or from time to time (such as the amount of traffic at a given intersection at different hours of the day or on different days of the week), and the relationships that occur between different features (such as that between climate, soil type, and vegetation), respectively. For example, precipitation is usually measured at widely spaced locations (see Chapter 8), whereas measurements of land surface elevation made during a ground survey, or generated by photogrammetric and other remote sensing techniques are much more closely spaced (see Chapter 10), but these discrete point measurements must be interpolated to create a continuous surface.

Irrespective of what is being measured, information or **metadata** (that is, data about data) about location, time and attribution should accompany each measurement. This information is necessary to fix the measurement's position in a coordinate system by specifying, for example, the latitude, longitude and time it was made, and provide relevant details about particular characteristics of a feature, instrument or observing practice that may be required to remove inconsistencies that occur between measurements (Longley et al. 2005; Chang 2010). If metadata are lacking it may limit how a measurement can be used. Crucially, then, for a measurement to have general utility it is necessary to fix its position in space and time. When researchers collect primary data they should always note and record metadata so that other people and/or agencies can use the data with confidence.

## Secondary Data Sources

Data may be obtained through either through primary data collection or accessing secondary data sources. Researchers exert control over primary data collection, and primary data collection is often done for a specific purpose or project (see discussion above about samples). Secondary data sources include data that have already been acquired, processed, and/or analyzed in some fashion (see Chapter 11). These kinds of sources could be in the form of existing maps (digital or analog), databases, charts, and tables. Secondary sources are normally used when the cost of doing primary data collection becomes prohibitive or when the data are historical. While it is possible for a single researcher or a team of researchers to conduct a local survey or undertake local field work, a large cross-sectional census, a panel study, or a longitudinal data collection effort are typically funded by government sources and made available for numerous types of analysis. Generally, these kinds of data sources were not acquired for the specific purpose of a given study, so care must be taken in how these kinds of data are implemented and used. In every case, metadata should describe the genesis, processing, and analysis of the data. Metadata are extremely important with secondary data sources because they describe why the data were collected (e.g., for what purpose), how the data were measured (e.g., instruments used, calibration of the instruments, what kind of remote sensing classification), and how the data were processed and/or analyzed (e.g., thematic overlay, database analysis, statistical interpolation, radiometric calibration, etc.). In addition, any error estimations or accuracy assessments performed on the data are generally provided. Each of these data characteristics is extremely important and will help to determine if the data are appropriate to use in a given study.

When no metadata are provided, extreme care must be taken because of the lack of knowledge as to how the data were acquired, processed, and analyzed. Usually one can be relatively safe using secondary data produced by governmental organizations (such as the United States Geological Survey, United States Bureau of the Census, etc.) because these data must conform to uniform accuracy standards. More care must be taken when the secondary data were provided by agencies or companies that do not adhere to any accuracy standard.

## Data Standards

Data standards are important to maintain uniformity within a single and across multiple datasets. Many agencies and governmental organizations have set standards that data must conform to before being used for spatial analysis or distribution to the public. Ideally, data should be acquired in standardized ways so that the data can be used by the largest number of organizations or people. These standards ensure that spatial data have been checked for accuracy and can be used with confidence. In addition, standard values for land cover classes or other mapped phenomena should be used. For example, the United States Geological Survey (USGS) has a set standard of land cover classes at multiple scales. When classifying land via remote sensing techniques it is often best to adhere to these standard values because it allows a larger number of people or agencies to use and more easily interpret the data for comparisons over time and space.

It is not uncommon for different agencies or companies to replicate the same or very similar datasets, but this redundancy can waste time and money. Unfortunately, even though two datasets seem very similar, it is often not possible to share the data because their sampling and data collection protocols were not the same or were inadequate for another agency's needs. For example, a local planning agency may need a transportation infrastructure database. A state transportation department may need the exact same database. However, each of the agencies may digitize the roads with different protocols or different definitions of roads. Further, each agency may collect slightly different road attributes. Ideally, these two agencies could collaborate and determine a common set of rules and protocols so that each could benefit from their combined efforts. If multiple agencies, offices, or researchers require the same data it is often useful to share the data as this adds value to the data and prevents the wasting of time and money. For example, the USGS Land-Use classification system is used by many local and state governments as the base classes for any classification (Jensen 2005). By following standard regimes – like the USGS classification system – data become more transferable between agencies or organizations. This is also true when collecting attributes for features. Another example of data sharing and data requirements is described in Box 6.1.

---

### Box 6.1    Polls and Margins of Error

Polls are common during election years or to gauge public opinion at any time on virtually any issue. Margins of error are provided with these polls to determine whether there is an actual statistical difference between the groups. For example, assume that a recent poll declares that 46% of 500 people polled are planning to vote for the Democratic Party presidential candidate and 43% are planning to vote for the Republican candidate (with 11% undecided). However, the poll's margin of error is calculated to be 4.3% (95% confidence). Therefore, the two percentages of people planning to vote for the two candidates are not statistically different. Often, television news anchors and others declare that a poll like this one describes a "statistical dead heat." The first step to calculate a poll's margin of error is to determine its standard error using equation 6.5,

$$\sqrt{\frac{p(1-p)}{n}} \qquad\qquad (6.5)$$

where $p$ is usually conservatively 50%. Then, the level of confidence is determined (e.g., 90%, 95%, 99%, etc.), and a z-score is looked up in the table of area under a standard normal curve. This value (e.g., 1.645 for 90%; 1.96 for 95%; 2.575 for 99%) is then multiplied by the standard error to determine the margin of error. The equation can be simplified to the following for $90\% = \dfrac{0.82}{\sqrt{n}}$, $95\% = \dfrac{0.98}{\sqrt{n}}$, and $99\% = \dfrac{1.29}{\sqrt{n}}$. A poll's margin of error decreases as the number of respondents increases, and increases as the confidence level increases. Most polls use the 95% confidence level when determining margin of error.

## Errors and Accuracy

All data, whether collected by the researcher for a specific purpose (primary data) or acquired by the researcher from a separate data source (secondary data) usually contain some error. The accurate manipulation of data requires an assessment of the error and the level of uncertainty associated with the data. The types of errors and the degree of uncertainty vary with the type of data collected, how they were collected, and the original purpose for collecting the data. For example, when collecting Global Positioning System data one must take into account the location error associated with this data and the quality of the GPS unit. Without understanding the limitations of the data used in research, the likelihood of making incorrect conclusions increases.

With any type of measurement there are two principal and related questions that must be asked: first, how *valid* is our measurement or measuring instrument; and second, how *reliable* or accurate is it? Validity basically asks: "are we measuring what we think we are measuring?" And reliability (or accuracy) asks: "what is the relationship between our measurement and reality?" (Also see Chapter 4.) If a sensor on a satellite is incorrectly calibrated, then the data it streams back may be invalid and not reliable. Another issue with sample data is whether the sample adequately represents the population from which it is drawn. Let's assume that we want to measure the impact of religion on presidential voting patterns for the entire US and, for some odd quirk in how the sample data was collected, rural areas are over-sampled. Will our generalization to the entire population be accurate?

Error is introduced into data in two ways: **inherent error** and **operational error** (Kitchen and Tate 2000). Inherent error is associated with the collection process, such as in incorrectly calibrated instruments, poorly worded and structured surveys, operator error during digitization or geocoding, incorrectly entering attribute data into a database, or imprecise locational measurements. Operational error results from the manipulation of the data once they have been collected.

## Conclusion

Sampling is an important step in the process of doing geographic research. Whether research starts out as inductive or deductive, at some point we want to link together our ideas (theories) of how the world works with what we observe. Sampling is one of the crucial steps in this process. How we sample our data is also important because it influences how we interpret our observations, and what type of mathematical manipulations and statistical analyses are appropriate. Finally, sampling is a key component of the research process that moves us beyond the particular to the general because sampling enables us to set up standards for comparisons.

## References

Agresti, A., and Finlay, B. (2009) *Statistical Methods for the Social Sciences* (4th edn). Upper Saddle River, NJ: Prentice Hall.

Burt, J. E., and Barber, G. M. (1996) *Elementary Statistics for Geographers* (2nd edn). New York: Guilford Press.

Chang, K. (2010). *Introduction to Geographic Information Systems* (5th edn). New York: McGraw Hill.

Gersmehl, P. J., and Brown, D. (1992) Observation. In *Geography's Inner Worlds*, Abler, R., Marcus, M., and Olson, J., eds. New Brunswick, NJ: Rutgers University Press, 77–98.

Goodchild, M. F. (1995) Attribute accuracy in elements of spatial data quality. In *Elements of Spatial Data*, Goodchild, M. F., and Gopal, S. eds. New York: Elsevier Science Ltd, 191–206.

IFAS (2008) *Determining Sample Size*. University of Florida, Institute of Food and Agricultural Sciences. (http://edis.ifas.ufl.edu/PD006#TABLE_1) [accessed August 17, 2009].

Jensen, J. R. (2005) *Introductory Digital Image Processing: A Remote Sensing Perspective* (3rd edn). Upper Saddle River, NJ: Prentice Hall.

Kitchin R., and Tate, J. N. (2000) *Conducting Research into Human Geography: Theory, Methodology and Practice*. Harlow UK: Prentice Hall.

Longley, P. A., Goodchild, M. F., Maguire, D. J., and Rhind, D. W. (2005) *Geographic Information Systems and Science* (2nd edn). New York: John Wiley and Sons.

NSGIC (2006) *Digital Imagery for the Nation*. National States Geographic Information Council, Maryland. (http://www.fgdc.gov/participation/steering-committee/meeting-minutes/january-2006/Imagery_for_the_Nation_Flyer_NSGIC_011606_V14.pdf) [accessed December 14, 2009].

## Additional Resources

Ebdon, D. (1985) *Statistics in Geography*. Oxford: Blackwell. Another good introductory statistics book for geographers with additional information on sampling.

Gatrell, J. D., Bierly, G. D., and Jensen, R. R. (2005). *Research Design and Proposal Writing in Spatial Science*. Berlin: Springer. Spatial research project design and proposal writing are addressed in this book. In addition, several sample proposals are provided.

Manly, B. J. (1992) *The Design and Analysis of Research Studies*. Cambridge: Cambridge University Press. A more advanced book that describes many applied research studies.

Patten, M. L. (2007) *Understanding Research Methods: An Overview of Essentials* (6th edn). Glendale, CA: Pyrczak Publishing. A book of short chapters that describe basic statistics and simple research processes. The book includes a discussion on sampling.

Ripley. B. D. (2004) *Spatial Statistics*. Hoboken, NJ: Wiley. This advanced book focuses on spatial statistics – especially those used to generate surfaces.

Webster, R., and Oliver, M. A. (2001) *Geostatistics for Environmental Scientists*. New York: Wiley. A very good reference to more advanced geostatistics and geostatistical analysis – including spatial sampling.

## Exercise 6.1   Common Spatial Sampling

You will probably be called on to perform sampling throughout your college and professional careers. Below are two examples of the kinds of sampling that you may need.

1   Open Google Earth (or another mapping program) and type in the following coordinates: 40.096414 N 111.613927 W and then zoom to these coordinates (click on the magnifying glass icon). Assume that the agricultural field to the north of this subdivision is being considered by the City Council to be rezoned as Commercial in prepara-

tion for a big box retailer. You have been assigned to determine how people living in this neighborhood feel about this rezoning, and the Council has designed a survey to help you with your job. Of course, budgets are limited, and you do not have enough money and time for everyone living in the subdivision to complete the survey. How would you sample the people in this subdivision to determine how they feel about the potential rezoning?

2  Type in the following coordinates: 39.381851 N 87.430233 W and then click the magnifying glass. Assume that you have been charged with sampling soil pH throughout this field. What would be the best way to sample the field? What are some of the limitations of using other sampling strategies?

# Part II
# Collecting Data

# Part II

# Collecting Data

# Chapter 7

# Physical Landscapes

*Michael J. Crozier, Ulrike Hardenbicker,*
*and Basil Gomez*

- Introduction
- Form and Pattern
- Process–Form Relationships
- Conclusion

Keywords

| | |
|---|---|
| Angle of friction | Exogenic processes |
| Angle of repose | Landslide |
| Boundary shear stress | Magnitude–frequency relationships |
| Catena | Regolith |
| Complex behavior | Relaxation time |
| Denudation | Return period (recurrence interval) |
| Dynamic equilibrium | Shear strength |
| Endogenic processes | Threshold |
| Ergodic transfer | |

## Introduction

The physical landscape is the framework of hills, mountains, plains, valleys, rivers, coasts, and other land surface features which humans now inhabit and gain their livelihood from. Research into the physical landscape primarily is concerned with the evolution of these landforms (or components of the physical landscape); how they change both naturally and as a result of human activity, the rate at which those changes occur and the factors that control them. Investigations may focus on the geometry of the landforms themselves, the material(s) they are composed of, and/or the processes that shape them. The processes involved ultimately are driven by climate and tectonic forces. But these fundamental forces typically are modified by other (local) factors, such as vegetation cover and human activity, so that changes to landforms occur at different rates in space and time. Instruments and observations may provide direct insight into the rate and cause of recent change but long term change, which occurs over centuries or millennia, must be deciphered indirectly, by dating sediment deposits or erosion surfaces, or by deriving information from models that portray components of the physical landscape and the processes that change them. Increasingly, because of concern about the way in which human activities are impacting climate, attention is being directed to the prediction of change that might occur to the physical landscape in the future, as well as to understanding the changes that have occurred in the past. In either case, however, knowledge of the relationships that exist between form and process, and the factors that cause them to vary is required to explain how the physical landscape changes through time.

Using examples drawn primarily from studies of hillslopes and rivers, which are omni-present features of the physical landscape, this chapter introduces some approaches and techniques for obtaining and analyzing data about landforms and the processes that create them. First, ways of representing the geometry of the physical landscape, its forms, features, and patterns are discussed. Some of the physics that underpins geomorphological processes is then outlined (geomorphology is the science of the study of landforms). This provides a signpost to those parameters that must be measured in order to understand how different geomorphological processes function to create landforms. We shall see that, when observed through time, these processes often operate episodically as they overcome thresholds, and leave their imprint on diverse components of the physical landscape that, for example, range in scale from the fabric of sediment deposits to entire river systems.

## Form and Pattern

In its most elemental sense, the physical landscape can quantitatively be represented by fundamental parameters such as altitude (the variation of altitude across a given area constitutes relief), slope angle and aspect (the direction a slope faces), and latitude. Location on the Earth's surface is determined by "$x$" (northing or latitude), "$y$" (easting or longi-tude) and "$z$" (altitude) co-ordinates (see Chapter 19). Depending on the scale of the investigation and the resources available, various methods may be used to collect data on these fundamental "geospatial" parameters. They include "field surveys" undertaken using, for example, a theodolite, an electronic distance measuring (EDM) device, or a global positioning system (GPS), and remote sensing techniques (see Chapter 10) involving aerial and satellite imagery. Topographic maps and Digital Elevation Models (DEMs) are digests or representations of geospatial data that provide the basic information required for char-acterizing and classifying the morphology of the physical landscape in two or three dimen-sions. Information derived from a topographic map or DEM can be used to identify patterns that can be used to characterize and classify different types of terrain, and may also reveal the processes responsible for creating them, as is demonstrated by the uses to which the globally consistent digital elevation data collected by the Shuttle Radar Topography Mission (SRTM) have been put. Such terrain classifications may be based on detailed information about distinctive attributes of the physical landscape obtained through field survey (such as recurring patterns of topography, soils and vegetation), but more usually they rely on fundamental morphological information, such as whether a hillslope is convex, concave or rectilinear and the rate of change of plan or profile curvature. These attributes are important because they control the concentration and dispersal of water and sediment across the physical landscape (for example, the plan shape of hillslopes favors convergence of water and sediment in hollows and divergence on spurs) and influence the relationship between erosion processes, such as landsliding, and landscape evolution (Montgomery et al. 1998). Other topographic attributes of the physical landscape that are important to the movement and distribution of water include primary characteristics, such as the catchment or contributing area (the area from which rivers and streams collect and convey water towards a given point). These attributes influence the volume of water flow (runoff) from the land surface, and can be represented by indices, such as the topographic wetness index ($TWI = \ln(a \mathbin{/} b \tan \beta$ where: ln is the natural logarithm; $a$ is the local upslope contributing area; $b$ is the unit contour length; and $\beta$ is slope), that is a measure of the degree to which water can accumulate at a site; sites where the index is higher, due to a large specific catchment area ($a \mathbin{/} b$) or low slope, are more likely to be saturated with water than dry.

Geomorphological maps provide a representation of the features and forms that make up the physical landscape. Most geomorphological maps are developed for small areas at quite large scales (typically between 1:5,000 and 1:50,000), and they commonly contain information about terrain conditions that can be used in land use planning or for environmental impact, resource, or natural hazard assessments. Ground shape (morphology) is recorded through field survey, which is often supplemented by information derived from remote sensing (see Chapter 10), and the land surface is subdivided into planar units partitioned by gradual changes or sharp breaks in slope delineated by decorated lines and other symbols. The initial morphological map (see Figure 7.1) may be embellished with information about the origin and age of individual landforms, the nature of the materials they are composed of and the processes acting on them, so that the resulting geomorphological map can also be used to record information about natural resources and hazards and landform genesis. One way in which this can be done is to employ an interpretative scheme which uses key features of the physical landscape to predict other conditions and processes. Soil landscape modeling, for example, relates the "soil **catena**" (the sequence of different soils that occur down a hillslope) to slope form and process, while the "nine-unit land surface model" associates the topographic position of different components of the physical landscape, from drainage divide to river bed (see Figure 7.2), with the dominant processes that are presumed to operate in those positions (Dalrymple et al. 1968). Process domains that identify the areas in which certain processes can operate may also be characterized by terrain conditions and identified using digital elevation data (Montgomery and Dietrich 1992).

## Processes

If landforms represent the anatomy of the physical landscape, processes correspond to the physiology. Each depends on the other. The processes affecting the physical landscape are both endogenic and exogenic (driven by energy from within or received at Earth's surface,

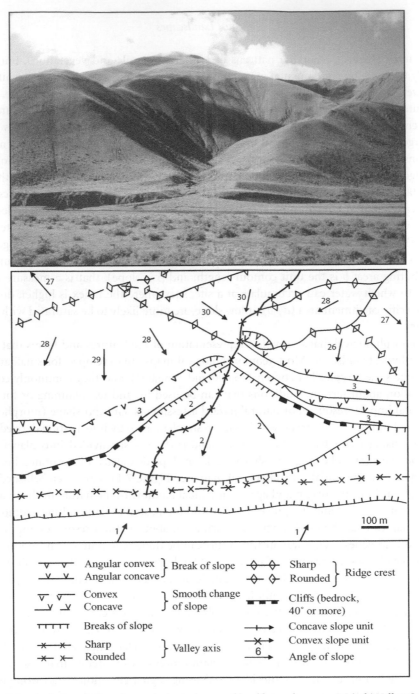

**Figure 7.1** A photograph and morphological map of landform elements, Waitaki Valley, South Island, New Zealand. In the foreground of the photograph is a gully and alluvial fan complex. The modern alluvial shingle fan is treated as one element and is separated from higher fan remnants by breaks of slope and cliffs. The higher and older fan remnants form terraces separated by risers demarcated by concave and convex breaks of slope. The steeper hillslope units are defined by summit ridges and spurs and concave breaks or changes in slope at their base. Mapping form elements in this way helps in identifying the history of landform development as, for example, indicated by fan remnants suggesting that through time conditions have alternated between those favoring fan formation and those favoring degradation. In this example fan degradation can be related both to periods of climatically driven reduction in forest cover and episodic tectonic uplift.

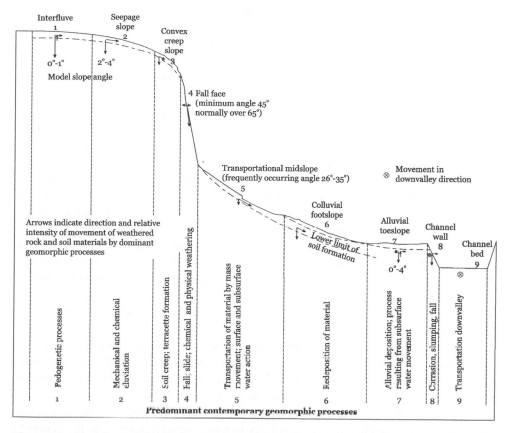

**Figure 7.2**   The nine unit landsurface model (after Dalrymple et al. 1968). The figure shows the hypothetical range of slope process-units and their sequential position within the slope altitudinal profile. Not all units will be represented in every landscape. Slope units are distinguished by the dominant processes, form and direction of water and sediment flux (whether into the slope, downslope or along the valley axis), and because they are also distinguishable on the basis of slope angle they can also be represented on morphological maps.

respectively). **Endogenic processes** cause large-scale deformation of Earth's crust by orogenesis (mountain building, which occurs in association with faulting, folding, metamorphism and plutonism along the boundaries of Earth's tectonic plates) and epeirogenesis (uplift or depression of Earth's crust that occurs in the absence of folding or faulting and, for example, produces plateaus and ocean basins). In combination with diagenesis and lithification (the chemical, physical, or biological changes sediment experiences after deposition, as it is compacted and becomes solid rock) these processes create the geological framework that resists exogenic processes. **Exogenic processes** encompass weathering (the *in situ* breakdown or disintegration of rock by physical and chemical processes); gravitational mass movements (such as rockfalls, **landslides**, and earthflows) and the processes of erosion (which involves both entrainment and transport of material by wind, liquid water and ice) that operate in different (for example, coastal, desert, glacial, periglacial and riverine) environments. The "effectiveness" of these processes (that is their ability to shape the physical landscape) depends on the resistance of the geological framework (as modified by weathering) and the "power" of the event concerned, and is governed by how much material is moved (magnitude) and at the rate this happens (frequency). Weathering itself involves the *in situ* breakdown of rock by mechanical processes, such as freeze–thaw activity, wetting and drying, heating and cooling and salt crustal growth (physical weathering), chemical decomposition, involving chemical reactions such as solution and oxidation (chemical weathering), and biological activity which may be mechanical (such as root growth) or chemical (organic acid production) in nature. Rock properties and climate are the principal influences on weathering, which acts on exposed surfaces to produce debris of all sizes, from boulders to clays and solutes. Physical and chemical weathering dominate in cold and warm, moist climates respectively. Weathering rates can be determined either by measuring the amount of alteration that has occurred or the amount of material that has been removed by mass movements and erosion processes. The development of the physical landscape can be viewed in terms of whether or not the rate of weathering exceeds the rate of material transport, that is whether the overall rate of **denudation** by the processes that remove relief is limited by the rate of rock weathering or by the rate at which the weathered material is removed (where the capacity to remove material exceeds weathering rates bare rock surfaces might be expected to predominate, whereas in the reverse situation the weathered material can accumulate on hillslopes and soil formation can take place).

A driving force is required to move material, be it the soil and rock fragments that mantle hillslopes or individual sediment particles on a stream bed. On hillslopes, this force or "shear stress" is due to the weight of the material that is directed down the hillslope, and is the product of the material's mass, the acceleration due to gravity and the sine of the slope angle. No movement occurs unless the resistance generated by the material in question is overcome. On hillslopes, resistance to movement, represented by the material's **shear strength**, is determined by the slope angle (and increases as the slope angle decreases), inter particle contacts (friction) and cohesion (due to molecular attraction). The steepest angle that cohesionless material can attain and still remain stable is termed the **angle of repose** or **angle of friction** which, depending on the size, shape, and consolidation of the material involved, varies between 30° and 35° for dry, unconsolidated sand and gravel, respectively. Mass movements occur when the slope is too steep for the material on it to resist the pull of gravity. Typically this occurs not because the slope angle changes, but because the material's resistance to movement is affected, for example, by the deterioration in root strength that occurs after vegetation is destroyed by humans or wildfires, or the increase in moisture content that occurs during intense rain storms. In the case of the flow of water in streams, the driving force may either be represented by the **boundary shear stress**, which is the product of the density of water, the acceleration due to gravity, water depth, and the slope of the water surface, or "unit stream power" (the amount of energy expended per unit area of the stream bed), which is the product of boundary shear stress and water velocity. Slopes are usually small in rivers, so the resisting force is simply a function of a sand or gravel particle's weight and the friction due to inter-particle contacts. There are similar expressions for the driving force due to the flow of ice and wind over the land surface and wave action on beaches.

Some of the parameters required to quantify the driving and resisting forces can be derived from high resolution DEMs, others are easy to measure in the "field," where data

**Box 7.1** Particle Size Analysis

Size is a fundamental physical property of sediment which can be used to describe, compare, or interpret different types of particulate matter (for example, the nomenclature for soil texture depends on particle size http://soils.usda.gov/technical/aids/ investigations/texture/). The techniques used for particle size analysis depend on the type of material being analyzed; thus, for example, the size of coarse gravel particles can be determined by direct measurement, that of fine gravel and sand by dry sieving, and the size of silt and clay by sedimentation analysis.

The technique usually applied to gravel particles involves measuring the length of the intermediate (*b*-) axis of a total of 100 particles, selected because they lie beneath the intersections of a sampling grid with a fixed spacing or are touched by the foot at each step taken in a pass across the sampling site. Dry sieving is a commonplace mechanical method of particle size analysis in which an air or oven dried sample (typically about 2 kg in weight) is placed on a nest of wire mesh screens, the openings in which decrease in size from top to bottom, stacked above a pan. The nest of sieves is agitated in a mechanical shaker for a few tens of minutes, and the weight of sediment retained on each sieve and in the pan recorded. Sedimentation analysis involves determining the diameter of a particle from its settling velocity (which according to Stokes' Law depends on the particle density and diameter and the fluid density and viscosity) in a water-filled column. The "pipette method" is an inexpensive but time consuming way (because clay particles take several tens of hours to settle) of doing this. Typically 20 g of sediment is suspended in 1000 ml of distilled water and allowed to settle in a column. Repeated sampling at a fixed time interval and depth below the water surface will entrap finer and finer particles, the diameter of which can be determined from the temporal variations in particle concentration in the samples. The process may be accelerated if changes in suspended sediment concentration at different depths down the column can be monitored as the particles settle and this is the principle instruments such as the "SediGraph" employ. Detailed descriptions of the standard laboratory procedures for particle size analysis can be found at http://pubs.usgs.gov/twri/twri5c1/pdf/TWRI_5-C1.pdf.

These techniques may be implemented individually or sequentially after a sample is split. They yield information about the size of individual particles from which statistics, such as the median particle size, that characterize the sample may be derived. Particle size data are presented as a histogram or cumulative frequency curve, the construction of which may be simplified if an equal-increment size scale, such as the phi, $\varphi$, scale (where: $\varphi = -\log_2 D$ (mm) $= -3.3219 \log_{10} D$ (mm), and $D$ is the specified particle diameter) is employed for the $x$-axis.

collection typically is accomplished using standard procedures, such as those developed by the US Geological Survey for measuring the discharge of a river (http://pubs.usgs.gov/twri/ twri3a8/pdf/TWRI_3-A8.pdf), or from "samples" analyzed using standard laboratory and computational procedures (see Box 7.1). It remains, however, that a considerable amount of time and effort must be expended to obtain such information, and for this reason many fundamental data are only available for specific points in the physical landscape. To deter-mine discharge (the volume of water flowing past a given location in a river system in a given time), for example, measurements must be made of the width, depth, and velocity of the water at many horizontal and vertical locations across the river and, because the water level (stage) changes over time, the discharge must be measured at many different stages before a relationship (rating curve) between stage and discharge can be established that permits the discharge at that location at any point in time to be calculated. The rating curve must also be updated periodically because, as the flowing water interacts with the material forming the channel boundaries, erosion and deposition of sediment changes the channel profile.

## Thresholds

Research to determine critical values (**thresholds**) for movement is common to all process studies over a wide range of scales and levels of abstraction. For example, rainfall intensity and duration can be analyzed to determine the values that trigger shallow landslides on soil mantled hillslopes. This may require empirical data, for example, information about when and where landslides occurred and the rainfall conditions at the time, so that condi-tions which produced no response can be separated from those that initiated landslides (see Figure 7.3). Alternatively, the force required to initiate motion can be deduced from physical principles. Thus, for example, the topographic wetness index can be modified to determine the relative saturation of the soil profile (the ratio of the thickness of the satu-rated soil, $h$, to the total thickness of the soil profile, $z$), which is a function of the steady-state (constant) rainfall, $R$, specific catchment area, soil transmissivity, $T$ (the rate at which water passes through the soil), and the local slope, $\theta$, such that $h \, / \, z = R \, a \, / \, bT \, \sin\theta$ (Montgomery et al. 1998). In the case of translational failures (slips) that occur along a single plane parallel to the soil surface, where the ratio of depth of the failure plane below the soil surface to the length of the failure plane is small ($\leq 10\%$), the criterion for slope failure is $\rho_s \, g \, z \sin\theta \cos\theta = C' + [\rho_s - (h \, / \, z) \, \rho_w] \, g \, z \cos^2\theta \tan\varphi$ where: $\rho_s$ is the density of the soil; $C'$ is the cohesion of the soil, including root strength; $\rho_w$ the density of water; and $\varphi$ is the angle of friction of the soil. Combining and rearranging the expressions for the relative saturation of the soil profile and the criterion for slope failure permits the critical steady-state rainfall, $R_{crit}$, required to initiate landslides to be specified:

$$R_{crit} = \frac{T \sin\theta}{(a/b)} \left[ \frac{C'}{\rho_w gz \cos^2\theta \tan\varphi} + \frac{\rho_s}{\rho_w} \left( 1 - \frac{\tan\theta}{\tan\varphi} \right) \right] \qquad (7.1)$$

Such thresholds are referred to as "extrinsic thresholds," that is, they are produced by the application of an external force. Note, however, because factors (such as the local geology) influence landsliding that neither the empirical data nor model framework account for, landslides can and do occur at times and in locations where the threshold for slope failure should not ordinarily be exceeded and *vice versa*. Such a situation may also arise because an "intrinsic threshold" has been crossed, that is, a change in the phenomenon of interest occurs in the absence of any change in the external forces. There are two categories of intrinsic threshold. An "internal threshold" may be exceeded when, for example, at some

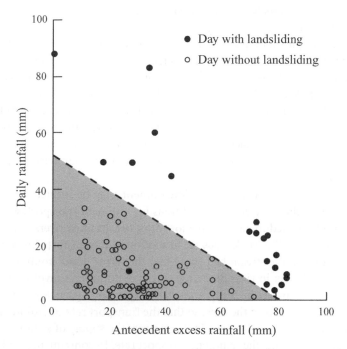

**Figure 7.3** Occurrence of landslides in Wellington, New Zealand, 1974. As a slope gets wetter the water content may become sufficient to trigger landslides (critical water content). This figure separates water content by source – rainfall on a given day (*y* – axis) and water that has accumulated in the slope of over a preceding time period (*x* – axis). In this case, accumulated water is represented by summing effective rainfall (rainfall minus evapotranspiration) that has occurred over ten days preceding a daily rainfall event (plotted as antecedent excess rainfall). When antecedent excess rainfall is high, relatively small rainfall events are capable of triggering landslides. One outlier on the graph is an exception to the distinct threshold separating landslide conditions from stable conditions and may represent a situation where leaking water pipes contributed to the water content of the slope (after Crozier 1999).

point in time a slope becomes unstable because weathering has reduced the shear strength of the materials it is composed of. Alternatively a "geomorphic threshold" can be exceeded if a change in the morphology of a landform leads to instability. For example, a cut-slope failure may be initiated when erosion undermines the base of a sea cliff or river bank and the overhang collapses. Human activities, such as road and building construction may also undercut slopes as may the retreat of glacial ice, on a longer timescale.

It is also possible to characterize thresholds on the basis of when and how processes operate (a temporal threshold), such as by determining the critical rainfall required to initiate landslides, or where processes operate and where they cannot (a spatial threshold). In the latter case a domain of process operation may be determined by specifying "preconditions". Thus, for example, a hillslope may be considered unstable if $\tan \theta \geq \tan \varphi = [C' / (\rho_s\, g\, z \cos^2 \theta)]$ and stable if $\tan \theta < \tan \varphi = [C' / (\rho_s\, g\, z \cos^2 \theta)] + [(1 - (\rho_w / \rho_s))]\, \tan \varphi$. Montgomery et al. (1998) show how the critical rainfall can be computed for locations with slopes between these two bounding preconditions using topographic data derived from digital elevation models and site-specific field measurements of soil properties (a description of the physically-based digital terrain model can be found at http://calm.geo. berkeley.edu/geomorph/shalstab/index.htm). Preconditions may also be determined by comparing the distribution of mapped landslides with the distribution of other factors, such as slope angle or lithology, that reflect the forces acting on or resistance of the landscape. Spatial thresholds are also recognized for fluvial erosion. For example, Montgomery and Dietrich (1992) showed that stream channels usually only occur on hillslopes when certain combinations of drainage area and valley gradient length are met (the smaller the drainage area the steeper the valley gradient must be for a stream channel to be initiated), and that there is therefore a topographic threshold between channeled and unchanneled regions of the landscape.

We have already seen how landscape development may be influenced by the rate of weathering. In rivers there are also preconditions for sediment transport, which are set by the size and amount of sediment available for transport. In most rivers, for example, only during large floods are all particle sizes set in motion and transported according to their relative proportions on the stream bed (note that the term "flood" routinely is applied to any storm-generated peak in river discharge and may not always involve overbank flow). During smaller events, by virtue of their greater weight, large particles may remain stable and shield small particles from the flow, so that the transport rate of sediment depends on both absolute and relative particle size. In many rivers the supply of sediment that is available for transport also limits the sediment transport rate. In consequence, only when there are no constraints on either the supply or amount of sediment available in the channel can the amount of coarse sediment moved by saltation or traction (as bed load) be directly related to the available stream power. Supply also exerts a strong influence on the amount of fine sediment that rivers transport in suspension. The suspended sediment concentration (mass of sediment per unit volume of water) fundamentally is determined by the river's ability to disperse the sediment through turbulence and the settling velocity of the sediment (which, for water of specified density and viscosity, is a function of particle size and shape). Suspended sediment concentration often is determined by sediment gaugings made at the same location (gauging station) as water discharge is measured, using a sampling device that is lowered down from the water surface to the bed and back up to the surface (descriptions of different sampling devices can be found at http://pubs.usgs.gov/ of/2005/1087/pdf/OFR_2005-1087.pdf). To obtain the suspended sediment discharge the concentration of sediment collected by the sampler (determined by filtering the sediment from the water, and measuring the volume of the former and the mass of the latter) is multiplied by the water discharge. Suspended sediment concentration may be expressed as a continuous function of water discharge by using concurrent measurements of suspended sediment concentration and water discharge made over a number of years to derive a sediment rating, which when it is combined with the water discharge record, can be used to determine a river's long-term suspended sediment yield. Like other relations that involve large variable ranges, the rating is usually derived by plotting concurrent measurements of suspended sediment concentration, $C_s$, against water discharge, $Q$, on a log–log graph (see Figure 7.4), and the underlying relation often exhibits a simple power form ($Cs = aQ^b$; where $a$ and $b$ are empirical coefficients). Scatter in the data is homoscedastic (that is, it is independent of discharge) and is a product of variations in sediment supply that in the short term may, for example, be attributed to differences in the rate at which sediment is

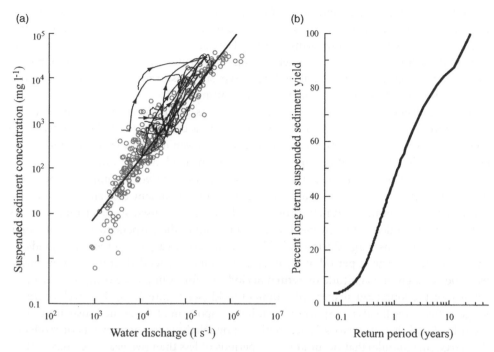

**Figure 7.4** **A** Relation between suspended sediment concentration and water discharge for the Waipaoa River, New Zealand. Open circles are individual gaugings obtained over a 40 year period, the thick solid line is a rating model with the simple power form $C_s = 0.001Q^{1.3}$, and the thin solid lines track the relation through individual runoff events. **B** Relation between the cumulative long-term suspended sediment yield and flood frequency. Compiled for a 23 year period of record. About 50% of the suspended sediment load of the Waipaoa River is transported by floods that recur at least once a year, and 86% is transported during floods with a return period of less than 10 years (after Hicks et al. 2004).

produced on hillslopes and delivered to stream channels during floods or throughout the year and, in the long term, to variations in climate and tectonics. Thus, continuous relations between suspended sediment concentration and water discharge for individual floods are highly variable and often characterized by hysteresis (that is two or more values of suspended sediment concentration are associated with a given discharge measured during the rising and falling stage of the flood, respectively). A discussion of the relationship between suspended sediment and water discharge and the methods used to characterize it can be found at http://water.usgs.gov/osw/techniques/OFR_87_218.pdf.

## Magnitude and frequency

Like most other geomorphological processes, the processes that produce sediment on hillslopes and deliver it to stream channels (including sheet and stream bank erosion and landsliding) do not operate continuously. Instead they operate episodically when given conditions occur and/or thresholds are crossed, and are associated with events of specific magnitude and frequency. Landsliding, for example, is a highly discontinuous and inherently episodic process that usually only occurs during large (high magnitude), comparatively rare (low frequency) rainstorms. Sheet and stream bank erosion, by contrast, typically contribute sediment to stream channels during rainstorms of all frequencies and magnitude (small, unexceptional as well as large, atypical rainstorms), and these ubiquitous erosion processes thus appear to operate more continuously.

The effectiveness of an event, or sequence of events, can be assessed not only in terms of the amount of change but also the amount of work accomplished. In the long term, the amount of work undertaken by a series of events of a given magnitude is a product of the effect that a discrete event of that magnitude has on the landscape and the frequency with which it recurs. Consider, for example, the amount of work a river accomplishes by transporting suspended sediment. Recall that the suspended sediment concentration may be expressed as a continuous function of water discharge and used to determine a river's long-term suspended sediment yield and, by integrating the suspended sediment rating curve with water discharge during individual floods, event suspended sediment yields. In this manner, over a given period of record, the amount of suspended sediment transported by floods of a given magnitude or **return period** can, for example, be computed as a percentage of the long term suspended sediment yield (see Figure 7.4B). Research on rivers suggests that large floods transport only a minor proportion of the annual suspended sediment load and that most work is accomplished by relatively frequent events of moderate discharge (magnitude) that occur at return periods of less than five years (see Box 7.2). In turn, the most effective discharge, that is the discharge that over a period of years transports the largest fraction of the annual suspended sediment load, has a relatively short (1.2 to 2.0) year recurrence interval and is also the flow responsible for determining the size and shape (morphology) of river channels. This may not always be the case though in situations where a threshold must be crossed before sediment is released, for example, by landsliding from hillslopes and delivered to stream channels, or in high-capacity stream channels that have an overabundance of suspended sediment.

In addition to using observational records, **magnitude–frequency** behavior may also be evaluated from inventories of events, such as landslides, complied from aerial photographs, or the record of hillslope erosion and large magnitude events preserved in sediment deposits. Studies performed using such data, which because they span longer periods of time than the instrumental record incorporate many more events, reinforce the notion that large systems with many components may to evolve into a poised, critical state, where minor disturbances induce events of all sizes. Evolution to the critical state occurs solely because of internal dynamical interactions among individual system components; that is, the critical state is self-organized (a system may be viewed as a structured set of objects and/or attributes, for example, a hillslope is an object and hillslope gradient an attribute). The classic example is that of a sand pile, which maintains a critical angle of repose and on which avalanches (events) of all sizes occur. For systems in a self-organized critical state, model calculations predict a power-law behavior. Landslide area mapped from aerial photographs has a power-law magnitude versus frequency distribution, as do magnitude–frequency distributions of lacustrine sediments attributed to landsliding and turbidites (sediment deposits created by underwater avalanches) that were deposited over thousands and millions of years, respectively. Self-organized criticality is an appealing frame of reference, because for systems to evolve into a self-organized critical dynamical state there must

## Box 7.2   Flood Frequency and Probability

Observations of discharge are required to estimate how often a river floods. To calculate the annual return period of a flood it is necessary to ascertain what the largest discharge was during a given year. The annual peak discharge may be estimated or measured directly, but it is more usually obtained by from a stage–discharge relationship (rating curve). Data from many years of record (shorter data records give rise to more uncertainty) are then used to derive a flood frequency curve, which is constructed by ranking the peak discharges in descending order of their magnitude (a ranking of 1 being assigned to the largest discharge). The recurrence interval, $T$, for a given peak discharge is $(n + 1) / m$: where $n$ is the number of years of observations, and $m$ is the magnitude ranking of the discharge. If it is assumed that the data are log-normally distributed (see Chapters 17 and 18), the annual peak discharge can be plotted against its recurrence interval on log–normal graph paper and a trend line added to highlight the relationship (detailed instructions for producing flood frequency curves are to be found at http://pubs.usgs.gov/twri/twri4a2/pdf/twri_4-A2_a.pdf). As it is projected from the trend line, the stage for a flood with a given recurrence interval such as the "hundred year flood," a term that is often used by planners and in the news media, can be back-calculated from the stage–discharge relationship, and the depth of inundation on the flood plain determined by transferring the elevation data to a topographic map or DEM (a flood hazard map can be created by interpolating the flood profile between gauging stations). The recurrence interval of past events can also be used to forecast the probability of an event occurring in the future. The probability, $p$, that a flood with a given discharge will occur is $1 / T$; thus, for example, a flood with a recurrence interval of 100 years has a 0.01 (1 in 100) probability or 1% chance of occurring in any given year, and the probability of an event being equaled or exceeded, $P_T$, at least once in the next, $n$ years, is $1 - p^n$. The probability of a flood occurring or being equaled or exceeded does not, however, depend on past occurrences and so does not imply that the "hundred year flood" will only occur once every hundred years, or that such an event is "overdue" if the recurrence interval has been exceeded. While this statement is correct in principle, it is based on the assumption that the conditions that produced the existing record of floods will continue into the future. However, in a world of rapidly changing environmental conditions, including deforestation, climate change and urbanization, frequency–magnitude relationships derived from the historical record may have little bearing on the future frequency–magnitude behavior of process systems. The concept of an event's return period and its associated probability of occurrence can also be applied to earthquakes, severe weather, wave height, drought, and wind speed among other event-based processes. Earthquakes provide one exception to converting return periods to constant annual probabilities. For certain types of earthquakes, stress release (earthquake magnitude) is a function of time since the last earthquake. Stress accumulates through time and thus, the annual probability of a hundred-year return period earthquake, for example, increases each year, with time since the last event.

be a separation of time scales (the external process responsible for driving the system is necessarily much slower than the internal relaxation process), a requirement that presupposes the existence of a threshold, the angle of repose of the sand pile, for example. Other frames of reference also can be used to describe **complex behavior** in systems where the outputs are not proportional to the inputs across the entire range of inputs. These include the notion of complex response and chaos theory; in the former case a stimulus may give rise to multiple outcomes and in the latter initial differences or minor disturbances tend to persist and grow over time. In some complex systems different initial conditions evolve to similar endpoints (equifinal or attractor states). The source of these nonlinearities includes controls and relationships, such as thresholds effects and hysteresis, which can be readily observed (Phillips 2003).

## Process–Form Relationships

Relationships between the form (morphology) of the physical landscape and the processes that created it are inevitably complex, because the landforms of which it is comprised reflect the imprint of processes that operate over a range of spatial and temporal scales. Thus, for example, Church and Slaymaker (1989) linked the increase in suspended sediment yield in drainage basins up to $3 \times 10^4$ km$^2$ in area in British Columbia to the remobilization of sediments deposited during the Quaternary (the time span covering Earth's recent glaciations). This is contrary to the decline in suspended sediment yield that is normally observed as basin area increases and relief, which sets the overall denudation rate, declines. Historical studies that focus on changes in landform characteristics over time spanning thousands to millions of years cannot rely on direct measurement of the processes that produce changes, but instead seek to reconstruct or retrodict past forms in order to establish how and at what rate change occurred. In some instances the succession of forms through which the physical landscape evolves can be derived from an **ergodic** sequence, in which landforms with different ages found at different locations are arranged chronologically to create a time (or topographic) series. However care needs to be exercised when using this approach because it assumes that differences in form are simply a function of time. In the time period represented there may have been significant changes in one or more of the controlling variables; that is, in climate, tectonics, sea-level, etc.

The way in which processes affect the physical landscape over time depends on the degree of stability (equilibrium) of the system, thus it may either be assumed that the processes and the forms they give rise to are in equilibrium, or that they are not. An equilibrium hillslope, for example, is one in which the driving and resisting forces acting on it compensate one another, such as in the case where slope angle equates with the angle of repose, or "rock mass strength," of the material the slope is composed of. In the former case, relationships (where changes in the patterns and processes involved sometimes scale with variations in size, that is they are "allometric relations") can be established between the characteristics of contemporary landforms in different locations that may help explain the changes that occur through time. Whereas, in the latter case field observations permit different "evolutionary" stages to be identified, and similar landforms of different ages that occur in different locations to be arranged in chronological order; a case in point being the time sequence of landform development (the Geographical cycle) derived by William

Morris Davis (1850–1934). Although as we have seen in Chapter 3, care must be taken to ensure that observations are rigorously evaluated and not simply interpreted in a way that is designed to lend support to a pre-existing theory.

Tectonic processes, which operate over timescales of millions of years, are responsible for creating a wide range of landforms including very large distinctive terrains such as the ranges of fold mountains that have developed in response to the horizontal (compressive) motion between two adjacent tectonic plates and thickening of Earth's crust; the elevation of which through the principle of isostasy (buoyancy) is determined by the thickness and rigidity of the crust, as well as by spatial variations in erosion processes, such as fluvial incision and landsliding. The rate of rock uplift often appears to be balanced by the rate of incision, which thus regulates relief; whereas hillslope angles either may be proportional to the rate of river incision (higher rates of incision give rise to steeper hillslopes), or maintain a threshold angle of stability and be independent of the rate of river incision (Burbank et al. 1996). In the latter case high sustained rates of tectonic activity (note that a rate of rock uplift of a few millimeters per year translates into a vertical displacement of several kilometers per million years) may give rise to a **dynamic equilibrium** topography; that is, a situation where although local relief may be increasing or decreasing, overall relief remains essentially constant over time. This comes about because erosion rates increase with increasing elevation, so that the overall increase in the elevation of a mountain range slows over time to the point where the long term rate of river incision (or glacial erosion) matches the rate of rock uplift. On a somewhat smaller scale the linear, elongate valleys (graben) bounded by mountain ranges (horst) that give rise to the topography of the Basin and Range province of the southwestern United States are, like rift valleys, a product of extension, thinning and faulting of the crust (horst and graben are produced by parallel faulting, and half-graben by vertical movement along a single fault). The basins were down-faulted as the ranges were uplifted, eroding as they rose. Rivers disperse the sediment which, because of the closed (interior) drainage pattern, accumulates in the adjacent valleys. The end result is a series of small isolated mountains (inselbergs) that rise abruptly from a surrounding, almost level plain.

Climate is also influenced by tectonic processes. For example, uplift of the Himalayas and Tibetan Plateau is thought to have caused the Indian and Asian monsoon to develop some 8 or 9 million years ago, and orographic precipitation may cause high erosion rates on the windward side of mountain ranges. It is also possible that high rates of precipitation and erosion may drive active faulting in fold mountain ranges, so that climate also influences tectonics (Wobus et al. 2005). For this reason it is not always clear whether tectonics or climate regulates long-term erosion rates. Nevertheless, in the last 2 to 4 million years, as the climate changed from a non-cyclic to a cyclic mode characterized by rapid oscillations between glacial and interglacial conditions, the global sedimentation rate (a proxy for erosion rate) appears to have increased. The rapid (Quaternary) changes in climate that occurred over a few hundred thousand years may also have had a significant impact on unglaciated landscapes. In regions where there is active uplift, topography appears to reflect the current climate and any relict features the wettest conditions previously experienced, whereas in stable regions the overall pattern of climate change has the potential to affect the contemporary geomorphology (Rinaldo et al. 1995). Elsewhere glaciation has had a significant impact on topography. Entire landscapes have been shaped by glacial deposition and massive water discharges that may also have influenced climate by switching the deep-ocean density driven, thermohaline circulation between its glacial and interglacial modes. Glaciers are also highly efficient erosive agents, to the extent that they may exert the ultimate control on the height of tectonically active mountain ranges; that is they function as a "buzzsaw," so that regardless of the rate of rock uplift, the equilibrium line (snowline) altitude in many glaciated mountain ranges parallels the summit elevations of the ranges, and the overall landscape looks the same.

The future effect the mega drivers of erosion (climate and human activity) may have is likely to depend on terrain conditions and the scale at which they are most effective, but there already exists compelling evidence that, on a global scale and over long time periods, human activity has caused a sharp increase in erosion from a long term natural rate of $16\,\mathrm{Gt}\,\mathrm{year}^{-1}$ during the Pliocene to current losses of $75\,\mathrm{Gt}\,\mathrm{year}^{-1}$ from cropland (Wilkinson and McElroy 2007), and that the accumulation of post-settlement alluvium on large floodplains represents the greatest geomorphic change currently being experienced on Earth (far exceeding former inputs from Pleistocene glaciations or from current alpine erosion). So clearly the physical landscape is in the process of adjusting to significant anthropogenically induced changes to the forces and buffers that control landform process and form.

To study and model process–form interactions over long (geological) time periods, a combination of age information derived from a variety of different dating methods (see Box 7.3), topographic information obtained from DEMs, and information about rates of earth surface sediment transport gained from theory or observation are required. Laboratory and field observations are important, not only because they provide fundamental information that can be used to derive "flux laws" that place limits on the amount of material flowing through a unit area in a given time and can be incorporated in numerical models of landscape evolution, but also because they provide the basis for linking process and form in the modern world. If **relaxation times** are short, links between process and form may be manifest through a morphological relationship, such that between the gradient of a slope and the size of the debris of which it is composed. The action of some processes also creates distinctive topography; for example, in mountainous terrain, fluvial and glacial erosion produce V- and U-shaped valleys, respectively. More commonly, as discussed previously, processes are treated as shear stress that act on the different earth materials and, when the resultant force exceeds the resisting force, promote erosion, transport and deposition. In other words, geomorphologists use Newtonian physics to describe the behavior of material at Earth's surface; they also accept that, as the laws of thermodynamics imply, geomorphological processes use and transform energy (which is conserved) and landforms evolve towards an equilibrium condition. This approach to characterizing relationships between process and form is exemplified by the work of Grove Karl Gilbert (1834–1918), and Arthur N. Strahler (1918–2002). For example, a century ago G. K. Gilbert showed how the convex–concave form of hillslopes (see Figure 7.1) reflects the gradual transition in process dominance (from soil creep to soil wash) as the distance from the drainage divide increases, and some 40 years later A. N. Strahler demonstrated how, in an area of uniform climate, geology, and vegetation, hillslopes maintained a characteristic mean maximum angle (time-independent form) that was adapted to maintain the processes of erosion and transport in a steady state.

The assumption that process and form are in equilibrium has also permitted relationships between different aspects of river morphology (such as width and depth) and the quantity of water or the amount and type of sediment that the channel conveys to be

## Box 7.3   Analytical Dating Methods

To understand how the physical landscape evolves geomorphologists require information about long-term erosion rates or when events occurred. Terrestrial *in situ* cosmogenic nuclides can, depending on the nuclide involved, be used to determine the ages of bedrock and sedimentary surfaces and quantify erosion rates over time-frames of 200 to 8 million years. Cosmogenic nuclides are formed by the interaction of cosmic rays produced by the sun with nuclei of atoms at the surface of mineral grains in rock and sediment that is exposed to the atmosphere. The interaction produces nuclides that are either stable (noble gases) or unstable and subject to decay, such as $^{10}$Be, $^{14}$C, $^{26}$Al and $^{36}$Cl. The dating techniques rely on knowing the rates of nuclide accumulation and decay, which vary with the altitude and latitude of the sampling site. Information about the basic principles involved and examples of how cosmogenic nuclides are used to date surfaces and sediment and determine erosion rates can be found at http://cnef.earthsciences.dal.ca/NewFiles/tcnweb/index.html

Fallout radionuclides such as $^{7}$Be, $^{137}$Cs and $^{210}$Pb have much shorter half-lives, ranging from 53 days to 30 years and can be used to date and source sediment and quantify sediment transport processes over time periods of one to two hundred years. $^{7}$Be is formed by cosmic ray spallation of nitrogen and oxygen within Earth's atmosphere, and is rapidly incorporated into sediment after it is washed out of the atmosphere and transferred to the terrestrial environment. $^{210}$Pb is an isotope in the $^{238}$U series that is derived from the decay of $^{222}$Rn (a gas, the daughter of $^{226}$Ra which occurs naturally in soils). It is introduced into the atmosphere by the diffusion of $^{222}$Rn from the soil. Fallout from the atmosphere creates an excess or "unsupported" amount of $^{210}$Pb that cannot be accounted for by decay of the *in situ* parent $^{226}$Ra. $^{137}$Cs is a thermonuclear by-product. Fallout from the atmosphere is fixed in the upper portions of the soil profile and is dispersed by soil erosion. Information about the use that can be made of fallout radionuclide measurements can be found at http://www.fao.org/docrep/X5313E/x5313e00.htm#Contents

Radiocarbon dating is a well known and widely used dating method that can be applied over about the past sixty thousand years. In Earth's upper atmosphere nitrogen is bombarded by cosmic radiation and broken down into an unstable isotope of carbon, $^{14}$C; which when it is transferred to the ground surface behaves like $^{12}$C and $^{13}$C and, through photosynthesis, becomes fixed in the biosphere. During the lifetime of an organism the ratio of $^{12}$C to $^{14}$C remains constant, but when it dies the ratio gradually decreases (the half-life of $^{14}$C is 5,730 years). Methods used to measure residual $^{14}$C activity include Gas Proportional and Liquid Scintillation Counting and Accelerator Mass Spectrometry: http://www.c14dating.com/meths.html

defined. For example, in alluvial rivers, the morphology of which is governed by the size and quantity of the material they transport, "hydraulic geometry" relations describe how characteristic properties of the channel and flow change as discharge increases downstream (typically, the width and depth of the channel increase while velocity remains constant or declines). The appeal of such relations, which often show remarkable consistency on a global scale, is that they can be used to characterize the adjustments that occur in terms of simple (power) functions that, inasmuch as they rely on surrogates for process (such as discharge), are convenient approximations of a much more complex reality. Indeed the apparent universality of some empirical relations has led to their adoption as "laws" that govern the way the physical landscape evolves. One such example is "Hack's Law", the relationship between mainstream channel length, L, and drainage basin area, $A$ (in this case area is used as a surrogate for discharge), $L \propto A^h$ where: the exponent $h$ varies between 0.5 and 0.6 (Montgomery and Dietrich 1992).

# Conclusion

The two fundamental prerequisites for an understanding of the physical landscape are: an ability to analyze and represent its diverse forms and features; and a methodology for measuring, recording, and analyzing the behavior of land-forming processes. Research into the temporal and spatial relationships between form and process helps us understand those factors that control the landscape system, how the physical landscape works, and how it evolves. That knowledge, in turn, may allow us to model and predict both the behavior of processes and the response of the physical landscape to environmental change that has occurred in the past and may occur in the future, not the least of which are the changes attendant upon rapid global changes in population pressures and climate. Transferring this knowledge to other settings also allows us to understand how, for example, "landscapes" on the sea bed or Mars evolve (Perron et al. 2003, Straub et al. 2007).

# References

Burbank D. W., Leland, J., Fielding, E., Anderson, R. S., Brozovic, N., Reid, M. R., and Duncan, C. (1996) Bedrock incision, rock uplift, and threshold hillslopes in the northwestern Himalaya. *Nature* 379: 505–10.

Church, M., and Slaymaker, O. (1989) Holocene disequilibrium of sediment yield in British Columbia. *Nature* 337: 452–4.

Crozier, M. J. (1999) Prediction of rainfall-triggered landslides: a test of the Antecedent Water Status model. *Earth Surface Processes and Landforms* 24: 825–33.

Dalrymple, J. B., Blong, R. J., and Conacher, A. J. (1968) A hypothetical nine-unit landsurface model. *Zeitschrift für Geomorphologie* 12: 60–76.

Hicks, D. M., Gomez, B., and Trustrum, N. A. (2004) Event suspended sediment characteristics and the generation of hyperpycnal plumes at river mouths: East Coast Continental Margin, North Island, New Zealand. *The Journal of Geology* 112, 4: 471–85.

Montgomery, D. R., and Dietrich, W. E. (1992) Channel initiation and the problem of landscape scale. *Science* 255: 826–30.

Montgomery, D. R., Sullivan, K., and Greenberg, H. M. (1998) Regional test of a model for shallow landsliding. *Hydrological Processes* 12: 943–955.

Perron, J. T., Dietrich, W. E., Howard, A. D., McKean, J. A., and Pettinga, J. (2003) Ice-driven creep on Martian debris slopes. *Geophysical Research Letters* 30: 1747.

Phillips, J. D. (2003) Sources of nonlinearity and complexity in geomorphic systems. *Progress in Physical Geography* 27: 1–23.

Rinaldo, A., Dietrich, W. E., Rigon, R., Vogel, G. K., and Rodriguez-Iturbe, I. (1995) Geomorphological signatures of climate. *Nature* 374: 632–5.

Straub, K., Jerolmack, D. J., Morhig, D., and Rothman, D. H. (2007) Channel network scaling laws in submarine basins. *Geophysical Research Letters* 34 L12613, doi:10.1029/2007GL030089.

Wilkinson, B. H., and McElroy, B. J. (2007) The impacts of humans on continental erosion and sedimentation. *Geological Society of America Bulletin* 119: 140–56.

Wobus, C., Heimsath, A., Whipple, K., and Hodges, K. (2005) Active out-of-sequence thrust faulting in the central Nepalese Himalaya. *Nature* 434: 1008–11.

## Additional Resources

Goudie, A., Anderson, M., Burt, T., Lewin, J., Richards, K., Whalley, B., and Worsley, P. eds. (1990) *Geomorphological Techniques* (2nd edn). London: Routledge. Geomorphology is a wide-ranging discipline that deals with all the features found and processes operating on Earth's surface. The core techniques geomorphologists use to study patterns and processes are described here.

Kondolf, G. M., and Piégay, H. eds. (2003) *Tools in Fluvial Geomorphology*. Chichester: Wiley. Describes the techniques used by fluvial geomorphologists to study patterns and processes.

Mitchell, C. (1991) *Terrain Evaluation*. Longman: New York. Describes the many methods of terrain evaluation.

Farr, T. (with seventeen others). (2007) The Shuttle Radar Topography Mission. *Reviews of Geophysics* 45 RG2004: 1–33. Provides information about the Shuttle Radar Topography Mission (SRTM) data base. Additional online information can be found at http://www2.jpl.nasa.gov/srtm/

Schumm, S. A. (1979) Geomorphic thresholds: the concept and its applications. *Transactions of the Institute of British Geographers* 4: 485–515. A classic paper on the concept of thresholds.

Burbank, D. W., and Anderson, R. S. (2001) *Tectonic Geomorphology*. Oxford: Blackwell.

Selby, M. J. (1993) *Hillslope Materials and Processes* (2nd edn). Oxford: Blackwell.

Gregory, K. J., and Walling, D. E. (1973) *Drainage Basin Form and Process*. London: Edward Arnold.

The above three books offer comprehensive coverage not only of the core techniques used to study the processes and forms referred to in this chapter, but also of the interplay between tectonic and surface processes and the relationships that exist between form and process in rivers and on hillslopes.

Bierman, P. R., and Nichols, K. K. (2004) Rock to sediment – slope to sea with [10]Be – rates of landscape change. *Annual Review of Earth Science* 32: 215–55.

Currie, L. A. (2004) The remarkable metrological history of radiocarbon dating II. *Journal of Research of the National Institute of Standards and Technology* 109: 185–217.

Muzikar, P., Elmore, D., and Granger, D. E. (2003) Accelerator mass spectrometry in geologic research. *Geological Society of America Bulletin* 115: 643–54.

The above three references showcase various dating methods and their applications.

## Exercise 7.1    Frequency Analysis

Frequency analysis is used to estimate the probability of the occurrence of a given event, such as a rainstorm or flood. A recurrence interval (or return period) is the probability that a given event will be equaled or exceeded in any given year. The following equation is used to determine the recurrence interval $(R_i)$ when there is a magnitude associated with the data (such as the discharge of a riverine flood):

$$R_i = (n+1)/m \tag{7.2}$$

where $n$ is the number of years of record and $m$ is the magnitude ranking (determined by sorting the discharge record in descending order). The probability $(P)$ of an event with recurrence interval $(R_i)$ is:

$$P = 1/R_i \tag{7.3}$$

and the probability $(P_T)$ that a given event will be equaled or exceeded at least once in the next $r$-years is:

$$P_T = 1 - Pr \tag{7.4}$$

(Probabilities can be expressed as a percentage, ratio, fraction or decimal.)

Construct a flood frequency curve using the tabulated record of the peak discharge of the Waipaoa River, New Zealand (Table 7.1) (plot the data points on log–linear graph paper and fit a regression line to them). Use the relation to determine, for example, the:

1   Return period of a discharge of $1500\,\mathrm{m^3\,s^{-1}}$ and $3000\,\mathrm{m^3\,s^{-1}}$
2   Discharges that have returns periods of 5 yr and 50 yr.
3   A "100-year flood" is an event in which the peak discharge is attained on average once every 100 years. What discharge would be associated with a 100-year flood in the Waipaoa River?
4   What are the implications of using this method to determine the recurrence interval of the 100 yr flood?

**Table 7.1**   Annual maximum (peak) discharge ($\mathrm{m^3\,s^{-1}}$ [cubic meters per second]) of the Waipaoa River, at Kanakanaia

|      | 0     | 1     | 2     | 3     | 4     | 5     | 6     | 7     | 8     | 9     |
|------|-------|-------|-------|-------|-------|-------|-------|-------|-------|-------|
| 1930 |       |       |       |       |       |       |       |       | 1,520 | 1,728 |
| 1940 | 814   | 580   | 758   | 1,430 | 2,240 | 386   | 1,280 | 1,210 | 3,600 | 1,161 |
| 1950 | 2,800 | 1,120 | 1,200 | 1,210 | 1,860 | 1,640 | 1,560 | 1,120 | 510   | 770   |
| 1960 | 2,410 | 1,140 | 1,270 | 440   | 430   | 1,500 | 890   | 670   | 980   | 480   |
| 1970 | 970   | 1,290 | 680   | 1,010 | 890   | 720   | 1,680 | 970   | 830   | 450   |
| 1980 | 2,580 | 870   | 2,240 | 530   | 1,830 | 1,490 | 1,080 | 1,100 | 4,000 | 1,210 |
| 1990 | 1,350 | 380   | 760   | 560   | 830   | 650   | 2,030 | 1,470 | 546   | 415   |
| 2000 | 621   | 565   | 1,872 | 811   | 747   | 3,446 | 1,245 | 317   | 1,596 | 1,606 |

5  The probability of such a flood occurring in the next (or any) year is 1 in 100 or 1%, but what is the probability of a 100-year flood occurring in the next 100 years?
6  What cannot be determined about the next 1 in 100 year flood?

## Exercise 7.2  Landslide Triggering Rainfall

By correlating climate and landslide occurrence it is possible to identify the conditions which produced landslides and those that did not. The threshold separating these conditions (Figure 7.3) is referred to as a "maximum probability threshold," above which rain events always produce landslides. Climate is represented by two parameters: antecedent excess rainfall ($AER$) and daily rainfall ($R$). To compute the $AER$ and discriminate between those conditions that cause landslide occurrence and those that do not, we also need to know the potential evapotranspiration ($Ev$), the maximum amount of water that can be held in the **regolith** (120 mm for Wellington, New Zealand), the amount of water in the regolith ($M$), and the drainage rate from surplus precipitation (which it is assumed declines exponentially from a value of 0.84). On a day to day basis, the amount of water in the regolith is:

$$M_{today} = M_{yesterday} + (R_{today} - Ev_{today}) \qquad (7.5)$$

where $R - Ev$ can be either negative (in which case $M_{today} < M_{yesterday}$) or positive (so that $M_{today} > M_{yesterday}$), and $M$ can *never exceed*, the maximum value of 120 mm. The excess rainfall ($ER$) is:

$$ER_{today} = (M_{yesterday} + (R_{today} - E_{today})) - 120 \qquad (7.6)$$

Negative values make 0 (zero) contribution to the AER, which is computed as the sum of the excess rainfall (positive values) multiplied by the drainage rate from surplus precipitation on each of the previous ten days:

$$AER_{today} = 0.87 ER_1 + 0.84^2 ER_2 + 0.84^3 ER_3 \ldots + 0.84^{10} ER_{10} \qquad (7.7)$$

where $ER_1$ = yesterday, $ER_2$ = the day before yesterday …

1  Use the data in Table 7.2 from the Kelburn climate station in Wellington, New Zealand, to determine the $AER$ on June 4 through 9, 1996 (the amount of water in the regolith on May 24 was 117.7 mm) – insert the tabulated values in a spreadsheet; add five additional columns (headed $R - Ev$, $M$, $ER$, positive $ER$, and $AER$); and calculate $R - Ev$, etc. for every day from 25 May onwards.
2  On the basis of the threshold shown in Figure 7.3, are landslides likely to have occurred in Wellington on any of the six days in question?

Figure 7.3 can also be used to quantify the amount of rain required to generate landslides on any given day. The threshold line bisects the $y$-axis at ~50 mm and the $x$-axis at ~80 mm. Its slope is therefore 50/80 (0.625) and so, for any given $AER$, the amount of rain required to generate landslides is $(80 - AER) * 0.625$. Table 7.3 gives the precipitation,

**Table 7.2** Precipitation and evapotranspiration at Kelburn, Wellington

| Date | Rainfall (R, mm) | Evaporation (E, mm) |
|---|---|---|
| 24-May-96 | | |
| 25-May-96 | 0 | 1.1 |
| 26-May-96 | 26.3 | 3.2 |
| 27-May-96 | 1.0 | 2.2 |
| 28-May-96 | 0.8 | 0.6 |
| 29-May-96 | 0.4 | 1.3 |
| 30-May-96 | 0 | 1.3 |
| 31-May-96 | 0 | 2.0 |
| 1-Jun-96 | 0 | 1.9 |
| 2-Jun-96 | 26.3 | 0.7 |
| 3-Jun-96 | 1.8 | 0 |
| 4-Jun-96 | 0 | 0.5 |
| 5-Jun-96 | 0.9 | 0 |
| 6-Jun-96 | 0 | 1.2 |
| 7-Jun-96 | 0.4 | 0.1 |
| 8-Jun-96 | 0.2 | 0.6 |
| 9-Jun-96 | 6.3 | 0 |

**Table 7.3** Precipitation at Kelburn, number of landslides in Wellington and antecedent excess rainfall

| Date | Rainfall (mm) | No. landslides | AER (mm) |
|---|---|---|---|
| 1-Jul-96 | 0.2 | 0 | 4.37 |
| 2-Jul-96 | 0 | 0 | 3.67 |
| 3-Jul-96 | 6.6 | 0 | 3.08 |
| 4-Jul-96 | 10.0 | 0 | 2.59 |
| 5-Jul-96 | 10.5 | 2 | 7.97 |
| 6-Jul-96 | 27.0 | 4 | 15.52 |
| 7-Jul-96 | 11.3 | 1 | 35.71 |
| 8-Jul-96 | 0.2 | 2 | 39.49 |
| 9-Jul-96 | 1.0 | 0 | 33.17 |
| 10-Jul-96 | 0 | 1 | 27.87 |
| 11-Jul-96 | 25.3 | 11 | 23.41 |
| 12-Jul-96 | 59.2 | 21 | 40.91 |
| 13-Jul-96 | 7.1 | 4 | 84.10 |
| 14-Jul-96 | 0 | 2 | 70.64 |
| 15-Jul-96 | 9.5 | 7 | 59.34 |
| 16-Jul-96 | 3.9 | 7 | 56.40 |
| 17-Jul-96 | 0 | 1 | 50.14 |
| 18-Jul-96 | 2.3 | 2 | 42.12 |
| 19-Jul-96 | 0.5 | 1 | 36.89 |
| 20-Jul-96 | 2.3 | 0 | 30.99 |
| 21-Jul-96 | 0.3 | 0 | 27.96 |
| 22-Jul-96 | 5.9 | 3 | 23.74 |
| 23-Jul-96 | 2.8 | 7 | 24.90 |
| 24-Jul-96 | 43.7 | 27 | 23.10 |
| 25-Jul-96 | 0 | 10 | 54.60 |
| 26-Jul-96 | 16.3 | 12 | 45.86 |
| 27-Jul-96 | 9.2 | 4 | 51.80 |
| 28-Jul-96 | 4.1 | 3 | 49.89 |
| 29-Jul-96 | 0 | 7 | 44.85 |
| 30-Jul-96 | 0.4 | 0 | 37.67 |
| 31-Jul-96 | 2.0 | 3 | 31.65 |

antecedent excess rainfall and number of landslides that occurred in Wellington in July 1996.

3 Copy these data into a spreadsheet and calculate the amount of rain required to initiate landslides in Wellington on each day in July 1996.
4 Using appropriate graphs to illustrate your answer and bearing in mind the accuracy of input parameters, how well does Figure 7.3 define the threshold for landslide occurrence in Wellington?
5 What other factors might affect the occurrence of landslides in the city of Wellington?

# Chapter 8

# Climates

*Julie A. Winkler*

- Introduction
- Climate Data
- Inhomogeneities in Archived Data
- Sources of Archived Data
- Research Methods in Climatology
- Detecting Climate Change: An Example of the Analysis of Climate Data
- Conclusion

Keywords

Anomaly
AOGCM
Climate change
Climate classification
Climate model
Ensemble forecasts
ENSO (El Niño – Southern Oscillation)
Greenhouse gases
Inhomogeneity
*In situ*

Intergovernmental Panel on Climate Change (IPCC)
Metadata
Multivariate statistics
NAO (North Atlantic Oscillation)
Remotely-sensed observations
Resultant
Scalar
Troposphere
Vector

# Introduction

Historically, climatology was claimed as a sub-discipline of both geography and atmospheric science. Current interest in climatology extends well beyond these two disciplines to all branches of the physical and human sciences, in part because of increased societal awareness of the potential impact of, and Earth's vulnerability to, climate variability and change. There is no universally-accepted definition of climatology, and given the wide-ranging interests of climatologists it seems unlikely that one will emerge in the near future. However, the American Meteorological Society's *Glossary of Meteorology* defines climate as "The slowly varying aspects of the atmosphere–hydrosphere–land surface system" and climatology as "The description and scientific study of climate" (American Meteorological Society 2000). These definitions go beyond the often popular conception that "climate is the average of weather" and that climatology is simply a descriptive endeavor. Rather, climatology is also concerned with the physical processes responsible for climate and its variability and with the impacts of climate on natural and human systems.

The different philosophical paradigms outlined in earlier chapters have not been extensively debated by climatologists, who for the most part work within a positivist framework and use the scientific method (see Chapter 3). For this reason, this chapter focuses on the data and methods that are used in climatological research and highlights some of the major data sources and fundamental issues that need to be considered when climate data are analyzed.

# Climate Data

The many facets of Earth's climate are described using a wide variety of variables (also referred to as "climate elements" or "climate parameters"). Three variables, maximum temperature, minimum temperature, and precipitation accumulation, are at the core of many climatological analyses. Air pressure, humidity, wind direction and speed, visibility, etc. are other examples of climate variables. Often these primary variables are used to derive additional variables that are useful for specific applications. One example is "growing degree days", which is a temperature-dependent measure of heat accumulation that can be related to plant, insect, and disease development. Information about specific atmospheric phenomena such as cloud type, the frequency of fog or severe weather (including strong winds, large hail, and tornadoes), or the tracks followed by tropical or mid-latitude cyclones may also be analyzed. An essential point, however, is that because the goal of much climatological research is to understand better the linkages and interactions within Earth's climate system, investigations are rarely limited to a single climate variable or atmospheric phenomenon.

## General considerations when taking or interpreting climate measurements

Climate variables and atmospheric phenomena can be measured directly (referred to as *in situ* observations), or they can be observed using remote sensing techniques. Most remote sensing techniques monitor the properties of electromagnetic waves emitted or reflected by objects, such as gas molecules or water droplets in the atmosphere. The amount of precipitation recorded using a standard rain gauge is an example of an *in situ* observation, whereas a precipitation estimate made on the basis of the quantity of microwave energy emitted by radar and reflected back to the source by precipitation-sized droplets or ice crystals in the atmosphere constitutes a **remotely-sensed observation**. Researchers may sometimes acquire information about particular climate variables or atmospheric phenomena by designing a field experiment specifically for that purpose, an example being the Atmospheric Boundary Layer Experiments (ABLE) in southern Kansas (USA) whose broad goal is to better understand processes in the lower atmosphere (see http://www. atmos.anl.gov/ABLE/ for more information). In other cases a researcher may rely on archives containing historical observations of climate variables that national or international meteorological organizations, such as the National Oceanographic and Atmospheric Administration (NOAA), National Climatic Data Center, United Kingdom Met Office, or World Meteorological Organization (WMO), have compiled from routine measurements made at one or a number of observation locations over a period of years or decades (for examples of climate information available in national archives see http://www.metoffice. gov.uk/climate/uk/ and www.ncdc.noaa.gov). In fact, the extensive use of archived observations is a hallmark of climatological research.

In all instances, careful consideration must be given to the research design (see Chapter 5). The specifics will inevitably vary between different projects, but issues relating to the type of data involved and the methods used to analyze them must be addressed by researchers who design their own field experiments as well as by those using archived data. Typically the relevant climate variables and spatial and temporal scales of investigation must be specified. For example, if the objective is to evaluate the risk to plants of damage by frost, observations of daily minimum temperature alone may be required, whereas if it is to evaluate the risk of heat stress to humans more frequent observations of both temperature and humidity, as well as other parameters that affect human comfort, such as wind speed and levels of solar radiation, are likely to be required. Some climate variables can also be expressed in different ways. Consider, for example, atmospheric humidity which may be expressed as "relative humidity" (a measure of how close the atmosphere is to saturation), "absolute humidity" (the amount of water vapor present in a unit volume of air), "specific humidity" (the ratio of the mass of water vapor present to the total mass of air), or as "mixing ratio" (the ratio of the mass of the water vapor present to the mass of dry air). Human comfort may be adequately described by the first term (relative humidity), whereas to determine the altitude at which condensation begins (a prerequisite for cloud formation) it is useful to know the specific humidity or mixing ratio of an air parcel.

The horizontal and vertical dimensions of a phenomenon affect the density and areal coverage of the observations needed to investigate it. Orlanski (1975) suggested that atmospheric phenomena may be classified as microscale (less than 100 m), local scale (100 m to 3 km), mesoscale (3 to 100 km), synoptic scale (100 to 3,000 km) and planetary scale (larger than 3,000 km). One carefully placed instrument may be sufficient for measuring a microscale phenomenon, such as the heat flux from an urban roof top, whereas hundreds of instruments are required to study a planetary-scale phenomenon such as **ENSO (El Niño – Southern Oscillation)**. The sensitivity and response time of the instrument must also be considered. Microscale atmospheric phenomena with short time scales usually require more frequent observations compared to larger scale phenomena. Thus, for example, wind gusts on city streets may need to be monitored at time scales of seconds or fractions of a second, whereas hourly wind observations may be sufficient when investigating mesoscale low-level wind maxima such as the jets that are frequent in the Great Plains region of the United States. Other considerations when taking atmospheric observations are outlined in Box 8.1.

**Metadata** (data about data) are essential if observations are to be shared with or used by others. The WMO recommends, that at a minimum, metadata should include a description of the type and condition of an instrument; its surroundings, exposure, height above the ground surface and the degree of interference from other instruments or objects; changes that have occurred to it and/or its replacement(s); and observation protocols.

## Inhomogeneities in Archived Data

Despite the apparent widespread availability of archived data, it is not always clear whether climatology is "data rich" or "data poor" as the archived climate records have a number of severe limitations. One limitation is that the spatial and temporal resolution of archived observations are not detailed enough to address many contemporary research questions. Coverage of Earth's surface is uneven; comparatively few observations have been made over tropical areas, polar regions, or the oceans. Also, records often do not extend far enough back in time for climate variability and change to be accurately detected. The length of the climate record varies by climate variable and location. Quasi-global observations of surface temperature and precipitation roughly date to the mid 1800s, whereas routine upper-level observations began around 1940 (Winkler 2004).

Another limitation of archived climate measurements is that most observational networks were initially designed for short-range weather prediction rather than for climate monitoring. The requirements for these two applications are quite different; for example, accuracy and precision are important for weather forecasting, whereas temporal and spatial consistency is a primary concern for climate monitoring. The installation of new instruments with greater accuracy and precision, while very useful for weather prediction, can pose problems for climate monitoring as the changes in instrumentation introduce inconsistencies (usually referred to as **inhomogeneities** in the climatology literature) into the climate record. Ideally, there is a long (a minimum of one year) overlap period when measurements are taken with both the new and old instrumentation so that correction factors can be calculated, but in practice this often does not happen.

Multiple other sources of inhomogeneities in climate records also exist, ignorance of which can lead to erroneous interpretations of climate observations. Studies have suggested that changes in station location, even a change on the order of only a few hundred meters, may introduce considerably more bias than that associated with instrument changes (Guttman and Baker 1996). Changes in instrument exposure, such as increased urbanization or even the growth of a nearby tree, also can introduce artificial trends into the observations.

Differences in observation practices also need to be considered when analyzing archived climate data, particularly a well-known, but difficult to correct for, inhomogeneity known

**Box 8.1**   Climate Research Checklist

All climatological research requires that careful attention be paid to experimental design. A "checklist" (Table 8.1) is provided below to assist students and others new to climatological analysis in taking measurements of the atmosphere. Although primarily designed for field measurements, the checklist is also broadly relevant when utilizing archived climate observations. The checklist is, of course, not exhaustive, and considerations specific to your research project should be added.

**Table 8.1**   Climate research checklist

| Question | Explanation | Why important? |
|---|---|---|
| What is the research objective(s)? | Research objectives need to be clearly stated at the beginning of every project. | The research objectives guide the type and number of climate variables included in the analysis. |
| What is the spatial scale of the climate phenomenon being studied? | Climate phenomena occur at scales ranging from microscale to planetary scale (see text). | The spatial scale determines the necessary density and areal coverage of observations. Remember that spatial scale refers to both horizontal and vertical dimensions. |
| What is the temporal scale of the climate phenomenon under study? | Temporal and spatial scale are often related as microscale and local scale climate phenomena tend to have shorter durations compared to synoptic and planetary scale phenomena. | The temporal scale determines the frequency of observations. |
| What instruments can be used to measure a particular climate variable, and how does the sensitivity of alternative instruments differ? | Many different instruments have been developed to measure a given climate variable. Sensitivity is the response of an instrument to a signal. | Less sensitive instruments may not provide the accuracy needed to address the research objectives, but highly sensitive instruments generally require more frequent calibration and are more costly. |
| What is the instrument's "footprint"? | An instrument "sees" only a portion of the surrounding environment. The area sensed, or footprint, depends on the characteristics of the instrument such as its field of view and the elevation at which it is installed. | It is necessary to know the footprint in order to correctly situate the instrument and to interpret the measurements. |

**Table 8.1** *Continued*

| Question | Explanation | Why important? |
|---|---|---|
| What is the instrument response time? | Response time is the time taken by an instrument such as a thermometer to respond to a change in the variable being measured (e.g., temperature). | If the instrument response time is too slow, important details of the climate phenomenon may be missed and the measurements may be misleading. If the response time is too fast, more detail is recorded than is needed. |
| How will the instrument be calibrated? | Calibration is the process of relating the magnitude of the output of a measurement instrument to the magnitude of the input (American Meteorological Society 2000). A standard needs to be established against which all instruments in an observing network are calibrated. Calibration usually occurs both in the laboratory and in the field. | Uncalibrated or mis-calibrated instruments lead to inaccurate measurements. Measurements from instruments at different locations cannot be compared if they are not calibrated to the same standard. |
| How might the measurements degrade with time? | Instrument drift is a gradual and unintended change in the reference value of an instrument with time. Drift is usually caused by deterioration of instrument parts or when an instrument becomes dirty. | Instruments need to be routinely maintained and frequently recalibrated. |
| What types of systematic measurement errors can be expected? | Systemic error is the part of the inaccuracy of a measuring instrument that is due to a single cause or small number of causes (American Meteorological Society 2000). | It may be possible to calculate bias factors to adjust for systematic error. |
| What is the geographical representativeness of the measurement site? | A measurement is usually considered to be representative of the surrounding area, although the size of that area varies. | To increase the area that a measurement is representative, it is necessary to site the instrument such that microclimate variations are minimized, usually done by placing instruments in flat open areas with uniform surface characteristics. |

**Table 8.1**   *Continued*

| Question | Explanation | Why important? |
|---|---|---|
| What is the instrument exposure? | Instrument exposure is simply the physical location of the instrument; the surrounding environment can have a large effect on the representativeness of the measurement (American Meteorological Society 2000). | Instruments should be installed away from obstacles such as tall trees or nearby buildings that may interfere with the measurement. Exposure may change with time. |
| What post-processing of measurements is required? | Many of today's instruments are electronic and measurements are recorded automatically every fraction of a second. These measurements often need to be reduced and summarized. | The appropriate averaging period for the measurements depends on the research objectives. |
| How will the data be stored? | The quantity of atmospheric observations is voluminous even for short field experiments. | Procedures for data storage should be determined at the onset of a research project. |
| What metadata need to be recorded? | Metadata are data about data. | Metadata are essential, especially if observations are to be shared by others. |

as "time of observation bias." Many temperature measurements are taken by volunteer observers who select an observation time that is convenient to them. Most often they find early morning or late afternoon times (before or after their work day) to be most convenient, and they record the maximum and minimum temperature that occurred during the 24-hour period ending at the observation time. Unfortunately, early morning and late afternoon correspond to the typical time of day of the lowest and highest temperatures, respectively. If an observer records maximum temperature at 4 pm local time, for example, the warm afternoon temperatures may be the highest temperature not only for that "observational day" but also for the following 24-hour period. Thus, locations with afternoon observation times tend to have warm biases compared to stations with a midnight-to-midnight observation time, whereas those with morning observation times tend to have cold biases (Baker 1975). In addition to time of observation bias, different observing practices in different countries introduce additional complications especially for continental or global scale climate analyses; for example, in most European countries temperature and humidity observations are taken at a height of 2 m above the ground surface, whereas in the United States they are made at 1.5 m. Although this difference may seem small, temperature typically changes rapidly with height near the Earth's surface, and consequently small variations in the elevation of the sensor can have a substantial influence on the observed temperature.

Given the potential inhomogeneities in climate records, an essential first step of any climatological analysis is to carefully inspect the observations for homogeneity. While never easy, this task is simplified somewhat if metadata are available. In this situation, the common approach is to compare the period of record when changes (such as an instrument change) are known to occur to that at a nearby station with no change, and to calculate an adjustment factor. Peterson et al. (1998) provide an excellent review of the many different methods used to detect inhomogeneities in the climate record.

## Sources of Archived Data

Finding the most appropriate retrospective climate data set for a particular research question may require considerable effort. Some, but not all countries, have data centers that archive and distribute climate data sets, although not all climatological data sets are freely available. An example of a national data set is the surface observations of the United States Historical Climatology Network (USHCN) http://www.ncdc.noaa.gov/oa/climate/research/ushcn/ushcn.html. The network is comprised of approximately 1,200 stations selected on the basis of their long period of record, small percentage of missing data, and modest number of changes in station location, instrumentation, and observing time. Another well-known data set is the central England temperature record (HadCET) http://hadobs.metoffice.com/hadcet/. Observations from multiple stations were combined to produce composite monthly average temperatures extending from 1659 to present (composite daily values begin in 1772). Two global data sets of surface temperature for land areas that are widely used in climate studies are CRUTEM3 http://www.cru.uea.ac.uk/cru/data/temperature/ and the Global Historical Climatology Network (GHCN) http://www.ncdc.noaa.gov/oa/climate/ghcn-monthly/index.php. The data are gridded monthly temperature anomalies, calculated as deviations from 1961–1990 mean values, with a 5° latitude by 5° longitude resolution. The HadCRUT3 data set http://hadobs.metoffice.com/hadcrut3/diagnostics/global/nh+sh/, by contrast, incorporates both sea-surface and land-based observations. Data sets that combine *in situ* observations with remotely-sensed observations provide even greater spatial coverage; an example being the Global Precipitation Climatology Project (GPCP) http://cics.umd.edu/~yin/GPCP/main.html, which provides estimates of monthly mean precipitation from 1979 onwards at 2.5° latitude by 2.5° longitude resolution.

These data sets are only a few examples of archived climatological observations. Whatever the type or source of archived climate data, it is important to remember that it is the user's obligation to inspect the data for completeness, inhomogeneities, and possible errors. Failure to do so may result in meaningless or misleading analyses.

## Research Methods in Climatology

Research methods range from the simple to the complex. The methods used in climate research are constrained only by the researcher's knowledge and imagination. Two common approaches in climate research – statistical analysis and numerical modeling – are briefly discussed.

## Statistical methods

Zwiers and von Storch (2004: 666) summarized the many ways that statistical procedures have been employed in climatological research. In their words, "the use of statistics is pervasive in the climate sciences not only for the extraction and quality control of data but also for the synthesis of knowledge and information from that data." Applications of statistics include: data retrieval and post-processing of remotely-sensed observations such as radar measurements; estimating climate parameters from proxy records of climate dependent phenomena such as the growth rings of a tree (see Chapter 9); quality control of observational data including tests for inhomogeneities; spatial and temporal interpolation of atmospheric observations; medium and long-range climate forecasts; forecast verification; the identification of modes of variability in the climate record such as the **North Atlantic Oscillation** (NAO) (the fluctuation in sea-level pressure between the low pressure typically found in the vicinity of Iceland and the subtropical high pressure known as the Azores High); the detection of **climate change**; synthesis of observations and/or model output; and hypothesis testing.

The statistical methods used by climatologists extend from the simple descriptive statistics used to summarize data (see Chapter 17) to the techniques, some of which are very sophisticated, that are used to make inferences about the characteristics or relationships of the climate of a location or region (see Chapter 18). Regardless of the simplicity or sophistication of a statistical method, researchers must be aware of the assumptions underlying the methods themselves and the nature of the climate data to which the methods are applied. For this reason it is worthwhile to highlight some idiosyncrasies of climate data that can complicate the application of statistical techniques and/or may violate their underlying assumptions. For example, some climate elements are **vector** quantities. A vector quantity, such as wind, has both magnitude and direction, and is consequently more complicated to analyze than **scalar** quantities, such as temperature, that only possess a magnitude. For example, it is not possible to calculate a simple arithmetic mean of a vector quantity. Of course the mean wind speed can be specified, but this is an incomplete description of the wind as it contains no information about wind direction. An alternative is to separate the horizontal wind into two components ($u$ and $v$ where $u$ is the zonal, west-east component of the wind and $v$ is the meridional, north-south component). These components can then be used to calculate the **resultant** wind, or vector mean, of a series of wind observations. The magnitude of the resultant vector, or the vector average wind speed, is not the same as the simple average of the wind speeds, or scalar average wind speed. Nor is the direction of the resultant vector the same as the prevailing wind direction (the direction from which the wind most frequently blows). When analyzing wind observations or for that matter any vector quantity, researchers need to carefully consider, in light of their research objective(s), which measure is most relevant.

Any statistical analysis of climate observations must consider the area for which a measurement is representative. Measurements of atmospheric phenomena rarely are considered to be point estimates, even though the field of view of the instrument taking the observation may be only a few square centimeters or meters. The outer cylinder of a standard rain gauge, for example, has a small diameter (127mm in the United Kingdom and 203mm in the United States), but the representative area assigned to a gauge may, depending on the surrounding terrain, extend several hundred kilometers. In some situations, this can lead to misinterpretation as precipitation events can begin and end abruptly, and large spatial gradients exist because some locations experience precipitation while other nearby locations do not. Not surprisingly, spatial interpolation (filling in the space between measurements to create a continuous data set) is a complex undertaking in climatology since observation locations are rarely distributed uniformly in space and some climate variables, such as precipitation, are not as spatially and temporally continuous as others.

A number of statistical methods assume that variables are normally distributed. Although some climate variables, such as temperature, closely approximate the symmetrical Gaussian (normal) distribution, others, such as precipitation, are not normally distributed. Precipitation amounts, especially daily totals, tend to be highly skewed with a large number of days with no precipitation and only a few days with heavy precipitation. Often precipitation is better represented with Gamma or similar distributions. Another common assumption in statistics is stationarity. The mean and variance of a stationary process do not change with time or position. Climate data often exhibit temporal trends that violate this assumption. This means, for example, that defining climate "normals," which are an estimate of the "expected value" of a climate variable and are often used for design, planning, and decision making purposes, may be more difficult than one might initially assume. A climate normal is defined by the WMO as a 30-year average updated every 10 years. Recent studies, such as that of Huang et al. (1996), recommend that because of trends in climate variables a shorter averaging period should be used and climate normals should be updated more frequently. Climate data are also correlated in space and time and the "degrees of freedom" (usually calculated as the number of data elements, $N$, minus the number of statistical parameters, for example $N - 1$; see Chapter 18) for statistical testing need to be reduced to account for this correlation. Not adjusting the degrees of freedom for spatial and/or temporal autocorrelation increases the likelihood of rejecting a null hypothesis when, in fact, it is true. Although a number of methods have been proposed for dealing with what is often referred to as the "multiplicity problem" (Wilks 2006), unfortunately many researchers often fail to make the necessary adjustment.

Statistical analyses, as already noted, have been used for a variety of applications but one area worth highlighting is **climate classification**, a historically important component of climatology. Climate classification has two primary aims. The first is to group together locations with similar climates, the rationale being to summarize spatial variations in climate and suggest causes for the observed differences; probably the most well known example is the Köppen-Geiger classification (http://koeppen-geiger.vu-wien.ac.at/) which is based on observations of average annual and monthly temperature and precipitation and the seasonality of precipitation. The second aim is to group together similar atmospheric circulation patterns in order to forecast weather and climate and study changes in circulation with time, an example being Lamb's catalogue of daily weather patterns for the British Isles which can be accessed at http://www.cru.uea.ac.uk/cru/data/lwt.htm. In recent years, subjective methods of climate classification have for the most part been replaced by computer-assisted approaches that rely heavily on multivariate statistics to group locations with similar climates or days/months with similar circulation patterns. An example is the automated catalog of mid-**tropospheric** airflow developed by Huth (2001) for Europe. Principal components analysis, a multivariate statistical technique frequently used in both physical and human geography, was applied to daily observations of the elevation of the 500 hPa pressure surface (which represents the half way point in the atmosphere in terms of density) to isolate the primary modes of the circulation. Component scores were then computed that measure the association between each case (or day in the case of Huth's study) and each component (circulation mode). The scores were grouped using another common multivariate technique known as cluster analysis and the final clusters represent the primary mid-tropospheric airflow patterns over Europe. Similar classifications have been developed for other regions and/or other climate variables such as sea-level pressure.

## Numerical modeling

Numerical models are physically-based models developed using the principles of conservation, the first law of thermodynamics, and the laws of motion. That is, numerical models use physical laws to govern their behavior, and they play an extremely important role in meteorology and climatology (the history of numerical weather forecasting is briefly summarized at http://celebrating200years.noaa.gov/foundations/numerical_wx_pred/welcome.html#ahead). Recently, **ensemble modeling** has become an important approach in numerical modeling. In the past, atmospheric scientists assumed that there was a "best" prediction based on a "best model" and accurate and complete data. Ensemble modeling, by contrast, recognizes the chaotic nature of the atmosphere, the imperfect nature of numerical models, and the inability to measure the atmosphere both completely and accurately. Multiple runs of numerical models (i.e. an "ensemble") are performed with slightly different initial conditions, and the range of projections provides an indication of uncertainty in the forecast (see Gneiting and Raftery (2005) and http://www.hpc.ncep.noaa.gov/ensembletraining/ for more information).

A group of **climate models** that has received considerable attention in recent years is the complex, three-dimensional global Atmospheric General Circulation Models (AGCMs) developed to study climate processes, natural climate variability, and the climatic response to anthropogenic forcing such as increased **greenhouse gas** emissions. Often these models are coupled with Ocean General Circulation Models (OGCMs) to create coupled **Atmospheric Ocean General Circulation Models (AOGCMs)** (see McGuffie and Henderson-Sellars (2005) for more information on climate models). Most of today's AGCMs and AOGCMs are run in what is referred to as transient mode, in that greenhouse gases in the atmosphere are allowed to increase with time. Selected output from several AOGCMs are available from the **Intergovernmental Panel on Climate Change** (IPCC) Data Distribution Center at http://www.ipcc-data.org/. Box 8.2 provides an example of how a group of geographers have used numerical models to derive climate change scenarios for different locations in Michigan, USA.

**Box 8.2**   Multiple Data Sources for Studying Climate Change

A recent investigation of the potential impacts of climate variability and change on agriculture and tourism in Michigan provides an example of the use of climate data from multiple sources. One component of this research project was to create a suite of future climate scenarios at the local scale. These scenarios were developed using a process referred to as statistical downscaling. The first step in the downscaling process was to use historical climate observations to generate statistical relationships, referred to transfer functions, between the large-scale atmospheric circulation and local climate. The primary predictor variables representing large-scale circulation were the height of the 500 hPa surface, specific humidity at 850 hPa, and mean sea-level pressure. These variables were obtained from a relatively new data set known as the NCEP/NCAR Reanalysis. Local temperature and precipitation were obtained from the volunteer United States Cooperative Observing Network. The 15 stations for which scenarios were developed were chosen based on the quality of their observational records. Prior to the analysis, the Michigan Office of the State Climatologist evaluated the daily time series at these stations for homogeneity including (a) inspecting the station histories for changes in location, time of observation, and instrumentation, (b) filtering the series for obvious errors, and c) applying a homogeneity test to the time series.

Multiple transfer functions were developed using several regression-based techniques. After the transfer functions were evaluated against observations reserved for a testing period, they were applied to coarse-scale simulations for 1990–2100 from four AOGCMs. The models used were CGCM2, HadCM3, ECHAM4, and NCAR CSM 1.2, developed in Canada, Great Britain, Germany, and the United States, respectively. For each AOGCM, two simulations were available that reflect different increases in greenhouse gases by the end of the century. The end result was a large suite of climate scenarios for each location.

Figure 8.1 shows the projected increase in the median value of growing degree days (GDDs) for 20-year overlapping periods for one of the locations (Eau Claire, Michigan). GDDs are a derived variable calculated from subtracting a base temperature from the daily mean temperature. For this example, the base temperature is 5°C (41°F), which is often used to calculate growing degree days for small grain crops (e.g., wheat). All the scenarios project that the annual heat accumulation will increase during the twenty-first century, suggesting that the growth stages of agricultural crops will occur earlier in the growing season and that crops will reach maturity earlier. Also, the growth and development of many insect pests are temperature dependent, and extra heat accumulation may increase of the number of generations per season of an insect pest. The uncertainty surrounding the projected change increases in the latter part of the century, as can be seen from the greater spread of the scenario ensemble.

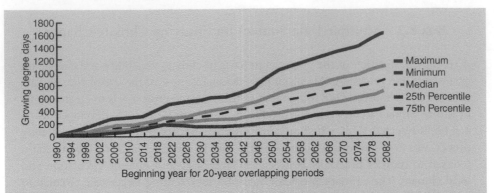

**Figure 8.1** Projected change in the median value of the number of growing degree days (base 5°C) per year for Eau Claire, Michigan. The heavy lines show the largest and smallest changes as projected by a 64-member ensemble of climate scenarios. The projected changes of 50 percent of the scenarios lie between the 25th and 75th percentile. The use of a scenario ensemble provides an estimate of the uncertainty surrounding future climate change. Example provided by the Pileus Project, Michigan State University. For more information see www.pileus.msu.edu

# Detecting Climate Change: An Example of the Analysis of Climate Data

The Intergovernmental Panel on Climate Change (IPCC) was established in 1988 by the WMO and by the United Nations Environment Programme (UNEP) to provide impartial information about climate change (it does not conduct any research or monitor climate). Its fourth (2007) assessment report stated that "warming of the climate system is unequivocal" and "most of the observed increase in globally averaged temperatures since the mid-20th century is very likely due to the observed increase in anthropogenic greenhouse gas concentrations" (http://www.ipcc.ch/pdf/assessment-report/ar4/wg1/ar4-wg1-spm.pdf). Thus climate change resulting from increased greenhouse gases may already be impacting Earth's environment and human activity (http://www.epa.gov/climatechange/science/stateofknowledge.html#ref), and this important scientific and environmental issue, and the uncertainty and controversy surrounding it, highlights both the importance and the limitations of climate observations.

Several key data sets have been used to detect climate change; one of the most important being the series of measurements of atmospheric carbon dioxide that have been made at Mauna Loa Observatory, Hawaii, since 1958 (Figure 8.2). These observations, characterized by a marked increase in carbon dioxide concentrations since measurements began, are one indicator of the effect human beings can have on the planet they inhabit. Anthropogenic emissions of so-called "greenhouse gases," including carbon dioxide but also other gases such as methane and nitrous oxide, are believed to be the primary reason for the recent observed increase in globally-averaged temperatures.

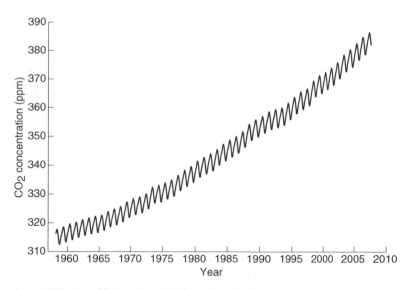

**Figure 8.2**  Trend in monthly average of carbon dioxide ($CO_2$) concentration in air samples obtained at Manua Loa Observatory, Hawaii. The data used to construct the graph and supporting information can be found at (http://scrippsco2.ucsd.edu/home/index.php)

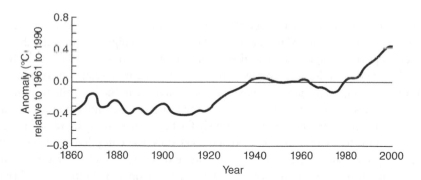

**Figure 8.3**  The Northern Hemisphere temperature curve constructed by Folland et al. 2001

Compilations of mean surface (land and/or ocean) air temperature were assembled in the early 1980s and have been regularly updated since that time. One of the best known of these data sets is the Northern Hemisphere temperature curve developed by climatologists at the University of East Anglia (UK) (Jones et al. 1982). Annual mean temperature is expressed in terms of a departure (or anomaly) from the mean temperature for a fixed time period (see Figure 8.3). The data themselves and the methods used to analyze them have been criticized; nevertheless, this and similar data sets, which have been expanded to include the land and ocean surfaces of the entire globe, reveal an increase in global mean temperature of 0.74°C ± 0.18°C over the last 100 years and almost double that rate of warming during the past 50 years (http://www.ipcc.ch/pdf/assessment-report/ar4/wg1/ar4-wg1-chapter3.pdf).

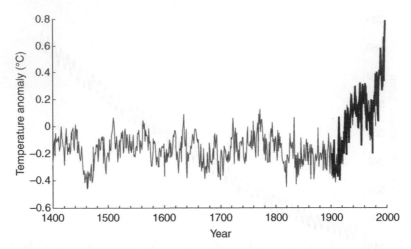

**Figure 8.4**  Estimates of annual Northern Hemisphere mean temperature are plotted as anomalies from the 1961–90 average temperature after Mann et al. (1999). Temperature estimates based on reconstructions from proxy data are shown with a thin line and those from the instrument record are shown by a thick line. The data as well as information about the methods used to derive them are available at http://www.ncdc.noaa.gov/paleo/ei/ei_cover.html

A somewhat controversial temperature series is Mann et al.s (1998) "hockey stick" graph, an expanded version (Mann et al. 1999) of which was highlighted in the third IPCC assessment report. The significance of this graph is that it suggests that the relatively recent trend in Northern Hemisphere average temperature is highly unusual in as much as the rate of warming is greater than that experienced at anytime in the last millennium (Figure 8.4). The graph, which received its name from the sharp discontinuity that occurs around 1900 (prior to which time the trend is relatively flat) and the ensuing increase in temperature that combine to resemble the shape of a hockey stick, was compiled using instrumental records and proxy data derived from a wide range of sources, such as tree rings and ice cores, obtained from around the globe (see Chapter 9). Concern about the validity of the "hockey stick" graph revolves around the issues of data quality and the methods used to merge proxy data derived from different sources. However, a report published in 2006 on "*Surface Temperature Reconstructions for the Last 2,000 Years*", authored by the special committee established by the National Research Council (USA), concluded that Northern Hemisphere surface temperature reconstructions for the past thousand years yield generally consistent results (http://books.nap.edu/openbook.php?record_id=11676&page=1).

A well-known time series that serves to illustrate the uncertainty surrounding climate observations has been compiled from global satellite temperature measurements from 1979 to the present (Spencer and Christy 1990). The data were obtained by Microwave Sounding Units (MSUs), mounted on polar orbiting satellites, that measure radiance (the energy passing through or being emitted by a given area) from which temperature estimates for different layers in the lower atmosphere are derived. Information about the data may be found at http://www.ncdc.noaa.gov/oa/climate/research/msu.html. These measurements have a broad spatial coverage, but the process of constructing a temperature record is a complex undertaking, not least because of changes in local observing time and a decay of orbital height caused by drag on the satellite by the atmosphere. There have been numerous recalculations of the satellite-derived temperature series, and although earlier versions suggested that the recent warming of the lower atmosphere was not as rapid as surface temperature observations suggest, more recent research reveals a better agreement between the satellite measurements and surface-based temperature observations (Mears and Wentz 2005).

# Conclusion

The use of climate observations by geographers can be expected to expand along with their interest in the linkages between climate and other physical and human processes. Taking field observations can be challenging, and attention must be paid to the research requirements, the nature of the climate phenomena being studied, and the specifics of instrument and site characteristics. Using and interpreting archived climate data can be equally taxing because inhomogeneities occur in the data, and their spatial and temporal resolution as well as the period of record are not always appropriate for the questions that are being posed. Regardless of how obtained, when analyzing climate data researchers also must be aware of the inherent assumptions and limitations of the statistical methods and models they use. Nevertheless, it is only through the careful treatment and analysis of climate observations, along with reconstructions of climate variables from proxy sources and the results derived from statistical and numerical models, that an understanding can be gained of the way in which Earth's complex climate system has operated in the past and the way it may function in the future.

# References

American Meteorological Society (2000) *Glossary of Meteorology* (2nd edn). New York: Allen Press.

Baker, D. G. (1975) Effect of observation time on mean temperature estimation. *Journal of Applied Meteorology* 14: 471–6.

Folland, C. K., Rayner, N. A., Brown, S. J., Smith, T. M., Shen, S. S. P., Parker, D. E., Macadam I., Jones, P. D., Jones, R. N., Nicholls, N., and Sexton, D. M. H. (2001) Global temperature change and its uncertainties since 1861. *Geophysical Research Letters* 28: 2621–4.

Gneiting, T., and Raftery, A. E. (2005) Weather forecasting with ensemble methods. *Science* 310: 248–9.

Guttman, N. B., and Baker, C. B. (1996) Exploratory analysis of the difference between temperature observations recorded by ASOS and conventional methods. *Bulletin of the American Meteorological Society* 77: 2865–73.

Huang, J., van den Dool, H. M., and Barnston, A. G. (1996) Long-lead seasonal temperature prediction using optimal climate normals. *Journal of Climate* 9: 809–17.

Huth, R. (2001) Disaggregating climatic trends by classification of circulation patterns. *International Journal of Climatology* 21: 135–53.

Intergovernmental Panel on Climate Change (IPCC) (2001) *Climate Change 2001: The Scientific Basis.* Houghton, J. T., Ding, Y., Griggs, D. J., Noguer, M., van der Linden, P. J., Dia, X., Maskell, K., and Johnson, C. A. eds. Cambridge: Cambridge University Press.

Jones, P. D., Wigley, T. M. L., and Kelly, P. M. (1982) Variations in surface air temperatures: part 1. Northern hemisphere, 1881–1980. *Monthly Weather Review* 110: 59–70.

Mann, M. E., Bradley, R. S., and Hughes, M. K. (1998) Global-scale temperature patterns and climate forcing over the past six centuries. *Nature* 392: 779–87.

Mann, M. E., Bradley, R. S., and Hughes, M. K. (1999) Northern hemisphere temperatures during the past millennium: inferences, uncertainties, and limitations. *Geophysical Research Letters* 26: 759–62.

McGuffie, K., and Henderson-Sellars, A. (2005) *A Climate Modelling Primer* (3rd edn). Chichester, West Sussex; Hoboken, NJ: John Wiley & Sons.

Mears, C. A., and Wentz, F. J. (2005) The effect of diurnal correction on satellite-derived lower tropospheric temperature. *Science* 309: 1548–51.

Orlanski, A. (1975) Rational subdivision of scales for atmospheric processes. *Bulletin of the American Meteorolological Society* 56: 527–30.

Peterson, T. C., Easterling, D. R., Karl, T. R., Groisman, P., Nicholls, N., Plummer, N., Torok, S., Auer, I., Boehm, R., Gullett, D., Vincent, L., Heino, R., Tuomenvirta, H., Mestre, O., Szentimrey, T., Salinger, J., Forland, E. J., Hassen-Bauer, I., Alexandersson, H., Jones, P., and Parker, D., (1998) Homogeneity adjustments of in situ atmospheric climate data: a review. *International Journal of Climatology* 18: 1493–517.

Spencer, R. W., and Christy, J. R. (1990) Precise monitoring of global temperature trends from satellites. *Science* 247: 1558–62.

Wilks, D. S. (2006) *Statistical Methods in the Atmospheric Sciences, Volume 91* (2nd edn). Burlington, MA; London: Academic Press.

Winkler, J. A. (2004) The impact of technology upon in situ atmospheric observations and climate science. In *Geography and Technology*, Brunn, S. D., Cutter, S. L., Harrington Jr, J. W., eds. Norwell, MA: Kluwer Academic Publishers: 461–90.

Zwiers, F. W., and von Storch, H. (2004) On the role of statistics in climate research. *International Journal of Climatology* 24: 665–80.

## Additional Resources

Carleton, A. M. (1999) Methodology in climatology. *Annals of the Association of American Geographers* 89: 713–35. The methodologies employed by climatologists trained in geography are compared to those of climatologists trained in atmospheric science.

IPCC (2007) *Climate Change 2007: The Physical Science Basis. Contribution of Working Group I to the Fourth Assessment Report of the Intergovernmental Panel on Climate Change*. Solomon, S., Qin, D., Manning, M., Chen, Z., Marquis, M., Averyt, K. B., Tignor M., Miller, H. L. eds. Cambridge: Cambridge University Press. The first volume of the IPCC Fourth Assessment Report reviews the current scientific understanding of the climate system and its sensitivity to greenhouse gas emissions, particularly those aspects judged to be most relevant to policymakers.

IPCC (2007) *Climate Change 2007: Impacts, Adaptation and Vulnerability. Contribution of Working Group II to the Fourth Assessment Report of the Intergovernmental Panel on Climate Change*. Parry, M. L., Canziani, O. F., Palutikof, J. P., van der Linden, P. J., and Hanson, C. E. eds. Cambridge: Cambridge University Press. The second volume of the IPCC Fourth Assessment considers climate change impacts, vulnerabilities and adaptation options.

IPCC (2007) *Climate Change 2007: Mitigation. Contribution of Working Group III to the Fourth Assessment Report of the Intergovernmental Panel on Climate Change*. Metz, B., Davidson, O. R., Bosch, P. R., Dave, R., and Meyer, L. A., eds. Cambridge: Cambridge University Press. The third volume of the IPCC Fourth Assessment describes the mitigation options and opportunities for the various sectors contributing to greenhouse gas emissions along with the costs of these options.

Oliver, J. E. (1991) The history, status, and future of climatic classification. *Physical Geography* 12: 231–51. The historical evolution of classification in climatology is reviewed, and the future importance of classification to climate research and teaching is considered.

Oliver, J. E., and Hidore, J. J. (2002) *Climatology: An Atmospheric Science* (2nd edn). Upper Saddle River, New Jersey: Prentice Hall. This classic textbook in climatology provides an excellent introduction to the fundamentals of the atmosphere and regional variations in climate.

Parker, D., and Horton, B. (2005) Uncertainties in central England temperature 1878–2003 and some improvements to the maximum and minimum series. *International Journal of Climatology* 25: 1173–88. Use of the central England temperature record is complicated by its many inhomogeneities. The greatest contribution to uncertainty is shown to be the number of stations used to create the composite average with thermometer calibration the second largest source.

Rogers, J. C., Winkler, J. A., Legates, D. R., and Mearns, L. O. (2003) Climate. In *Geography in America at the Dawn of the 21st Century*, Gaile, G. L. and Willmott, C. J., eds. Oxford: Oxford University Press, 32–46. The major thematic areas currently being addressed by climatologists within the field of geography are identified as atmospheric circulation, surface-atmosphere interactions, hydroclimatology, and climate change.

Rohli, R. V., and Vega, A. J. (2008) *Climatology*. Sudbury, MA: Jones and Bartlett Publishers. This recent textbook includes a detailed discussion of the variation of climate with time and of the applied aspects of climatology.

World Meteorological Organization (2008) *Guide to Meteorological Instruments and Methods of Observation WMO-No. 8* (7th edn). Geneva, Switzerland: Secretariat of the World Meteorological Organization. Available at: http://www.como.int/pages/prog/www/IMOP/publications/CIMO-Guide/CIMO%20Guide%207th%20Edition,%202008/CIMO_Guide-7th_Edition-2008.pdf This publication provides extensive recommendations on good practice when taking meteorological measurements and observations with particular emphasis on quality assurance and instrument maintenance.

## Exercise 8.1 Statistical Characteristics of Temperature and Precipitation Observations

This exercise is intended to introduce you to the statistical characteristics of daily temperature and precipitation observations and draws on methods that you were introduced to in earlier chapters.

1 Obtain a time series of daily temperature and precipitation observations that is at least 20 years in length for a station located close to your university. Speak with your professor and/or a climatologist on campus about local sources of climate data. Also, check the national climate archive for your country for available data.

2 Find the available metadata for these temperature and precipitation time series. Also obtain any available photographs or aerial images of the site. [Hint: Google Earth is an excellent source of aerial imagery for present-day site conditions.] What types of instruments were used to measure temperature and precipitation? At what heights were the instruments installed? Has the type of instruments and their installation heights changed over the record period? What is the land cover of the area surrounding the instrument site? Have the site conditions changed with time?

3 Enter the data into a spreadsheet such as Excel. Carefully check the time series for missing observations. What percentage of observations are missing?

4   Create histograms of daily mean temperature and daily precipitation totals. If daily mean temperatures are not included in your data set, first calculate the daily mean for each day from the average of the maximum and minimum temperature reported for that day. You will need to experiment with bin sizes for the histogram. Keep the bins small enough so the details of the distributions are clearly evident.

5   Inspect the histograms for the two variables. Based on the histograms, do both of the variables appear to be normally distributed? Why or why not? If you have access to statistical software, calculate one or more of the better known measures for evaluating normality (e.g., the K-S test). Do the test results agree with your visual inspection?

6   Speculate on why these variables are, or are not, normally distributed. One factor to consider is whether the variable has upper or lower bounds.

7   Plot the monthly average temperature and the monthly average precipitation for the period of record. You can calculate the monthly means within the spreadsheet, or alternatively you can use the published climatological normals for the station (if available). Which of the two variables has a stronger annual cycle? Speculate on the reasons for any differences in the strength of the annual cycle.

8   Plot the daily mean temperatures for the last year of the record. Closely inspect the plot. What does the graph of daily mean temperature tell you about the temperature variation during that year?

9   Using the spreadsheet functions (or writing a small program), calculate the average temperature for the period of record for each day of the year. In other words, calculate the average temperature for January 1, January 2, etc. where the average is the sum of all the daily mean temperatures for that day divided by the number of years of data. Then calculate daily temperature **anomalies** by subtracting the daily average from each day's mean temperature. Plot the daily anomalies for the last year of the record. What does the graph of daily temperature anomalies tell you about the temperature variation during that year? Which graph did you find the most informative? The graph of daily mean temperature or that of daily temperature anomalies? Why?

## Exercise 8.2    Urbanization and Temperature Observations

Changes in land use and land cover are thought to be an important contributor to climate change, at least on the local and regional scale but likely at the hemispheric and global scales as well. This exercise explores the influences of urbanization on temperature observations.

1   Obtain temperature observations for two stations in and near a large urban area of your choice. One station should be located close to the urban core and the second should be located in the rural environment close to the periphery of the urban area. Both should have at least 20 years of maximum and minimum temperature observations. Carefully check the time series for missing observations.

2   What metadata are available for these stations? At a minimum, you should know the latitude, longitude and elevation of the stations and the period of record. See if you can also find out whether the stations have been moved or if the instrumentation has changed during the period of record. At what height have measurements been taken

and what is the land use/land cover of the measurement site? How have the site conditions, especially those for the urban site, changed with time?

3 For both stations, use a spreadsheet to calculate the average minimum temperature for each year of the observational record. Plot the time series of annual average minimum temperature for both stations on the same graph. Do you see any trends in the annual average minimum temperature? Are the trends the same for both stations? You may want to use the line-fitting function of the spreadsheet to help estimate trends.

4 Now plot the time series of annual average maximum temperature for both stations on the same plot. Do you see any trends in the annual average maximum temperature? Are the trends the same for both stations? Is there any difference in the trends for annual average maximum temperature compared to annual average minimum temperature? If so, speculate on the possible reasons for this.

5 Based on your observations above, is there any indication that urbanization may be impacting the temperature of the station near the urban core? If so, has the influence of urbanization increased with time? To help answer this last question, calculate for each year the difference in the average minimum temperatures at the two stations and plot the differences. An increase or decrease with time in the temperature difference between the stations would suggest changes with time under the influence of urbanization. Create a similar plot for annual average maximum temperature.

6 Do the plots above suggest that the differences in annual maximum and minimum temperature between the two stations (calculated above) increase or decrease fairly uniformly with time? Or are discontinuities or breakpoints seen in the plots? If the latter, do the discontinuities correspond to any of the known inhomogeneities in the data such as a location change of one of the stations?

7 Time permitting, perform the analyses above separately for the winter and summer seasons. Is the influence of urbanization on temperature larger for the winter season or for the summer season? Speculate on the reasons for any differences between seasons.

## Exercise 8.3 Isolating Climate Trends from Time Series

The goal of the following exercise is to illustrate the challenges of isolating climate trends from a time series with a high degree of variability.

1 Obtain a time series of temperature that is at least 50 years in length for a station of your choice. Speak with your professor and/or a climatologist on campus about local sources of climate data. Also, check the national climate archive for your country for available data.

2 Check the available metadata for the station for possible inhomogeneities including stations relocations or instrument changes. Also inspect the time series for missing observations.

3 Using a spreadsheet program, calculate the average temperature of each year of the time series from the daily temperature values. [Note: Depending on your data source, the annual means may already be available.] Plot the time series of annual mean temperature using a line graph. Comment on the degree of interannual variability. In other

words, does the average annual temperature vary considerably from one year to the next? Or is the amount of interannual variability relatively small?

4 Does there appear to be a trend in the time series of annual temperature? Or is any trend masked by the degree of interannual variability? If a trend is evident, what is the sign of the trend (i.e., positive and negative)? Is the trend uniform over the entire period of record?

5 Using the smoothing function or moving average options available in the spreadsheet, apply a three-point smoothing function to the time series data. A three-point smoother replaces the annual value for a particular year with the average of the temperature for that year and the two surrounding years. Are you now better able to identify trends in the temperature time series? Next try a 5-point smoother, a 7-point smoother, and so on. How does the degree of smoothing of the time series change your perception of the trends in the time series?

6 Using the smoothed time series that you feel best illuminates the trends in the time series, add a trend line to the plot (you can do this using the spreadsheet program). Is the overall trend positive, negative, or neutral?

7 Print a copy of the plot of the smoothed time series. Mark on the paper copy those subperiods of the record when average annual temperatures appear to have increased, those when average annual temperatures appear to have decreased, and those when average annual temperatures were relatively constant with time. Do any of the break-points (times when the trend changes sign) correspond to known inhomogeneities in the data series?

8 Compare the time series plot for your location to the Northern Hemisphere temperature curve shown in this chapter. Do the times of increasing (decreasing) temperature for your location correspond with the upward (downward) trends seen for the Northern Hemisphere temperature curve? What does this comparison suggest regarding local versus hemispheric temperature change?

# Chapter 9

# Vegetation

*Thomas W. Gillespie and Glen M. MacDonald*

- ■ Introduction
- ■ Quantifying Vegetation
- ■ Past Vegetation
- ■ Contemporary Vegetation
- ■ Conclusion

Keywords

Biomass
Dendrochronology
Density
Diameter at breast height
Environmental gradient
Floristic provinces

Macrofossils
Palynology
Scientific name
Species abundance
Species richness
Vegetation cover

## Introduction

Vegetation is defined as all the plant life in a specific place or time period (Barbour and Billings 2000). Vegetation can be described in terms of the species of plants that are found, the morphology, life history, and physiognomy of the dominant plants or the general structure of the vegetation. The type of vegetation growing in a specific location is related to factors such as climate, topography, geology, soil conditions, natural disturbance, and anthropogenic disturbance. Due to the importance of vegetation and its linkages with the environment and human activities, geographers have long been interested in the classification of vegetation and the processes that structure vegetation over global, regional, and local spatial scales (MacDonald 2003). The scientific classification of the earth into biomes, regions with similar vegetation and climate regimes, was of great interest to early biogeographers such as Alexander Von Humboldt (1769–1859) and Alfred Russell Wallace (1823–1913) 150 years ago. Today, there are a number of vegetation classification systems in use around the world, as well as many theories and models of how vegetation responds to disturbance and environmental change over different spatial scales. Any vegetation classification system, or consideration of how environmental change or disturbance impacts vegetation, must be based upon, supported with, or tested by, data on present or past distribution that are collected in the field in a systematic and scientific manner (see Chapter 3).

## Quantifying Vegetation

One way to quantify vegetation is to examine the present or past species composition of a vegetation type. Species composition is defined as a list of plants found in a particular region of the world or study site. Species composition and **abundance** can also be examined from a specific time period for historic studies of vegetation. Biogeographers have quantified the species composition of regions around the world by collecting and identifying plants in different vegetation types. At a global spatial scale, this has resulted in the identification of a number of biogeographic **floristic provinces**, which contain many endemic species that are unique to each region. Geographers also quantify species composition by adding up the number of different species encountered in a study area, often referred to as **species richness**, and biogeographers have long been interested in patterns of species richness across a number of **environmental gradients**.

A second way to quantify vegetation is to examine the structure or physiognomy of the dominant vegetation canopy. This is a two-dimensional view or profile of plants in an area. All terrestrial ecosystems can be classified based on structure into a few simple classes such as forest, woodland, shrub, herbaceous-grassland, and desert communities (Figure 9.1). Forest communities are tracts of land dominated by trees with at least 70% canopy closure like tropical rainforests or the deciduous forests of the eastern United States. Open forests with less than 70% canopy closure are referred to as woodlands, such as the oak woodlands in California and Texas or the eucalyptus woodlands in Australia. Shrub communities are dominated by multi-stem, woody plants that generally reach a height of 2 meters. Many

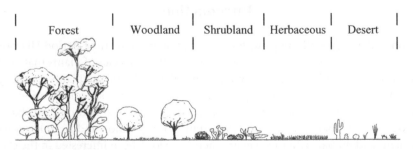

**Figure 9.1**   Profiles of forests, woodlands, shrub, herbaceous-grassland, and desert communities

shrub communities have closed canopies like the chaparral in California or open canopies with isolated shrubs and herbaceous ground cover, such as the mesquite shrub communities in Texas. Herbaceous communities are generally composed of relatively continuous cover of annual or perennial herbs, grasses, and grass-like species. Those dominated by species in the grass family are generally classified based on plant height, including the short grasslands in the drier regions of the western United States and the tall grasslands in the more humid regions of the eastern United States. Grasslands in the tropics are referred to as savannas. Finally, desert communities are found in arid regions, with less then 25 cm of annual rainfall, where shrubs and/or herbaceous plants cover 50% or less of the soil surface. These simple structural patterns for terrestrial vegetation occur throughout the world (Walter 1985).

There are a number of general questions and research approaches related to the past and present distribution of vegetation that are of widespread interest to geographers today. Although geographers address spatial questions and use a variety of field techniques to examine past and present vegetation patterns and processes, some basic terms and methods are widely used by all researchers.

## Past Vegetation

Research on past vegetation by geographers has generally focused on changes in vegetation types since the late Pleistocene (after the peak of the last great ice age about 20,000 years ago) and on changes in vegetation as it relates to past climatic conditions or disturbance regimes. Some geographers remove sediment cores from lakes and peatlands (highly organic material found in marshy areas) to reconstruct vegetation in the recent to more distant past (Edwards and MacDonald 1991; MacDonald and Edwards 1991). These cores can contain wood, leaves, seeds, charcoal, and microscopic pollen from plants that lived in the region thousands of years ago. Fossil pollen is one of the most important types of evidence used by biogeographers. The sub-field of fossil pollen analysis is called **palynology**. Because many plants depend on the wind to transport pollen and produce large quantities of airborne pollen, pollen grains and spores are often very resistant to decay, particularly in anaerobic environments, and can be preserved for millions of years in lake sediments, marine sediments, and peats. Many pollen grains are distinctive in shape, size, and other features and the palynologist can use these features to determine the family, genus, or species of the plant that produced the pollen (Figure 9.2). Studies have shown that different plant communities produce different and distinctive pollen assemblages (Roberts 1998). Lines of evidence such as large plant fossils or pollen can be used to reconstruct the species composition of the vegetation of an area through time. Other geographic studies collect cores and disks from living and dead trees in order to count the annual growth rings and reconstruct vegetation changes over the past few centuries. The term **dendrochronology** is applied to such studies. Many trees, such as most species of pines or oaks, grow in seasonally variable climates and produce annual growth rings. The rings are formed by couplets of large cells (earlywood) formed at the start of the growth season and small cells formed near the close of the growth season (latewood) (Figure 9.3). Each couplet represents one year of growth and counting the couplets allows one to calculate the age of the tree. Both sediment core and tree-ring studies provide evidence of past climate and

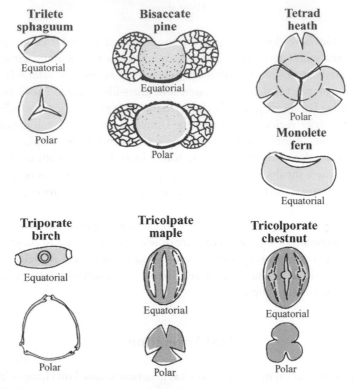

**Figure 9.2**  Different types of pollen grains showing typical shapes and arrangements of pores and openings

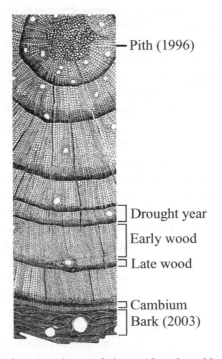

**Figure 9.3**  Cross sections of a tree with annual rings with early and late wood

other environmental conditions which are useful in understanding the environment in which vegetation grew in the past.

## Sampling and field equipment

The two basic approaches most commonly used by biogeographers to study past vegetation, analysis of sediments and analysis of tree-rings, require very different methods and equipment. Analyses of sediments are generally conducted on cores of sediment taken from existing lakes and peatlands. A number of different sediment coring devices are used to obtain samples from soft lake sediments and peat. Russian and Hiller samplers are pushed into the sediment and then rotated to capture a section of the sediment adjacent to the core barrel (Figure 9.4). The Livingstone piston-corer is pushed into the sediment and captures the sediment within the barrel as the corer descends; it can be used in situations where the depth of lake water and sediment does not exceed 20 meters. An airtight piston in the barrel creates an air seal as the core barrel penetrates the sediment, holding them in place while the corer is extracted. Coring devices used in deeper sites use the power of gravity to push the core barrel into the sediment to obtain a short core of the upper sediments. Frozen-finger corers are used to sample disaggregated sediments, particularly when there are fine structures such as annual layers (called varves) present that it is desirable to preserve. The core barrel is filled with dry ice and an agent such a trichloroethylene, and the super-cooled barrel is inserted or dropped into the sediment. The sediment freezes onto the outer surface of the corer and is retrieved in a frozen state.

Tree-ring samples are usually collected from living trees through the use of an increment borer. This device is essentially a thin metal tube with a threaded head that is screwed into the side of the tree at a 90-degree angle to the long axis of the trunk. A small cylindrical sample of wood is captured inside the corer (Figure 9.5). The wood sample is removed and stored in a plastic soda straw. If the corer is long enough to penetrate from the bark to the pith at the center of the trunk, it will capture all annual growth rings and other features formed since close to the time of the germination of the tree. For studies of paleoclimate, or when knowing the exact year of initial establishment of the tree is not important, the core is taken at about 1.5 meters from the ground. However, when studying issues such as tree recruitment, the core may be taken closer to the base of the trunk in order to sample rings formed when the tree was still young and short in stature. Dead trees and downed snags of dead wood are often sampled by removing disks of the trunk. These disks, which are generally 2 to 10 cm thick, are obtained using handsaws and powered chainsaws in the field.

## Site selection

The laboratory analysis of fossil plant material or tree-rings can take a great deal of time and therefore it is important to be careful in the selection of sampling sites. The exact lake chosen to core or stand of trees sampled will often depend on the research question being addressed. Different types of lakes or different stands of trees from the same region can tell us different things about past environment. In general, small lakes and peatlands

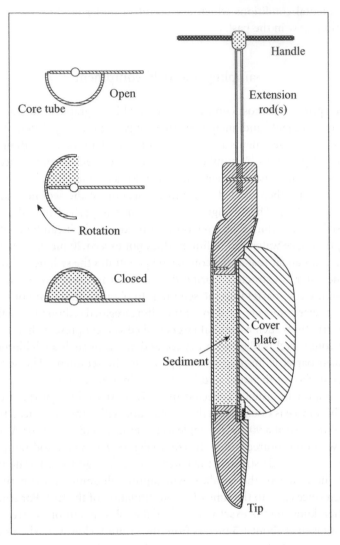

**Figure 9.4**   A sediment coring device (Russian corer) that can be used to obtain samples from lakes and peats http://www.epa.gov/ORD/SITE/reports/600r01010/600r01010.pdf

capture sediment and/or plant fossils from the local area surrounding them. This is even true for fine airborne material such as fossil pollen. In small lakes or forested peatlands the proportion of locally derived pollen in the sediments tends to be relatively high compared to pollen carried from some distance away. Studies which attempt to reconstruct past vegetation at a very local scale often obtain sediments from very small lakes of a few hectares or less. In some cases researchers obtain sediments from small damp hollows in forests that cover only a few meters. Large lakes capture a higher proportion of extra-local pollen and are more useful for reconstructing past regional vegetation. Lakes with large rivers flowing into them may contain fossils from very large source areas.

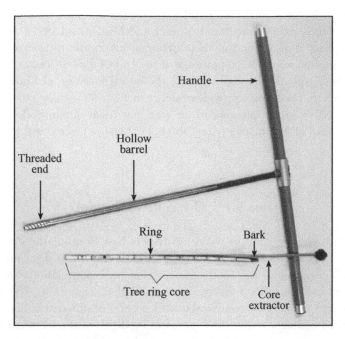

**Figure 9.5** Tree corer and tree ring core

Many tree-ring studies are conducted to examine the relationship of climatic change on tree recruitment or mortality. In these cases, the trees are sampled from stands at the forest edge near the tree species range limits. Such sites might include a high elevation alpine treeline if the relationship between temperature and forest vegetation is being studied. In other cases, trees might be sampled at dry sites at the edges of grasslands and deserts to study the influence of moisture stress on tree populations. For studies of tree to tree interactions, such as the impacts of root and canopy competition, one might sample trees growing in dense stands in the middle of the forest.

## Laboratory preparation and analyses

The analysis of plant **macrofossils** and pollen requires relatively sophisticated laboratory facilities for processing the sediments to isolate the fossils. Good quality stereo microscopes and/or compound microscopes (400x magnification minimum for pollen grains which are only about 10–200 µm in size) and extensive reference collections of plant and pollen samples are necessary to allow identification (see Additional Resources). The ages of the samples are usually determined by radiocarbon dating if the materials are older than 200 years and younger than 40,000 years. Chronologies for samples deposited over the past 200 years are often obtained by $^{210}$Pb dating of the surrounding sediment (see Chapter 7).

Once the plant macrofossils or pollen are isolated, identified, and dated they can be used to infer what the past vegetation cover in the vicinity of the site was like. For example, the bottom-most sediments from many small lakes in southern Ontario, Canada and adjacent New York State contain the fossil wood, cones, and needles of black and white spruce (*Picea glauca* and *Picea mariana*) (Szeicz and MacDonald 1991). Abundant fossil pollen from spruce is also found in these deposits. Radiocarbon dates show that these fossils were deposited some 13,000 years ago at the end of the last ice age. These trees do not grow in these regions today, but occur in the boreal forest located hundreds of kilometers to the north. Their fossils provide evidence that 13,000 years ago there was a boreal forest dominated by spruce growing in an area that today supports deciduous forests dominated by trees such as maples (*Acer*) and oaks (*Quercus*) (see Box 9.1).

## Box 9.1   The Pollen Diagram

An important consideration in pollen analysis is how to numerically analyze and present the raw pollen counts in order to provide a reasonable description of past vegetation. The main tool used for the presentation and interpretation of fossil pollen data from an individual site is the fossil pollen diagram (Figure 9.6). The basic fossil pollen diagram presents the relative abundance (%'s) of different pollen and spore types graphed against a vertical axis representing depth of the sediment section or age (years before present – BP). Qualitative or mathematical approaches are then used to divide the pollen record into stratigraphic zones. In each zone the pollen percentages for the different taxa are similar to each other and show marked differences from the pollen percentages from other portions of the core. The example given here is from a small lake in Ontario, Canada that shows an early boreal forest vegetation dominated by spruce being replaced by pine forest and then various types of mixed deciduous and conifer forest dominated by trees such as oak, maple, beech and hophornbeam and ironwood.

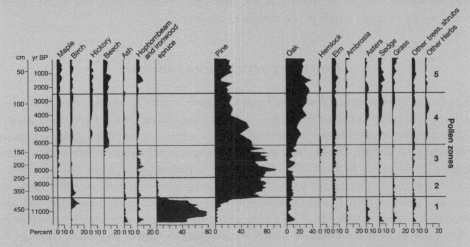

**Figure 9.6**   Pollen diagram from Decoy Lake, Ontario (after Szeicz and MacDonald, 1991)

When tree-ring samples are taken back to the laboratory, they are dried and then sanded to produce extremely smooth surfaces. The thin wood samples obtained from tree-ring cores are generally glued to wood supports to keep them from breaking during sanding and analysis. The rings are counted and their widths are measured using a stereomicroscope linked to a computerized sample holder. In some cases the samples are X-rayed and the variations in wood density analyzed and recorded.

Simply counting the rings from the inner side of the bark to the center of the tree can provide a rough indication of the age of the tree. However, in many cases the rings of trees from the same stand or region show common patterns of variations in width over time, and this variation can be used to verify the dating of the tree rings. A great many trees show a general pattern of large rings near the pith and smaller rings towards the trunk. This is usually a biological growth feature that is not related to the external environment and is statistically removed from the data. Following removal of the growth trend there are often patterns of individual years or groups of years that are represented by particularly large or small rings and these patterns can be found in many samples from trees in the same stand or region. Such variations found in many trees are usually the result of external factors such as climatic variations. For example, periods of extreme cold produce small rings in trees that grow near the arctic treeline. Dendrochronologists use such repeated patterns to verify the ages they assign to rings in individual samples. In addition, if the ring pattern in a piece of deadwood can be matched to the pattern in a sample from a living tree, the time at which the dead tree lived and formed rings can be ascertained. The dating of tree samples by comparing ring widths or densities is called cross-dating. The variations in ring widths are also often used to reconstruct past climate.

Once samples from living trees are aged by dendrochronological techniques the timing at which they became established can be estimated. If most of the samples were established at the same time, this can infer that the stand was created following a disturbance such as a fire or that it has become established during a period of particularly favorable climatic conditions. Similarly, if many pieces of deadwood found at a particular stand date to the same period, it can be inferred that they died due to some disturbance such as a bark beetle infestation, or due to a period of particularly unfavorable climate. Many studies by biogeographers have used tree rings to determine the timing and frequency of disturbance events or the sensitivity of tree populations to climate change. For example, a dendrochronological study of larch trees (*Larix sibirica*) growing near the arctic treeline in northern Siberia showed that many of the currently living trees became established during a prolonged period of warmer summer temperatures that commenced in the early twentieth century (MacDonald et al. 1998). In contrast, most of the dead trees in the stand died during a long period of intense cold in the mid-ninetenth century. This study highlighted the sensitivity of these tree populations, located at 70 degrees north latitude (they are some of the northern-most trees in the world), to even relatively small temperature variations.

## Contemporary Vegetation

Geographers are interested in processes such as how natural and human disturbances can change species composition and structure of vegetation communities. They also develop and deploy tools such as satellite imagery and geographic information systems (GIS) in

**Table 9.1**   Examples of species list from Florida shrub

| Family | Scientific name |
|---|---|
| Apocynaceae | *Asclepias curtissii* |
| Asteraceae | *Chrysopsis floridana* |
| Asteraceae | *Palafoxia feayi* |
| Caryophyllaceae | *Paronychia Americana* |
| Cistaceae | *Helianthemum corymbosum* |
| Cistaceae | *Lechea deckertii* |
| Clusiaceae | *Hypericum reductum* |
| Commelinaceae | *Callisia ornate* |
| Cyperaceae | *Bulbostylis ciliatafolia* |
| Ericaceae | *Lyonia fruticosa* |
| Fabaceae | *Dalea pinnata* |
| Fabaceae | *Galactia regularis* |
| Oleaceae | *Osmanthus megacarpus* |
| Pinaceae | *Pinus clausa* |
| Poaceae | *Aristida gyrans* |
| Polygonaceae | *Polygonella robusta* |
| Scrophulariaceae | *Seymeria pectinata* |
| Selaginellaceae | *Selaginella arenicola* |

efforts to map local, regional, and global vegetation patterns (MacDonald 2003). This is done in order to test theories concerning what environmental factors influence the distribution of contemporary vegetation. Geographers are also interested in measuring and modeling the rate of land-cover change to identify which plants may become endangered due to their inability to migrate fast enough to keep pace with anticipated future climate change or habitat destruction.

There are a number of protocols and terms that geographers should be aware of when dealing with species composition and structure of vegetation. All plants are reported in geographical research by their **scientific names** such as family, genus, and species. All plant family names end in "aceae" like the Poaceae (grass family), Orchidaceae (orchid family), or Pinaceae (pine family). Genus and species names are written in italics or underlined. For instance, the scientific name of the Black Oak is written as *Quercus nigra* or Q̲u̲e̲r̲c̲u̲s̲ n̲i̲g̲r̲a̲. Species lists are simple lists of the scientific names of species in a vegetation type. Most species lists are organized alphabetically by family name, then alphabetically by genus and species (Table 9.1).

Several general terms are used to classify vegetation structure. Individual plants can be classified into three common life-forms: trees, shrubs, and herbs. Trees refer to woody plants taller then 2 m with a single trunk. Shrubs, in contrast, are multi-stemmed plants that are generally less than 2 m tall. Herbs refer to small herbaceous (not woody) plants such as grasses and wildflowers. Sometimes grasses and grass-like plants are referred to as graminoids while wildflowers are referred to as forbs. There are a number of terms that apply to the community structure of vegetation. Species **density** is the number of individuals per unit area. However, it is sometimes hard to measure the absolute density of each individual plant. Therefore some shrub and herbaceous communities are sometimes quantified by the percent **vegetation cover** or the percentage of vegetation covering a ground area. **Biomass** is the amount of living matter in an area and vegetation biomass is the amount of above ground and below ground living plant matter in an area. The most precise way to measure biomass is to cut the vegetation, heat it in an oven to remove the water content, and weight it. This is very labor intensive so many geographers use other methods to approximate biomass. For forest communities, basal area is the most common way to assess biomass. Basal area is the total cross-section area of tree trunks per unit area. Basal area is generally quantified by measuring the **diameter at breast height** of trees (DBH) and expressed as the area of the trees per hectare $(m^2ha^{-1})$ or $(m^2/ha)$.

## Field equipment and materials

Several straightforward pieces of field equipment are needed to study present vegetation. Fifty-meter fiberglass tape measures are commonly used to delineate the boundaries of a study area. Flexible 2 m tape measures can be used to measure the circumference of trees and shrubs and also the height of plants. A special tape measure called a 'diameter tape' is calibrated to measure the diameter of trees instead of the circumference, and is commonly used in forestry research. Datasheets should be prepared before going into the field and should contain information on the location of the study area and other pertinent information about a site including climate, elevation, geology, soils, and disturbance data (see Chapters 7 and 8). Datasheets are generally formatted for the field method that will be employed so that a field researcher can easily record the vegetation characteristics that are being measured. The site characteristics are generally collected for descriptive purposes or for testing specific hypotheses. Finally, a notebook for voucher specimens should also be brought into the field. Voucher specimens are dried specimens of plants collected in the field to aid in the identification of species. When working in an area or region where there is high diversity or the possibility of identifying a new species, voucher specimens of plants are collected in the field and sent to a local or international herbarium where they are mounted on special herbarium paper and given to a taxonomic specialist for identification.

A variety of field methods can be used to study all the world's vegetation types (Bonham 1989). These include the line transect, belt transect, point-centered quarter method, and quadrat or plot. In particular, we examine how line transects, belt transects, and plots are established, what data is recorded with each method, and the advantages of each method.

## Line transect

A line transect entails placing a tape measure straight through an area and recording the species and structure of the plants encountered at selected intervals or intercept points along the transect (Figure 9.7). For instance, a 50 m tape measure can be placed in a straight

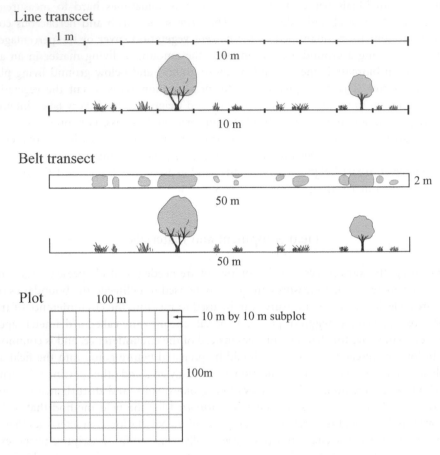

**Figure 9.7**    Examples of a line transect, belt transect, and quadrat or plot

line through a herbaceous community like a grassland (Bonham 1989). A small pole or stick is then placed directly over the interval location or intercept point on the tape measure. Data on the plant that is encountered or "hit" by the pole when it is lowered over the point are recorded. Along a short line transect of 10 m it is possible to record the plants encountered every 50 cm along the line transect, resulting in a total of 20 intercept points, whereas along a 50 m line transect one might record the number of plants encountered every 1 m, for a total of 50 intercept points along the line transect. If no plants are encountered at a point intercept along the transect, for instance the pole is lowered and hits bare soil or rocks, then no vegetation is recorded. The life-form or height of each plant encountered can also be recorded at each intercept point. The length of the line transect and the number of points recorded along the line transect depends on the amount of time one is able to spend in the field. Line transects are usually undertaken in herbaceous communities where it is often difficult to count the number of individuals in an area. This method is also ideal for shrub communities with plant heights less than one meter because it is easy to estimate vegetation cover.

## Belt transect

Belt transects entail the establishment of a narrow rectangle strip in a study area (Figure 9.7). For instance, a belt transect can be 1 m by 10 m or 2 m by 10 m or 2 m by 50 m. The size of the transect depends to a certain extent on the density of the vegetation, but belt transects of 2 m by 50 m or 20 m by 50 m are common (Whittaker 1975). Many vegetation surveys use multiple belt transects through a vegetation type in order to survey a larger sample area. For instance, ten belt transects that are each 2 m by 50 m can be established 20 m apart in one vegetation type. Beginning at one end of the transect data, plant species and location data are recorded for each plant rooted within the transect. For shrubs or trees within the belt transect, the diameter at breast height (DBH) of each tree or shrub can be recorded along with data on the plant height. Belt transects are ideal for creating vegetation profiles of an area by first drawing the location of each stem along a transect, then drawing the height and shape of each plant species. Belt transects are an excellent method for quantifying vegetation in shrub and forest communities where the individual stems for each plant are easy to identify. This method has the advantages of being easy to establish in one vegetation type and the linear shape also ensures that a large sample of a vegetation type is undertaken. The belt transect also provides a better estimate of species richness within a habitat type than other field methods such as square plots of equal size.

## Quadrat or plot

A quadrat or plot refers to a square with all sides of equal length (Figure 9.7). The term quadrat and plot are interchangeable, but quadrats are generally smaller than plots. Geographers usually examine a plot of 1 hectare, or 100 m by 100 m, in forests, plots of 10 m by 10 m in shrub communities, and smaller plots of 1 m by 1 m in herbaceous communities. When surveying large plots such as 1 hectare or larger, researchers generally subsample the plot, or divide it into smaller subplots (e.g., 10 m by 10 m) that are more manageable to survey. This insures that no species or individuals are missed when surveying a large area. Many researchers will establish "nested" or smaller plots within a larger plot to survey for different plant life forms. For instance, a 100 m by 100 m plot can be established to survey all trees with a DBH of 10 cm or greater and four small 10 m by 10 m plots can be established within the plot to survey the shrubs or herbs. This provides a sample of the understory without exhaustively surveying the entire 100 m by 100 m area. Plots have a number of advantages over line or belts transects. For forest communities, they can provide high-resolution data on species richness and the density of individuals within a sample area. Plots are most often used for long-term studies of vegetation and vegetation change because they can easily be resampled. In addition, certain spatial patterns of individual species like clumping are more easily identified in plots as compared to line or belt transects.

## Data analyses

Once the field measurements have been collected the results can easily be analyzed. Species richness is estimated as the total number of different plant species encountered at intercept

**Table 9.2**   Formulas for vegetation surveys

| Element | Formula |
| --- | --- |
| Basal area | $3.14 \times (\text{radius of tree})^2$ |
| Relative density | $\dfrac{\text{individuals of species} \times 100}{\text{total individuals of all species}}$ |
| Relative frequency | $\dfrac{\text{frequency of species} \times 100}{\text{sum of frequency of all species}}$ |
| Relative dominance | $\dfrac{\text{basal area of species} \times 100}{\text{total basal area of all species}}$ |
| Importance value | Relative density + Relative frequency + Relative dominance |
| Index of similarity | $\dfrac{2 \times \text{the no. of species that occur in both samples}}{\text{total species in sample 1} + \text{total species in sample 2}}$ |

points along the line transect or as the total number of species encountered in transects or plots. Density can be calculated as the total number of individuals encountered in the sample area. Species frequency is the number of times a species was encountered and is expressed as a percentage. For line transects, species frequency is quantified as the number of times a species was encountered divided by the total number of intercept points. For belt transects and plots, species frequency is quantified as the number of times a species is encountered divided by the total number of plants recorded in the sample. Basal area can be calculated as the total cross section area of the trees per unit area using the same formula to calculate the area of a circle (Table 9.2). Divide the diameter at breast height by two to calculate the radius, then square the radius, and multiply it by Pi (3.14). Species importance values can also be calculated from belt transects and plots. This is done using a simple formula that quantifies the relative density, relative frequency, and relative dominance (measured as basal area) of each species in a sample (Table 9.2). The importance value of each species in a sample can then be calculated by adding the three values. When two or more samples are taken it is possible to compare how similar two communities are with a similarity index. To calculate the similarity of two vegetation types it is only necessary to know the total number of species found at both sites and the number of species at each site (Table 9.2). This provides an estimate of the percentage similarity between two sites or vegetation types.

## Conclusion

The study of vegetation by biogeographers applies the perspectives of both space and time to understand the form and functioning of the earth's plants. Almost all life on the planet depends upon the thin cover of vegetation. Not only do plants capture solar energy through photosynthesis to power the food-chain, but they also provide habitat for animals and many resources for people, ranging from wood and fiber to pharmaceutical products. The state of the world's vegetation is a major concern today. Vegetation loss and fragmentation due to human land-use and climatic changes due to global warming place stresses upon vegetation and require careful and informed management strategies to mitigate such loses. Biogeographers, through the use of paleoecological techniques such as pollen analysis and tree-ring analysis, can help in understanding the natural dynamics of vegetation, responses to past climatic and environmental changes, and the prehistoric impacts of people. Through modern survey techniques, biogeographers document the present distributions of plants and vegetation formations, detect recent changes and provide critical information on the current distribution of the planet's biodiversity. Carefully conducted paleo-vegetation studies and modern vegetation surveys provide data that is not only useful today, but can be drawn upon by future biogeographers, ecologists, and resource professionals to help in their efforts to understand and conserve plant species and vegetation communities.

# References

Barbour, M. G., and Billings, W. D. (2000) *North American Terrestrial Vegetation*. Cambridge: Cambridge University Press.

Bonham, C. D. (1989) *Measurements for Terrestrial Vegetation*. New York: Wiley and Sons.

Edwards, K. J., and MacDonald, G. M. (1991) Holocene palynology II: human influence and vegetation change. *Progress in Physical Geography* 15: 364–91.

MacDonald, G. M. (2003) *Biogeography: Space Time and Life*. New York: John Wiley and Sons.

MacDonald, G. M., and Edwards, K. J. (1991) Holocene palynology I: principles palaeoecology and palaeoclimatology. *Progress in Physical Geography* 15: 261–89.

MacDonald, G. M., Case, R. A., and Szeicz, J. M. (1998) A 538-year record of climate and treeline dynamics from the lower Lena River region of northern Siberia Russia. *Arctic and Alpine Research* 30: 334–9.

Roberts, N. (1998) *The Holocene: An Environmental History*. Oxford: Blackwell.

Szeicz, J. M., and MacDonald, G. M. (1991) Postglacial vegetation history of oak savanna in southern Ontario. *Canadian Journal of Botany* 69: 1507–19.

Walter, H. (1985) *Vegetation of the Earth*. New York: Springer-Verlag.

Whittaker, R. H. (1975) *Communities and Ecosystems*. New York: Macmillan.

# Additional Resources

Bradley, R. S. (1985) *Quaternary Paleoclimatology: Methods of Paleoclimatic Reconstruction*. Boston: Unwin Hyman. An excellent book that covers the basics of Quaternary climate change causes, patterns, and reconstruction techniques. It has good sections on both tree-ring analysis and pollen analysis.

Fritts, F. H. (1976) *Tree Rings and Climate*. New York: Academic Press. A classic exposition on the use of tree-rings to reconstruct climate. Although it is over 30-years old, it remains a well-referenced source of information. Should be read by anyone who is interested in climatological applications of tree-ring analysis.

Gentry, A. H. (1988) Changes in plant community diversity and floristic composition on environmental and geographical gradients. *Annals of the Missouri Botanical Garden* 75: 1–34. This classic article provides an analysis of plot data from the Neotropics. Using over 70 plots, Gentry examines patterns of species richness, floristic composition, and structure in Central and South America.

Jensen, J. R. (2000) *Remote Sensing of the Environment: An Earth Resource Perspective.* Upper Saddle River, NJ: Prentice Hall. This is the best remote sensing textbook to date with a number of chapters that deal with the remote sensing of vegetation.

Kent, M., and Coker, P. (1994) *Vegetation Description and Analysis: A Practical Approach.* New York: John Wiley and Sons. This book examines methods that can be used to classify vegetation types. It also reviews standard methods to assess the biogeography of plant communities.

Mueller-Dombois, D., and Ellenberg, H. (1974) *Aims and Methods of Vegetation Ecology.* New York: John Wiley and Sons. This book provides an extensive review of field methods that can be undertaken in a diversity of ecosystems.

Pickett, S. T., and White, P. S. (1985) *The Ecology of Natural Disturbance and Patch Dynamics.* Orlando: Academic Press. This book provides an extensive review of the role of disturbance in forest ecosystems. It is still widely cited by biogeographers interested in natural and anthropogenic disturbance.

Schweingruber, F. H. (1988) *Tree Rings Basics and Applications of Dendrochronology.* Dordrecht: Kluwer Academic Press. A comprehensive and amply illustrated book on ecological applications of tree-ring analysis. A good reference for those interested in using tree-ring approaches for biogeographical studies.

Whittaker, R. J., Willis, K. J., and Field, R. (2001) Scale and species richness: towards a general hierarchical theory of species diversity. *Journal of Biogeography* 28: 453–70. This article examines the importance of spatial scale as it relates to patterns of species richness and diversity. It also provides a nice review of current theory related to patterns of species diversity.

## Exercise 9.1   Dendrochronology

Although modern research in techniques such as plant fossil analysis and tree-ring analysis requires sophisticated equipment and facilities, it is possible to undertake preliminary dendrochronological analysis using a scanned image of a tree core and a ruler with millimeter marks. A tree ring core sample and an enlarged subset of the core are provided in Figure 9.8. This tree ring core was extracted from a Big Cone Douglas Fir (*Pseudotsuga menziesii*) in the San Gabriel Mountains, California, in the winter of 1997. Note the difference in color between the latewood and earlywood. It can be assumed that the outermost ring represents the last year's growth season because it was sampled in the winter. Quantify each year towards the center of the enlarged tree core beginning at the last year of the core. Then measure the total radial distance from the outside of the bark to the center and measure the width of each ring in mm.

First, identify the age of the enlarged sample in years based on the number of rings. Then, calculate the average annual radial growth rate of the tree by dividing the total radial distance from the center to outer bark by the number of rings counted. Identify the years during which there might have been an extreme drought based on the number of small rings and the number of years when there might have been extreme amounts of rain. Finally, compare results of ring width with the annual variations in precipitation and summer temperatures in the region. Is there an association between ring width and precipitation? Are climate patterns such as pronounced changes during El Niño years (i.e., 1982–83) present?

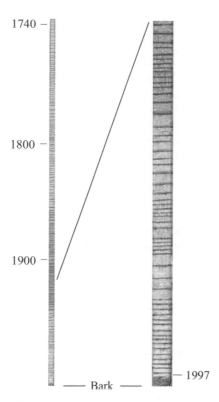

1740 —

1800 —

1900 —

— Bark —

— 1997

**Figure 9.8**  A tree ring core of Big Cone Douglas Fir (*Pseudotsuga menziesii*) collected during the winter of 1997 in the San Gabriel Mountains, California

## Exercise 9.2   Vegetation Surveys

We will use a vegetation survey to quantify the species richness, floristic composition, and structure of endangered tropical dry forests from Hawaii, and create an index of similarity between two fragments of forest. Data on one of the largest and best preserved tropical dry forests in Hawaii are provided in Table 9.3. The data from Manuka State Park on the island of Hawaii includes all trees within a 2 m × 50 m belt transect (100 m²) with a diameter at breast height ≥2.5 cm. The first column identifies the distance encounter along the belt transect and the second column is the scientific name of the species. The islands of Hawaii were historically known as the Sandwich Islands, which is why most species names end in *sandwicensis*. The third column is the diameter at breast height in cm of all stems in the transect. Note that there are multi-stems for a number of plants when measured at breast height. The last column is height of each tree in 2 m cohorts. A second 2 m × 50 m transect was collected from Kaluakauila Gulch, Oahu. This site contained five species (*Cordyline fruticosa, Diospyros sandwicensis, Leucaena leucocephala, Nestegis sandwicensis, Psydrax odoratum*).

To quantify the species richness, floristic composition, and structure of endangered tropical dry forest in Hawaii, you first summarize the level of species richness for the site. Second, create a species list of all plants encountered and the density of each individual

**Table 9.3**   Field data on tropical dry forest from a 2 m × 50 m belt transect for tree ≥2.5 cm DBH

| Meters | Species | DBH (cm) | Height (m) |
|---|---|---|---|
| 0 | *Nestegis sandwicensis* | 3.7 | 6 |
| 10 | *Aleurites moluccana* | 4.7 | 4 |
| 13 | *Aleurites moluccana* | 2.7 | 4 |
| 15 | *Diospyros sandwicensis* | 3.2 | 6 |
| 15 | *Nestegis sandwicensis* | 2.6 | 4 |
| 17 | *Nestegis sandwicensis* | 3.1 | 12 |
| | | 15.8 | |
| 17 | *Aleurites moluccana* | 4.1 | 4 |
| 18 | *Nestegis sandwicensis* | 2.9 | 4 |
| 27 | *Nestegis sandwicensis* | 12.2 | 10 |
| 28 | *Nestegis sandwicensis* | 8.9 | 10 |
| 31 | *Nestegis sandwicensis* | 3.8 | 10 |
| | | 4.0 | |
| | | 3.5 | |
| | | 16.4 | |
| 32 | *Diospyros sandwicensis* | 2.8 | 6 |
| 34 | *Nestegis sandwicensis* | 3.6 | 14 |
| | | 12.3 | |
| 36 | *Nestegis sandwicensis* | 19.6 | 14 |
| | | 2.7 | |
| | | 5.7 | |
| 36 | *Psychotria hawaiiensis* | 5.7 | 10 |
| 39 | *Psychotria hawaiiensis* | 3.8 | 6 |
| 40 | *Nestegis sandwicensis* | 5.9 | 6 |
| 42 | *Nestegis sandwicensis* | 4.2 | 6 |
| 43 | *Metrosideros polymorpha* | 20.1 | 14 |
| 44 | *Nestegis sandwicensis* | 6.5 | 8 |
| 45 | *Nestegis sandwicensis* | 2.5 | 4 |
| 46 | *Wikstroemia sandwicensis* | 6.5 | 6 |
| 47 | *Wikstroemia sandwicensis* | 3.8 | 4 |
| | | 2.8 | |
| 47 | *Nestegis sandwicensis* | 3.0 | 4 |
| 49 | *Nestegis sandwicensis* | 3.4 | 4 |

species per 100 m². Third, identify the number of multi-stem trees, total stand density, total number of tree ≥10 cm DBH, basal area in cm, and mean tree height. Draw a two dimensional profile of the vegetation based on the data provided. Finally, using the similarity index in Table 9.2, calculate the similarity between the two sites. Which trees are most abundant or rare? What are the largest trees? Is it possible to determine which trees are non-native/exotic to Hawaii? How similar are the two sites and which has the highest species richness?

# Chapter 10

# Remote Sensing

*Douglas A. Stow*

- Introduction
- Electromagnetic Radiation (EMR)
- Types and Sources of Remotely Sensed Imagery
- Acquiring Image Data
- Image Processing
- Interpreting Remotely Sensed Images
- Objectives of Image Analysis
- Calibration and Validation
- Conclusion

Keywords

| | |
|---|---|
| Absorption | Platform |
| Geo-referenced | Photogrammetry |
| Image interpretation | Radiometric |
| Infrared | Reconnaissance |
| Inventory | Reflectance |
| Irradiance | Remote sensing |
| Mapping | Scattering |
| Monitoring | Sensor |
| Optical | Spectral |
| Passive | |

# Introduction

Geographers often need to map the distribution of phenomena such as vegetation, landforms, structures, or patterns of urban and rural land use, in both the present and historical past. A viable means to accomplish this task is to utilize remote sensing technology, image processing, and interpretation procedures: areas which geographers have made substantial research contributions. The term **remote sensing** was coined in the 1960s by Evelyn Pruitt (1918–2000), a geographer who held a civil service appointment in the Office of Naval Research (ONR), which was then a major sponsor of geographic research (Jensen et al. 1989). Remote sensing of the environment involves collecting data about the features or phenomena that comprise the Earth's surface and atmospheric environments, without being in direct contact with the features or phenomena in question. This is achieved by capturing "electromagnetic radiation" (EMR) using sensors mounted on aircraft or satellites; it is these **platforms** that enable the remote (that is detached or separated) nature of the data collection. The EMR captured by the sensor ranges from very short wave ultraviolet rays, through the optical wavelengths, to longer microwave (radio) wavelengths. Passive remote sensing involves EMR that is reflected or emitted naturally, and may have originated from the sun and then been reflected off material or matter on Earth's surface or in the atmosphere. Active remote sensing makes use of EMR that is generated artificially (by radar or laser, for example) and reflected from natural or built environments. Normally, the EMR is sampled spatially, in a manner that enables a two-dimensional image (or picture) to be created, from which a researcher can extract geographic information. In some cases remotely sensed imagery, such as the aerial photographs or satellite images that are used widely in textbooks, the news media, and on the Internet, may simply provide a visual representation (or map) of an area of interest and the features contained in it. In others it can provide quantitative measurements of a locally specific phenomenon, such as surface temperature or elevation.

Although it can trace its beginning to 1839, when the camera was invented, remote sensing is a relatively modern technology. Systematic aerial photography became widely available during the 1950s and global data from Earth observation satellites began to be collected in the 1970s. Today, both spatially continuous and contiguous views of large areas of the Earth's surface and atmosphere are possible. Remotely sensed data are comprehensive, synoptic (that is many features can be observed simultaneously), and can be collected efficiently and non-invasively, even from areas that are inaccessible or inhospitable to ground-based observations. However, false information or artifacts may be introduced whether in data collections, image processing, or **image interpretation**. Moreover, remote observations can sometimes be less detailed and precise than similar measurements made on the ground. Historical information about geographic features or conditions can also only be limited by the quality and availability of archived imagery captured in the past.

This chapter emphasizes the utility of remote sensing for geographic research by providing an overview of the physical principles involved in remote sensing, the types and sources of images, and the fundamentals of image interpretation. Light detection and ranging (LIDAR) and interferometric synthetic aperture radar (IFSAR) technologies provide the basis for creating topographic maps and the digital terrain data sets that are used in a

Geographic Information Systems (GIS) environment (see Chapter 22). However, the scope of this chapter is limited to passive, optical and imaging remote sensing of the Earth's terrestrial (land surface) environment. Examples are drawn primarily from analyses of terrestrial phenomena. Some of the uses of remotely sensed data are discussed in Chapter 8. It is the author's hope that this chapter will spark student interest in what is a dynamic subfield of geography that experiences rapid change.

## Electromagnetic Radiation (EMR)

Since it is EMR that enables information about the Earth's surface to be extracted, it is helpful to introduce some information about its generation, transfer, and interaction. Radiation involves the transfer of energy, but unlike conductive and convective energy transfers, EMR transfer can occur in the absence of a material medium (that is, it can be transferred through the vacuum of outer space). EMR is made up of electric and magnetic fields that travel at the speed of light, and have a characteristic wavelength and frequency. What causes EMR to be emitted is a change in the electrical charge of matter, which can result from atomic-level changes in energy states or molecular-level motions such as vibration and rotation. The majority of geographic remote sensing is based on the passive imaging of solar reflective EMR, primarily in the wavelength range between 0.3 and 3.0 micrometers ($\mu$); that encompasses the longer portion of the ultraviolet (0.3–0.4$\mu$), the visible (0.4–0.7$\mu$), near **infrared** (0.7–1.4$\mu$) and shortwave infrared (1.4–3.0$\mu$) wavelength bands of the EMR spectrum (Jensen 2007).

**Passive optical** remote **sensors** on aircraft or satellites capture EMR that reflects off or is emitted from Earth surface materials and scatters off atmospheric constituents; the reflected or emitted surface radiation is the signal of interest and the scattered atmospheric radiation constitutes "noise." Noise occurs because EMR from the Sun passes through the atmosphere on its path to the surface and then, upon being reflected, passes through the atmosphere again before reaching the sensor. Atmospheric constituents, such as gas molecules, aerosols and particulates, as well as water in its liquid and solid forms (in clouds, for example,) have the ability to absorb (intercept) and/or scatter (redirect) EMR. **Scattering** impacts solar reflected radiation because it diffuses the direct solar irradiance, reduces the amount of reflected solar radiation traveling to a sensor, and redirects solar radiation into the view of the sensor. **Absorption** impacts long wave emitted radiation by reducing the magnitude of emitted radiation leaving a surface and limiting the regions of the EMR spectrum that can be used for sensing Earth surface properties.

The characteristic pattern of the amount of EMR that is reflected or emitted by a surface material or object at varying wavelengths is known as its "spectral signature." Extracting information about the composition or condition of surface features from the **spectral** signatures represented in remotely sensed data is further complicated because, although the amount of reflected EMR leaving a surface is primarily a function of the **reflectance** properties of the surface materials, it is also influenced by the magnitude and direction of the solar **irradiance** and the view direction of the remote sensor. The amount of emitted longwave radiation leaving a surface is primarily a function of the surface temperature and, to a lesser extent, the emissivity of the surface involved.

# Types and Sources of Remotely Sensed Imagery

A wide variety of types and sources of remotely sensed imagery are in the public domain, and the key to selecting the appropriate image type is to match the spatial, temporal and attribute aspects of the researcher's information requirements with the spatial, spectral, **radiometric**, and temporal characteristics of the available imagery and the system(s) used to acquire it. As with most geographic data, the important specifications associated with these characteristics pertain to two properties: (1) resolution or grain; and (2) extent or coverage. Specifically, spatial resolution is a measure of the amount of surface detail a sensor can detect; spectral resolution is determined by the number and location of wavelength bands in the EMR spectrum that a sensor measures and the width of those bands; radiometric resolution is the sensitivity a detector has to variations in reflectance; and temporal resolution refers to the frequency with which data can be collected, which is mostly determined by the platform's mobility. The number and location of wavelength bands in the EMR spectrum are also the determinants of spectral coverage, so one data property typically must be traded-off relative to the other and thus, for example, high spatial resolution usually comes at the expense of limited coverage, and *vice versa*. An overview of the information content of different types of imagery contain can be found at http://www.fas.org/irp/imint/niirs_ms/msiirs.htm.

Primary distinctions between the different types of imagery available are made on the basis of platform (airborne or satellite) and sensor type (photographic or digital). Radiometers and spectrometers, for example, quantitatively measure EMR over a wavelength band or discrete wavelength of the EMR spectrum, respectively, while cameras image an entire field of view. Both analog and digital sensors can operate from either airborne or satellite platforms, but the most commonly available imagery is obtained using particular platform-sensor combinations. Thus although airborne digital cameras and linear array sensors that measure reflected or emitted EMR are increasingly being utilized to capture high spatial resolution images of land areas, the most common airborne imagery is aerial photography captured by cameras mounted on fixed platforms, such as airplanes, helicopters, airships or balloons that operate at relatively low altitudes (usually between 1,000 and 21,000 m) and tend to be more mobile than satellites. Aerial photography is often of the highest possible spatial resolution, but normally has limited spatial coverage (aerial photographs typically cover an area of between 1.5 and 15 km$^2$).

Satellites, by contrast, provide coverage of much larger ground areas, but with a lower spatial resolution. Landsat images, for example, are 185 × 170 km and have a spatial resolution of between 80 and 15 m. Examples of Sun-synchronous (or polar orbiting) satellites, such as Landsat 7, tend to provide high spatial resolution images less frequently (such as once every 16 days), while geostationary (equatorial orbiting) satellites, such as NOAA's GOES satellites or the EUMETSAT (http://www.metoffice.gov.uk/satpics/latest_IR.html), provide much more frequent images (at hourly intervals, for example), but have a broader field of view and a coarser spatial resolution (Jackson 2009; Kramer 2002). Commercial satellite sensors provide very high spatial resolution images for smaller areal extents and at a substantial cost (Baker et al. 2001)

Photographic sensors on aircraft platforms are completely analog systems that capture imagery on film when it is exposed to particular wavelengths and processed as negatives

**Box 10.1**   State-of-the-Art Airborne Sensors

While large format film cameras have provided most of the aerial imagery for **photogrammetry** and GIS applications, large format multispectral imaging systems (LFMIS) provide a cost-effective alternative to generating high-fidelity and high spatial resolution imagery. Large format film cameras can provide large area coverage with high detail at a relatively low cost, although the radiometric range is more limited and the spectral quality is lower than LFMIS due to film characteristics and scanning.

LFMIS provide large format imagery with the benefits of direct digital image acquisition, which include: (1) higher radiometric resolution; (2) reduced operating costs per frame/image since no film purchase, processing, or scanning is required; (3) greater spectral coverage with blue, green, red, and near-infrared wavebands acquired simultaneously; and (4) direct digital processing flows which may include the use of GPS-AINS and direct geo-referencing. The availability of larger format CCD arrays, innovative image array technologies, larger hard drive capacities, and faster digital processing have made large format digital imaging more attractive.

Large format digital sensors are those that are capable of acquiring an image frame containing at least 36 megapixels. Commercially produced large format LFMIS employ either push-broom line or frame arrays. The ADS40, DMC, and UltraCam$_X$ have been on the market longest and are the most commonly available systems.

Most LFMIS are employed to generate orthoimagery or image maps, which means that sensor and terrain-related distortions have been corrected and that image conforms to an Earth coordinate system (i.e., pixels have geographic coordinates). The spatial resolution of such orthoimages tends to range between 0.1 to 1.0 m. Orthoimages are very useful for direct mapping of Earth surface features, as well as an image backdrop for GIS analyses. Orthoimages derived from LFMIS can be displayed as single waveband grey tone, true color, or false-color infrared images. The figure associated with Exercise 10.2 is an example of a LFMIS image captured by the DMC sensor on a fixed-wing aircraft that is shown in false-color infrared format.

and photographs. The height and the focal length of the camera lens determine the scale of the photo. Such images can have a very high spatial resolution and fidelity (that is, they can be accurately rendered without distortion or information loss). Their spectral range, however, is limited to the visible and near-infrared regions of the EMR spectrum and, although other film formats can be used, the standard image is 9 inches wide. Aerial photographic film, which is available in black and white panchromatic, black and white infrared, true color, and false color infrared formats, can be electronically scanned to produce digital images, but higher spectral and radiometric qualities are achieved with digital sensors that may use analog detectors to measure EMR levels, which are then converted to digital numbers that can be stored in a computer memory and transferred electronically as part of a raster data set (a two-dimensional array of cells comprising pixels, the smallest picture elements, of which each have a value in a pre-defined range). The resulting digital images are more readily corrected, enhanced, and converted into GIS compatible forms. Most satellite imagery is captured by digital imaging radiometers, since it is difficult to retrieve photographic film from a satellite.

## Acquiring Image Data

Although remote sensing can provide substantial cost effective benefits over conventional ground-based data collection methods, students should not harbor unrealistic expectations when it comes to obtaining imagery to meet specific needs. Often, it will not be possible to locate images that are freely available in the public domain for exactly the time(s) and location(s) of interest. That being said, the Internet is the first place one should start searching for appropriate imagery. For example, Google Earth and the NASA Zulu site provide global coverage of imagery that may be viewed and often downloaded for free over the Internet. A tutorial that describes how Landsat images from the latter site can be downloaded into a GIS format can be found at http://education.usgs.gov/common/lessons/ nasa_zulu.html. Another source for public domain imagery, including global coverage of satellite data and historical aerial photography for the United States, is the Earth Resources Observation and Science (EROS) Data Center of the US Geological Survey (USGS) http:// eros.usgs.gov/. In the UK, the Dundee Satellite Receiving Station maintains an archive of images from both polar and equatorial orbiting satellites http://www.sat.dundee.ac.uk/, and *Intute* draws together a variety of imagery and other resources http://www.intute.ac. uk/sciences/worldguide/. If particular spatial or spectral characteristics or current imagery is needed, data must typically be purchased from a commercial vendor.

EOSDIS (Box 10.2) is an excellent source of imagery and other geospatial Earth observation data sets. Most of these data sets are free and readily available through web-based portals such as EOS Data Gateway (EDG) http://redhook.gsfc.nasa.gov/~imswww/pub/ imswelcome/ and GloVis http://glovis.usgs.gov/. What is unique about the EOSDIS is that EOS data are freely available to students, educators, and researchers in a variety of formats, from raw image data, to geometrically and/or radiometrically corrected images, to maps of Earth science properties derived from sophisticated models.

NASA-funded science teams and scientists around the world are exploiting EOS to study the Earth by observing the atmosphere, oceans, land, ice, and snow, and their influence on climate and weather. A key to gaining a better understanding of the global environment is exploring how the Earth's systems of air, land, water, and life interact with each other. By using satellites and other tools to intensively study the Earth, NASA hopes to expand understanding of how natural processes affect us, and how we might be affecting them. As a bonus, EOS data and studies will yield improved weather forecasts, tools for managing agriculture and forests, and information for fishermen and local planners.

## Image Processing

Once acquired, it may be necessary or useful to process remotely sensed images prior to interpretation. Since aerial photographs are interpreted in hard copy form, processing is typically performed in a photo laboratory. By contrast, processing performed in the digital

## Box 10.2   NASA Earth Observing System

In the early 1980s, scientists and managers at the US National Aeronautics and Space Administration (NASA) initiated a comprehensive Earth observation system, called Mission to Planet Earth, that primarily relied on satellite sensors and ground receiving and processing centers. In the early 1990s NASA established the Earth Science Enterprise (ESE), a comprehensive program to study the Earth as an environmental system. The ESE has three main components: (1) a series of Earth-observing satellites called the Earth Observing System (EOS); (2) an advanced data system called the Earth Observing System Data and Information System (EOSDIS); and (3) teams of scientists who process, study, and synthesize EOS data.

Phase I of the ESE consisted of focused satellite and Space Shuttle missions, and airborne and ground-based campaigns. Phase II began in December 1999 with the launch of the first EOS satellite called Terra (formerly called AM-1). A coordinated series of polar- and equatorial- orbiting satellites have and will be launched to provide long-term global observations of the land surface, biosphere, solid Earth, atmosphere, and oceans (see Figure 10.1 and http://eospso.gsfc.nasa.gov/eos_homepage/mission_profiles/index.php)

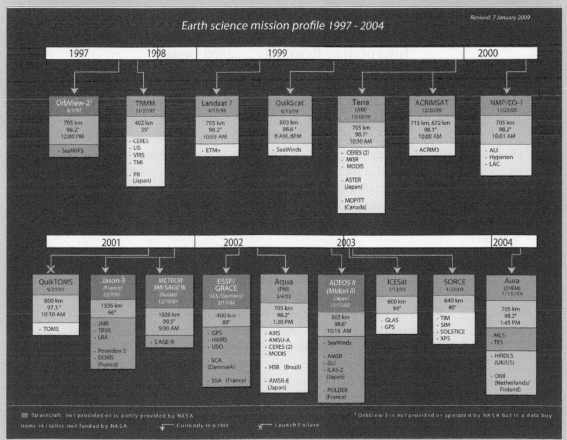

**Figure 10.1**   NASA's Earth Observing System mission profiles (1997–2004) http://eospso.gsfc.nasa.gov/eos_homepage/mission_profiles/docs/mission_profile.pdf

domain can be undertaken using commercially available software. The most important image processing requirement for many research projects involves the geometric correction and geographic referencing of an image (Jensen 2005). Geometric correction and **geo-referencing** are necessary because remotely sensed images have inherent geometric distortions, due to displacements that arise from variations in the terrain being viewed, platform orientation, and sensor imperfections, so that when they are acquired they are not always directly planimetric (or map-like) (McGlone et al. 2004). This means that to generate maps from images and/or to incorporate image-derived products into GIS databases, an image must be geo-referenced to an earth coordinate system. With an appropriate level of instruction, this may be accomplished using a GIS or image display software package, but it is also possible to obtain (sometimes at extra cost) digital orthophotography or orthorectified satellite image data from a public image archive or commercial source. Contrast or color enhancements that involve maximizing the grey level or color range within digital images can greatly facilitate visual interpretation, and spatial enhancements that sharpen or smooth digital images and bring out details or trends are also supported by many software packages. It is also possible to modify the scale of a digital image, and with access to the appropriate software, correct and enhance its spatial and radiometric characteristics.

## Interpreting Remotely Sensed Images

Image interpretation is the act of analyzing remotely sensed images and extracting useful information about a scene (the Earth surface area represented in an image) (Simonett et al. 1983). Interpreting images is both an art form and a science. Humans have innate interpretative abilities and there are at least eight elements of image interpretation that allow us to perform image interpretation (although it is not necessary to consciously employ or exploit them all in the process of interpreting an image). In increasing order of complexity, these interpretative elements are: tone or color; texture; shape; size; height; shadow; pattern; and context. In addition, there are four general types of image interpretative tasks that a student researcher may wish to accomplish. In order of complexity these tasks are:

- *Detection*: Locating the occurrence of a particular type of feature or phenomenon; a binary or dichotomous search process (e.g., detecting plant stress or landscape changes).
- *Identification*: Same as for detection, only determining the types of features or phenomena (e.g., identifying land use categories or vegetation types).
- *Measurement*: Quantifying the length, area, or number of occurrences of objects (e.g., measuring the length of a road segment or counting the number of dwellings in a neighborhood), once such objects have been detected or identified.
- *Analysis*: Examining spatial relationships and geographic attributes of a scene, often by incorporating information derived from detection, identification, and measurement of scene objects and phenomena (e.g., analyzing the socio-economic characteristics of a neighborhood or the soil erosion potential of a ranch).

The tools and strategies used for image interpretation vary, depending on whether the images are in analog (hardcopy) or digital form. The most common hardcopy images are aerial photographs, which are not normally geo-referenced. As for a map, the spatial scale of an aerial photograph is specified by a representative fraction (the ratio of the length of a feature on an image to its actual length on the ground), which is expressed as a dimensionless ratio per unit image length, such as 1:10,000-scale or 1/10,000-scale. However, it is common practice for local government agencies in the United States to use dimensioned scale equations in non-metric units, 1 inch = 2,000 feet for example. Dimensioned scale equations can be readily converted to a dimensionless representative fraction by converting the units of one side of the equation to match the units of the other side, so that 1 inch = 2,000 feet → 1 inch = 24,000 inches → 1:24,000-scale. Since the representative fraction increases when the denominator (the scale factor) decreases, a large-scale image will portray a smaller areal extent with greater spatial resolution than a small-scale image. There is no inherent representative fraction associated with a digital image that is displayed on a computer monitor, since the spatial scale changes as the operator zooms in and out.

Hardcopy images are usually interpreted with the aid of analog optical viewing tools and magnification devices that enhance the intricate spatial details recorded on aerial photographic film, and aerial photographs often have a 60% overlap that enables a pair of images to be viewed stereoscopically (Campbell 2002). A stereoscope viewer is a device that permits each of the interpreter's eyes to simultaneously view the overlapping portion of a pair of two-dimensional photographs, creating the illusion of depth, so that the interpreter "sees" the landscape in three dimensions. Being able to see topographic features and terrain variations in stereo is an effective way of visualizing landscape relationships and is often used to map terrain-controlled features such as landforms or vegetation. A zoom transfer scope is an optical device that allows analog images to be optically magnified and warped so that differences in scale and projection can be accounted for, and they can be aligned and superimposed directly on a planimetric base map; this obviates the need to geometrically correct and geo-reference aerial photographs. A digital GIS layer can also be generated from any analog map if it is digitally encoded by scanning or digitized by hand. Point, line, and polygon objects can also be counted, measured and/or digitized interactively, and the objects displayed in a graphics or overlay plane and/or saved to a digital file. Information attributes, such as the category names, of these spatial objects must be encoded manually, but if the image has been geo-referenced then the file of image-digitized features and their encoded attributes can be integrated into a GIS database.

Digital images can be displayed on a computer monitor or printed onto a variety of hardcopy media. Computer-based image processing and display systems enable interactive viewing, processing, and interpretation, in a very flexible and efficient manner known as "on-screen interpretation" (Jensen 2005). While the fairly novice student may not attempt to exploit semi-automatic image interpretation (e.g., image classification or object recognition) and quantification capabilities of digital image data and image processing software, "on-screen interpretation" can be very powerful, the purpose of more automated approaches is to generate and update geo-spatial data sets in a faster, cheaper, more repeatable, and more reliable manner (Landgrebe 2003). While the reliability of semi-automated approaches is improving, some form of manual, interactive editing, and quality checking of resultant products is required.

Up to three different wavelength bands, enhanced images, or dates of imagery can be displayed simultaneously in the three color planes (red, green, and blue) of a computer

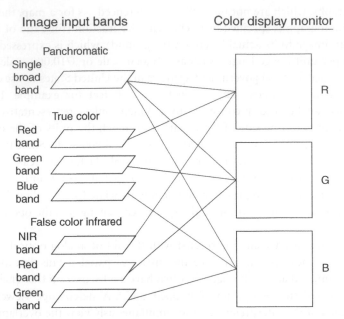

**Figure 10.2**   Digital image color display conventions. (R = red color plane, G = green color plane, B = blue color plane)

image display (Figure 10.2). A single band, panchromatic (i.e. broad visible wavelength band) satellite image can be displayed simultaneously in all three color planes to yield a grey tone (i.e., black & white) image. A true color image is displayed when red, green, and blue wavelength bands are displayed in red, green, and blue color planes, respectively. When other waveband-to-color plane combinations are displayed, the resultant images are called false color composites. The most common is the false color infrared composite, where near infrared, red, and green wavelengths are displayed in red, green, and blue color planes, respectively. This is the same color representation as color infrared aerial photographs, such that healthy green vegetation appears bright red and inorganic urban surfaces (e.g., concrete and roofing materials) are portrayed in blue-gray.

A multitude of image analysis and mapping tasks can be performed with digital images displayed on a computer monitor. Image displays are considered volatile, meaning that the image being displayed at any one time is temporary, unless it is saved and stored in a new computer file. An analyst can conveniently modify the scale and tone/color of a digital image, and with the particular software, correct and enhance spatial and radiometric characteristics. Since spatial scale changes by digitally zooming in and out, there is no inherent representative fraction associated with an image displayed on a computer monitor. Once an image is suitably enhanced and displayed, a student researcher can visually interpret the image on the computer monitor, as they would with a hard copy image. Point, line, and polygon objects can be counted, measured and/or digitized interactively by controlling the cursor with a computer mouse or trackball. These objects are displayed in the graphics or overlay plane of the computer display monitor, and can be saved to a digital file. Information attributes (e.g., category names) about these spatial objects must be encoded manually. If the image had been geo-referenced then a file of image-digitized features with encoded attributes is essentially a GIS layer that can be integrated into a GIS database. Similarly, a dated GIS file can be overlaid graphically with a more current geo-referenced image and then be updated interactively through this "heads-up digitizing" approach.

## Objectives of Image Analysis

Most image analysis is conducted in support of one or more of four general types of geo-spatial objectives: (1) **reconnaissance**; (2) **inventory**; (3) **mapping**; and (4) **monitoring** (Simonett et al 1983). Each of these objectives is explained below in the context of geographic research, with specific examples provided with the four application scenarios.

### Reconnaissance

This involves conducting an initial survey of a landscape, often to gain appreciation of the geographic setting or context. Remotely sensed images enable a student researcher to obtain a qualitative perspective of the biophysical environment or a "sense of a place" about a cultural landscape. This can be achieved for study areas that are poorly accessible or too far away to be conveniently visited on the ground. Image-based reconnaissance may support ground-level observations by enabling a different viewing perspective, or support direct observations from an aircraft, by providing documentation of what was observed in flight. Airborne videography is particularly useful for the latter case, since voice information can be recorded along with the video image data. Imaging requirements in support of reconnaissance are not usually stringent, due to the qualitative nature of interpretation and less challenging objectives. This means that readily available archive and/or on-line image sources, or taking hand-held airborne imagery are viable options. Examples of reconnaissance that might be conducted by a student researcher with the aid of remotely sensed imagery are: (1) assessing the extent of damage from a natural hazard event incurred by a city (see scenario no. 1 in Table 10.1); (2) determining the geomorphic setting of a watershed or region; and (3) understanding the distribution and characteristics of urban and rural settlements within a county, state, or country. The information product can be a map or descriptive text, that defines the extent of flooding and/or provides a qualitative analysis of damage to buildings and structures (see for example, http://www.fema.gov/hazard/flood/recoverydata/katrina/katrina_la_maps.shtm).

### Inventory

An inventory of a study area involves an accounting of its land surface features. Such an objective is quantitative in nature and may be achieved by means of statistical sampling. Remote sensing based inventories are commonly conducted for natural resources such as forest timber or agriculture crops (Lillesand and Kiefer 2007; Ustin 2004). Student researchers may also wish to inventory built or other cultural features such as buildings or roads with high spatial resolution imagery. Even human resources (i.e., population) can

**Table 10.1** Four application scenarios

| Project objective | Image data | Image interpretation | Information product |
|---|---|---|---|
| 1 Assess the geographic extent and spatial pattern of damage resulting from an earthquake in Mexico | Quickbird satellite imagery (true color with 0.6 m spatial resolution) provided pro bono through DigitalGlobe commercial imagery website | On-line download and computer display, with visual interpretation and qualitative analysis of damage to buildings | Descriptive text |
| 2 Analyze spatial and temporal patterns of beach recreational use and associated automobile transportation for a coastal county | Color digital images captured at very high spatial resolution with a hand-held digital camera on a low-altitude, high-wing aircraft; a sample of segments along the coastal strip that include beaches and adjacent parking lots are imaged | Hard copy prints of images are used to count beachgoers and automobiles in public parking lots | Statistical estimates of total number of beachgoers and automobiles for entire length of county's coastline |
| 3 Analyze forest fragmentation and possible impacts on endangered animal species for tropical forests covering a large watershed in Southeast Asia | Landsat Enhanced Thematic Mapper Plus (ETM+) digital image data obtained in geo-referenced format from researchers at the World Wildlife Fund | On-screen interpretation of false color composite image display of ETM+ wavebands and heads-up digitizing of forest and non-forest patches (e.g., clear cuts) | Digital GIS layer and hard copy map of forest and non-forest areas |
| 4 Examine wildfire frequency and extent within the western US | Terra MODIS multispectral digital imagery (250 and 500 m spatial resolution) captured annually since 2000 in early summer; ordered free of charge from the NASA Land Processes Distributed Active Archive Center | Internet download and computer display, on-screen interpretation of multi-temporal MODIS images (one year apart) and heads-up digitizing of fire scar boundaries | Maps and areal statistics on fire scars burned since previous year |

be estimated through image analysis (Ridd and Hipple 2005). Since inventories involve enumerative estimates or listing of objects or resources, which can never be exact when conducted over large areas, a complete or "wall-to-wall" sampling may not be required. Some type of spatial sampling scheme can be implemented, such as sporadic image coverage of the study area and the use of transects or dot grids overlaid on each image. Examples of inventories that might be conducted by a student researcher with the aid of remotely sensed imagery are: (1) counts of automobiles in parking lots for transportation studies (see scenario no. 2 in Table 10.1); (2) quantification of total stream length in a watershed; and (3) estimation of areal extent of land used for grazing in a state or country. In addition, the United State Environmental Protection Agency (EPA) has a variety of national, spatially contiguous environmental monitoring initiatives that use remotely sensed data both to inventory different properties across large regions and evaluate the status of large regions by sampling a subset of the total area, details of which can be found at http://www. epa.gov/cludygxb/html/natinv.htm.

## Mapping

This is a common objective of many studies that utilize remotely sensed imagery, but the imaging requirements in support of mapping can be stringent. Two essential requirements are that the imagery is geo-referenced imagery, or at least can be co-registered to a base map using, a zoom transfer scope for example, and that the coverage is continuous. These requirements are necessary to ensure that a map derived from remotely sensed imagery represents the positional and attribute characteristics of objects or phenomena in an accurate and precise manner (Congalton and Green 2008). Examples of mapping that might be conducted by a student researcher with the aid of remotely sensed imagery are: (1) delineation of forest and non-forest patches for habitat analyses (Ustin 2004) (see scenario no. 3 in Table 10.1); (2) identification of urban land uses; and (3) regional-scale portrayal of geologic structural features in support of tectonic studies (Rencz 1999). See for example, http://daac.gsfc.nasa.gov/geomorphology/.

## Monitoring

This entails time-sequential observations of land surface areas, normally with an interest in assessing landscape change. For remote sensing, this generally means that multiple images captured at different times are acquired and interpreted. Monitoring can be achieved with multi-temporal imagery from either a reconnaissance, inventory, or mapping perspective. The image requirements for monitoring are stringent; the multi-temporal images must be geo-referenced and co-registered (that is, spatially aligned in a relative positional sense). Examples of monitoring that might be conducted by a student researcher with the aid of multi-temporal imagery are: (1) determining the location and extent of wildfire scars (see scenario no. 4 in Table 10.1; also such monitoring can be conducted in real time – see, for example, http://www.ssd.noaa.gov/PS/FIRE/hms.html); (2) estimating the number of new residential dwellings built in a given time period; and (3) assessing changes in agricultural practices in a developing country.

## Calibration and Validation

The interpretation of remotely sensed imagery as well as information products, be they inventory statistics or maps derived through such interpretation, are most reliable when supported by ground-level observations. Such observations can be used to train an interpreter to detect or identify objects or phenomena in images (a form of calibration), and to test the reliability or determine the accuracy of an image-derived product (a form of validation) (Congalton and Green 2008). Whether they are used for calibration or validation, ground observations used in support of remote sensing are often referred to as "ground reference data" or "ground truth" (although the latter, popular term is a misnomer that should be avoided, since ground observations can also contain errors or uncertainties and do not necessarily represent true states or conditions). The collection of ground reference data should follow a spatial sampling scheme that minimizes samples and maximizes reliability and representativeness (see Chapter 6). Since this can be equally or more time consuming and as expensive as acquiring and interpreting remotely sensed data, very high spatial resolution imagery may be substituted for ground observations for training purposes, or for making assessments about the accuracy of information products derived from coarse resolution imagery.

## Conclusion

Remote sensing provides an efficient and exacting means for researchers to derive geo-spatial data and extract geographic information about the Earth's surface and atmosphere. With a modest amount of information about the different types and sources of imagery that are available, it is possible for any informed individual to access remotely sensed imagery in support of their research needs, and many fundamental geo-spatial data and information requirements can be met through the visual interpretation of remotely sensed images. Although this requires little advanced training, digital imagery and image processing software is becoming more accessible, and this allows for more efficient and reliable interpretation and digital encoding of geo-spatial data. It is also important to remember that remote sensing is a rapidly advancing technology and the availability, cost and characteristics of remotely sensed imagery change constantly. Most of these changes are positive; meaning that higher quality imagery is being made available, in more convenient, timely and cost-effective forms.

## References

Baker, J. C., O'Connell, K. M., and Williamson, R. A. (2001) *Commercial Observation Satellites: At the Leading Edge of Global Transparency*, Co-published by American Society for Photogrammetry and Remote Sensing and RAND Corporation.

Campbell, J. B. (2002) *Introduction to Remote Sensing* (3rd edn). New York: Guilford Publishing.

Congalton, R.G. and Green, K. (2008) *Assessing the Accuracy of Remotely Sensed Data: Principles and Practices* (2nd edn). CRC Press Publishers.

Jackson, M. (2009) *Manual of Remote Sensing, Volume 1.1: Earth Observing Platforms & Sensors*. American Society for Photogrammetry and Remote Sensing.

Jensen, J. R. (2005) *Introductory Digital Image Processing: A Remote Sensing Perspective* (3rd edn). New Jersey: Prentice-Hall.

Jensen, J.R. (2007) *Remote Sensing of the Environment: An Earth Resource Perspective* (2nd edn). New Jersey: Prentice-Hall.

Jensen, J., Campbell, J., Dozier, J., Estes, J., Hodgson, M., Lo, C., Lulla, K., Merchant, R., Smith, R., Stow, D., Strahler, A., and Welch, R. (1989) Remote sensing. In *Geography in America*, Gaile, G. and Willmott, C., eds. Columbus, OH: Merrill Publishers.

Kramer, H. J. (2002) *Observation of the Earth and its Environment: Survey of Missions and Sensors* (4th edn). New York: Springer-Verlag.

Landgrebe, D.A. (2003) *Signal Theory Methods in Multispectral Remote Sensing.* John Wiley & Sons Publisher.

Lillesand, T. M., and Kiefer, R. W. (2007) *Remote Sensing and Image Interpretation* (6th edn). New York: John Wiley & Sons.

McGlone, C., Mikhail, E., and Bethel, J. (2004) *Manual of Photogrammetry* (5th edn). American Society for Photogrammetry and Remote Sensing.

Philipson, W. (1997) *Manual of Photographic Interpretation* (2nd edn). American Society for Photogrammetry and Remote Sensing.

Rees, W. G. (2001) *Principles of Remote Sensing* (2nd edn). Cambridge, UK: Cambridge University Press.

Rencz, A.B. (1999) *Manual of Remote Sensing, Volume 3: Remote Sensing for the Earth Sciences.* John Wiley & Sons Publishers.

Ridd, M., and Hipple, J.D. (2005) *Manual of Remote Sensing, Volume 5: Remote Sensing of Human Settlements.* American Society for Photogrammetry and Remote Sensing.

Simonett, D., Ulaby, F., Estes, J., and Thorley, G., eds. (1983) *Manual Remote Sensing* (2nd edn), *Volumes I and II.* Falls Church, VA: American Society of Photogrammetry.

Ustin, S. (2004) *Manual of Remote Sensing, Volume 4: Remote Sensing for Natural Resource Management & Environmental Monitoring.* John Wiley & Sons Publishers.

## Additional Resources

Anderson, J. R., Hardy, E. E., Roach, J. T., and Witmer R. W. (1976) A land use and land cover classification system for use with remote sensor data. *US Geological Survey Professional Paper* 964. Most commonly cited remote sensing reference work. Provides classification systems and interpretation guidelines for mapping land use and land cover types from remote sensing imagery.

Beck, L. R., Lobitz, B. M., and Wood, B. L. (2000) Remote sensing and human health: new sensors and new opportunities. *Emerging Infectious Diseases* 6 (3): 217–27. Introduces public health and epidemiology practitioners to recent developments in remote sensing technology and practical applications.

Fox, J., Rindfuss, R. R., Walsh, S. J., and Mishra, V., eds. (2003) *People and the Environment: Approaches for Linking Household and Community Surveys to Remote Sensing and GIS.* Boston: Kluwer Academic Publishers. Provides an overview of how remote sensing and GIS can support demographic and socio-economic analyses of settlements.

Gates, D. M., Keegan, H. J., Schleter, J. C., and Weidner, V. R. (1965) Spectral properties of plants. *Applied Optics* 4 (1): 11–20. One of the original works that characterize the spectral reflectance properties of vegetation.

Graetz, R. D. (1990) Remote sensing of terrestrial ecosystem structure: an ecologist's pragmatic view. In *Remote Sensing of Biosphere Functioning*, Hobbs, R. J., and Mooney, H. A., eds. New York:

Springer-Verlag. Provides an ecosystem perspective to remote sensing by defining the spatial and temporal sampling attributes of remote sensing.

Lulla, K., Warner, T. A., Nellis, M. D., and Stow, D. A. (2004) Remote sensing, geospatial analysis, and geographers at the centennial of the Association of American Geographers. *Geocarto International* 19: 5–12. Summarizes the contributions that geographers have made to the field of remote sensing and how the discipline of geography benefits from remote sensing technology.

Myneni, R. B., Hall, F. G., Sellers, P. J., and Marshak, A. L. (1995) The interpretation of spectral vegetation indexes. *IEEE Transactions on Geoscience and Remote Sensing* 33(2): 481–6. Spectral vegetation indexes (indices) (SVIs) provide an effective measure of green plant matter and photosynthetic capacity of vegetation. This article compares and contrasts SVIs and explains their biophysical meaning.

Quattrochi, D., Walsh, S., Jensen, J., and Ridd, M. (2003) Remote sensing. In *Geography in America*, Gaile, G., and Willmott, C., eds. Oxford, UK: Oxford University Press.

Ridd, M. (1995) Exploring the V-I-S (Vegetation-Impervious-Soil) model for urban ecosystem analysis through remote sensing: comparative anatomy of cites. *International Journal of Remote Sensing* 16(12): 2165–85. Emphasizing urban remote sensing, a model is provided for characterizing urban ecosystems by the proportions of vegetation, impervious, and soil land covers.

Strahler, A. H., Woodcock, C. E., and Smith, J. A. (1986) On the nature of models in remote sensing. *Remote Sensing of Environment* 20: 121–39. This seminal article on spatial scale relationships and attributes of remote sensing data introduces the scene model, an important concept pertaining to the manner in which Earth surface phenomena or features are generalized.

## Exercise 10.1    Exploring Landsat Images

Go to this website for color Landsat images set aside for this exercise: http://geography. sdsu.edu/People/Pages/stow/book.html. The first two images are from a Landsat 5 Thematic Mapper (TM) image captured October 7, 2002 (top) and a Landsat 7 Enhanced Thematic Mapper Plus (ETM+) image captured October 7, 2005 (bottom). These images are displayed as false color infrared composites and have been geometrically and radiometrically processed. The image scene covers portions of eastern San Diego County, California and Baja California, Mexico. These anniversary date images were both captured in early autumn, which is normally the peak of the dry season in the annual cycle of this Mediterranean-type climate zone. Notice how color hues are not as red (meaning less vegetation cover) in the image from 2002, one of the driest years on record, relative to that of 2005, one of the wettest years on record. The topography of the scene is highly variable, as geologically, it falls within the Peninsular Range Batholith, which is composed mostly of granitic rocks.

Explore these Landsat TM/ETM+ images and particularly examine the following scene features described below (that are marked on the image by the corresponding letter symbols).

(a)    The US-Mexico international border is accentuated by the juxtaposition of land use types on either side of the border, as well as by the border fence and wide unpaved border road used for border security purposes.

(b) The cities of Tecate, Baja California (south of border) and town of Tecate, California (north of the border at the international port of entry) are seen mostly as a blue grey color. Parks and other areas of landscape vegetation are seen as red and pink patches. See Exercise 10.2 for more information about Tecate.

(c) A burn scar is evident on the 2005 ETM+ image, appearing as a mostly contiguous grey patch interspersed by a few unburned vegetation patches observed in pink. From the 2002 TM image we can see that the fire occurred in sparsely vegetated coastal sage scrub and grassland vegetation.

(d) A burn scar 2002 has begun to recover by 2005, through regrowth of shrub and herbaceous vegetation.

(e) Some of the non-vegetated and non-burned patches (i.e., small, bright, non-linear features that do not appear red or dark grey) are outcrops of granite rock.

(f) Tecate Peak has an elevation of around 1,200 m and a portion of its shrubland vegetation cover was burned (part of the burn scar seen in c.).

(g) The bright red sinuous and curvilinear features are riparian corridors, composed mostly of broadleaf trees, that follow the small ephemeral streams of this semi-arid region.

(h) Hauser Mountain is composed mostly of dense chaparral shrubland vegetation, which is portrayed in dense red color on the false color imagery (particularly in the wet 2005 year).

(i) A large stand of coastal sage scrub vegetation is seen on the images as a brown patch with mottled pink pattern. Also known as soft chaparral, coastal sage scrub vegetation is dominated by drought deciduous sub-shrubs that had limited leaf cover in October when these images were captured.

## Exercise 10.2 Remotely Sensed Built Environments

For this exercise, scroll further down the website (http://geography.sdsu.edu/People/Pages/stow/book.html) for two very high spatial resolution (ground sampling distance = 15 cm) orthoimages of Tecate, a small (population approximately 60,000) city in Baja California, Mexico that fronts the US–Mexico border. The multispectral image data from which these images were derived were captured with the Digital Mapping Camera airborne digital multispectral sensor by Digital Mapping, Inc. The images are subsets of the same DMC image portrayed in true color (top) and false color infrared (bottom). The small city of Tecate is seen as mostly blue grey color. The city is home to the Tecate Beer brewery and the historical Rancho la Puerta Spa and Resort. Notice the small town of Tecate, California just north of the border, at the international port of entry. The street pattern is regular and rectilinear. Landscape vegetation is mostly sparse. The following built features are found within the small enlarged subset images that are delineated on the larger subset image.

- *Left*: Spaced rural residential land use. A large ranch style residential structure is adjacent to a garage or parking structure. Irrigated landscape vegetation consisting of large lawns and trees surround these structures.
- *Center*: Multiple family residential land use. Large apartment type buildings surround a concrete patio area that is rimmed by trees. Smaller single family residential structures are seen below.

- *Right*: Single family residential land use with low- to middle-income housing. The dark linear feature with the rectilinear bend is the border fence. The neighborhood streets south of the border are both paved (concrete) and unpaved, while the border security enforcement road north of the border is completely unpaved. Irrigated landscape vegetation is evident but not ubiquitous.

Address the following questions:

1  What spectral-radiometric differences are most apparent between the built environment and the adjacent rural landscape?
2  Compare the true color and the false color infrared (CIR) images. What features can be better distinguished in false CIR rather than true color format?
3  Considering the types of buildings and the road network depicted, what are the main types of land uses found within this scene?
4  Based on housing density/type and the spatial distribution of vegetation, what socio-economic characteristics would best describe this urban area (e.g., low/high income, single family homes, etc.)?

# Chapter 11

# Secondary Data

*Kevin St Martin and Marianna Pavlovskaya*

- Introduction
- Many Kinds and Sources of Secondary Data
- Advantages of Secondary Data
- Limitations
- Working with Secondary Data
- Conclusion

Keywords

Data mining
Database querying
Ecological fallacy
Geovisualization
MAUP (Modifiable Areal Unit Problem)
Mixed methods

Qualitative data
Quantitative data
Relational database
Sample
Spatial exploratory data analysis

## Introduction

Secondary data are data that researchers do not create themselves but use in their research. Compared to primary data that are generated over the course of fieldwork (that involves, for example, measuring water quality or interviewing respondents), secondary data are already created by someone else. Secondary data providers include government agencies and private companies or such sources as published scientific studies, archives, or collections. Most commonly, the term secondary data refers to relatively large databases that individual researchers would not be able to gather themselves, such as census data, newspaper archives, resource inventories, or satellite imagery. Although called secondary, these data inform a great deal of academic work and are central to entire subdisciplines in the social and environmental sciences. Moreover, the importance of secondary data in research and policy development is likely to increase with time. This is because information technologies have facilitated an explosion of a wide range of both environmental and socio-economic digital information as well as methods for its analysis. Widely available and accepted as legitimate, secondary data has come to influence in important ways what kind of knowledge we produce and how. The ubiquity of secondary data, especially within the global north, demands that we carefully evaluate its potentials and limitations before integrating it into any research project or using it to answer specific research questions.

This chapter addresses some of the issues related to the use of secondary data by geographers. We point to the wide variety of secondary data and their many sources and discuss the important advantages and limitations of secondary data. We also address some issues of particular importance to geographers; namely, **ecological fallacy** and the **Modifiable Areal Unit Problem (MAUP)** as they relate to secondary data. Finally, we illustrate the need to engage creatively and critically with secondary data by focusing on non-standard approaches to analysis that use **mixed** research methods. In so doing we draw on examples from our own and our students' work in urban geography and resource management.

## Many Kinds and Sources of Secondary Data

Secondary data includes many different kinds of information about natural and human processes that is collected by various government agencies, non-government organizations, or corporations. Examples of such data include population census data, health statistics, school attainment scores, weather monitoring data, remotely sensed images, ocean surface temperature measurements, fish stock abundance calculations, quantities of hazardous materials released into the environment, results from public opinion polls and other population or business surveys, as well as data often presented in map form such as voting patterns, land use, or elevation.

In the United States, much secondary data is collected and distributed by government organizations such as the Census Bureau, Environmental Protection Agency (EPA), National Institute of Health (NIH), National Oceanic and Atmospheric Administration (NOAA), and United States Geological Survey (USGS) to name but a few. In addition, numerous private agencies collect and sell large amounts of data. They include real estate and environmental consulting firms, insurance and financial companies, marketing companies, and so on. Finally, a number of private agencies re-process government collected data, often doing much of the work that is required before such data can be effectively analyzed, or they operate as distributors for data products the government may not be interested to produce in great quantities.

Secondary data clearly encompasses disparate information that originates in a wide variety of sites. As such, one must assume that such data will vary greatly in terms of its form and type, its spatial or temporal coverage, and the categories or classifications through which it is organized. In many cases, these qualifications will determine the utility of a given dataset for a particular research project. In addition, each collection method, technique of recording and aggregating, and resultant dataset is embedded within the historical and social context of the agency or corporation that developed it. For example, National Marine Fisheries Service (NMFS) data are collected and recorded as a means to quantitatively assess fish stock abundance. This focus clearly emerges from the service's historic mandate to manage fisheries resources such that maximum yield can be obtained rather than, for example, maintaining fishing communities. This service's core dataset is thus primarily concerned with the quantity of different types of fish in the sea. Sea sampling of fish populations is done using a spatial grid with a resolution appropriate for statistical sampling, but which is too coarse for community level studies. The information gathered is both **quantitative** and **qualitative** (that is it consists of numerical measurements as well as reflecting differences in kind), but because the information is stored in a database, the latter information is limited to short string descriptors rather than, for example, the detailed text one might generate from fieldwork (that is contained in field notes). Also, while the temporal coverage of the core NMFS data sets is impressive (several decades), much of the data that would be useful to social scientists (such as the crew size on fishing vessels) has only been collected since 1994. Finally, NMFS data that might aid socio-economic analysis are organized by the category of fishing vessel rather than by individual industry participants. This makes socio-economic analysis at the level of the fisherman (concerning, for example, issues related to employment, job description, wages and benefits, work tenure) virtually impossible to accomplish. While NMFS is tasked with collecting data relevant to fisheries in the US, it is clear that the data collected are of limited use to social scientists interested in the scale of community, questions of employment, or socio-economic change over time.

Secondary data thus vary greatly, are produced by a wide range of organizations and reflect the idiosyncratic history of those organizations. Yet, there are many issues that are common across datasets. This is especially true insofar as information is increasingly stored within digital databases that share principles of organization, methods of query, and forms of reporting.

## From paper to digital databases

Just a few decades ago secondary data existed only on paper; all transformations and calculations were made by hand or using a calculator. Paper was the medium on which the data were stored and used to provide the results of any query, analytical operation, or interpretation. Today much secondary data, especially in post-industrial societies, is created, stored, analyzed, and distributed digitally. Digital spreadsheets and **relational databases** have come to replace printed tables. The implications are profound. For example, the volumes of data that are created and stored have increased dramatically, datasets can be accessed much faster, datasets residing in various locations can be remotely linked to act as a single database via the Internet, and very large databases can be easily imported, visualized, and analyzed with various software packages that include statistical analytical programs and geographic information systems (see Chapter 22).

Digital secondary data are most often structured in databases, organized as one or multiple tables which can be logically related according to shared attributes (termed a relational database). In these tables, rows represent individual cases such as, weather stations, land parcels, or census tracts, and columns (or fields) represent their quantitative or qualitative characteristics or variables. While there is much secondary data that are not organized in relational databases, there is clearly a movement in that direction even for those data not normally associated with a tabular form or even digital storage. For example, newspaper articles are now mostly organized in digital form and indexed as cases within a database. The same sort of search and query operations that could return, for example, all sea sample sites where a particular number of juvenile cod were observed (or not) by a NMFS scientist could, given a very different database, return all newspaper articles published in the last five years that mention the crisis in cod fisheries and the loss of local livelihoods. Even archives of visual information such as photographs and maps are being organized via relational tables and ruled by the same principles and logics.

Overall, the amount of digital information has grown dramatically in the last two decades and will continue to do so in the future. Despite being called secondary, these types of data are becoming the "primary" source for many research projects. As such, it is important to understand their advantages and fundamental limitations as well as the politics surrounding secondary data production, distribution, and use.

## Advantages of Secondary Data

Among the obvious advantages of secondary data, we will briefly consider are their scale and size, professional quality and accessibility, and their association with spatial referencing.

These attributes of secondary data provide opportunities for particular forms of analysis that simply would not otherwise exist. Yet, as many historians of science have made clear, the type of data collected, their scale and form, their categories and classification schemes will advance the interests of some but not all. For example, we may point to how NMFS datasets are aligned with the interests of a corporate and large scale fishing industry; indeed, the close relationship between economic power and state sponsored data collection is not uncommon (or undocumented). Yet, a close examination of any dataset can reveal its potential to do unintended or unimagined work, and variables within NMFS datasets can be re-interpreted by social scientists in new ways. For example, the number of crew members on a vessel is collected as an indicator of fishing pressure, an important variable in biological assessments of fish stocks, but that same variable could be re-interpreted and used as an indicator of employment and its change over time, despite its being buried within a table concerned with fish stocks rather than socio-economic analysis. The issue of to whom secondary data can be an advantage, whether it be corporate vessel owners, crew member or labor organizations, is never fixed; the advantage is that secondary data are open to those willing to spend the time to "get to know" the data and who can then take advantage of its scale, legitimacy, and accessibility.

### Scale

Most secondary data, because of their extensive spatial coverage and the amount of information collected, simply have no substitute. Individual researchers or even research teams could not possibly produce datasets of comparable size or scale. Government population censuses, for example, cover national territories and entire populations. They generate hundreds of variables for detailed spatial units and do so as often as every ten years. Real estate databases, too, describe housing stock in great detail and, in some countries, local real estate databases are integrated through the Internet such that hundreds of thousands of properties can be queried. Inventories of resources, such as fisheries, are not only nationally collected but integrated into international systems of data collection and reporting, such as the Food and Agriculture Organization (FAO) of the United Nations fisheries databases, that make global environmental analyses possible. In addition, some secondary datasets contain data that are nearly technically impossible to achieve without considerable government investment, including the products of remote sensing (see Chapter 10).

Data collection often begins with individual cases and small areas; data are then aggregated to include multiple cases and larger spatial units. In the past, when data were stored on paper, aggregations were fixed. For example, mapped census data would have been aggregated to one type of spatial unit, a census tract, such that it would literally only exist at that scale (in addition to the raw data). With digital databases, aggregation levels are no longer fixed and, in most cases, the data provider or researcher must, themselves, specify the appropriate and desired level of aggregation given the project at hand; thus it is possible to delimit census blocks and block groups in addition to census tracts. Furthermore, if the spatial scale of a secondary dataset is received at a fixed level or in static map form, such as census data at the state level (when one's project focuses on local communities), the agency that created the data may also make the data available at other scales, at a finer spatial resolution such as county, zip code, or census tract. Indeed data often exist in a form that can be output at a variety of scales that differ from the scale of standard data products.

## Legitimacy

Information contained in secondary datasets is usually organized consistently making it well suited for many types of quantitative or statistical analysis, which is often the very reason for collecting such data. In addition, secondary data usually are created by specially trained professionals who pre-test questions and verify categories in order to produce standard and comparable information, both across time and space, that can be used to examine trends or compare information across similar areal units, such as counties or provinces. The standardized format of secondary data also allows researchers to design data collection projects that add to or can be compared with existing secondary datasets.

Importantly, the professional systems of collection, assembly, storage, and retrieval that constitute secondary data confer a legitimacy that is widely recognized and works to empower secondary data, make it rhetorically convincing, and allow it to convince in ways other datasets cannot. For example, many datasets are derived from dubious information that is self-reported by businesses, individuals, or resource users. Yet, once aggregated in a consistent and organized form, such information, despite its origins, becomes the basis for formal scientific analyses. In fisheries, for example, log books are a form of self-reporting where vessel captains report fish catch, discards, trip location, and other variables to NMFS. While fishermen's individual stories are often derided as anecdotal or exaggeration, their log book entries are made believable via the technical systems within which they are embedded. Similarly, we observe that a great deal of the digital spatial data (map layers) currently available in secondary databases were digitized from paper maps that might be decades old, interpolated from sparse control points, or simply geocoded incorrectly (such as the location of the Chinese embassy in Belgrade). Yet, such layers, once in digital form, appear to exude accuracy and instill confidence in the analyses being performed.

## Accessibility

Importantly, the largest and most comprehensive datasets, such as census data, are often produced by public agencies and are publicly available at a low or no cost. This makes them accessible to academics but also analysts working for NGOs and grass-root organizations who can analyze these data with respect to their needs or political causes (Elwood 2006). Overall, such democratization of digital technologies and information serves to empower a variety of social actors beyond the state and corporations. The increasing accessibility of secondary data also facilitates their use as an exploratory first step in research projects that then focus on primary data collection. Widely available, affordable, and easy to use, secondary data can be used to more efficiently target costly and time-consuming primary data collection. Among other things, they are often used to identify places and/or populations for more in-depth qualitative or quantitative study. In one project, for example, we used census data to identify neighborhoods within New York City that contain large numbers of Spanish and Russian speakers. In addition, municipal level information (available from the New York Department of Education website) was examined to estimate the number of immigrant students attending the public schools within those same areas. Taken together, these data made it possible to identify neighborhoods where recent immigrants with young children reside. These populations then became the target of a major interview-based research project that focused on the multiple economic practices of immigrant households.

Nevertheless, while seemingly ubiquitous from the perspective of the global north, there are limits to the accessibility of secondary data. In particular, a large gap exists in the relative abilities of rich and poor countries to access, produce, utilize, and control digital information. As this gap reflects differences in economic and political power, the advanced post-industrial societies have obvious advantages. Countries of the global south, however, are increasingly conscious about the need to narrow the digital divide and, as digital technologies become more affordable and easy to use their governments are launching their own data collection projects. International corporations, too, fill their digital data banks with information about new resource, labor, and consumption markets in the global south. It would seem that, for better or worse, the digital coverage of the world is rapidly expanding and providing ever more sources of information for research.

As geographers, we note also that the growing accessibility of digital secondary data is closely linked to the growth of geomatic technologies such that access to secondary data increasingly implies access to geo-referenced data. Much of the data in secondary datasets is either collected by spatial units, such as census tracts or electoral districts, or includes other locational information such as street address or geographic coordinates. These data, therefore, can be visualized, explored, and analyzed using Geographic Information Systems (GIS).

Working with secondary data has many advantages; including those of scale(s) and magnitude, widespread legitimacy, and their ever growing accessibility, which make them an incomparable source of both social and environmental information to the geographer. And yet we cannot uncritically rely upon secondary data. It is important to remember that the advantages of secondary data should be evaluated relative to their limitations which can be, at times, severe.

# Limitations

Despite the many advantages of secondary data, their use may, ironically, narrow research opportunities and decrease the quality of findings. In this section, we will discuss the limitations of secondary data that can, without critical interrogation, hamper one's project. We will discuss how secondary data simply are not explicitly created for your particular project, how datasets may become internally inconsistent over time or across space, how what appears as full coverage may be based on sampling, how such data may not represent the population that you think it does, and how its precision must be balanced with issues of privacy, errors, and locational inaccuracy.

## Data created for which purposes?

Using secondary data means that we use the data created by someone else and for their own purposes to answer our specific research questions. Even large multiuse datasets are structured according to some original purpose. Census data, for example, is obtained for voting or taxation purposes and NMFS data is for the biological assessment of fish stocks. Embedded in the data, the initial design influences and limits our research questions, methods, and findings. For example, it would be very difficult to study some aspect of global climate change that has not been already incorporated into pre-existing global datasets. The latter compile many but certainly not all variables of interest to researchers of global climate change. Similarly, a social scientist researching poverty must rely on a particular definition of income that has been built into particular census categories. That is, a census will typically report a household's official monetary income but is unlikely to include other types of economic activity, such as informal and/or unpaid production of goods and services that may be important for coping with poverty. Domestic work, informal work for cash, in-home childcare, and exchanges between households and within a community are as important for social reproduction as formal wages, yet they are absent from census data.

As in the case of social scientists' use of NMFS databases, categories designed for one purpose may be creatively re-interpreted for another. This reinterpretation is, however, limited by the history and context of the agency or organization which is then reflected in the databases they create. Clearly, secondary datasets are initiated and maintained for particular purposes and, therefore, may only be useable if researchers can creatively reinterpret existing data or, as in all too many cases, modify their original research questions to fit the data. The use of secondary data has the capacity to limit analytical possibilities such that original research questions may, in the end, remain unanswered.

## Data collection practices change

Large-scale data collection practices do not stay constant and researchers who use secondary data have no control over these changes. Even in such a uniform and consistent data set as the United States census, analytical categories (variables) or the boundaries of spatial units such as census tracts may change from one decade to another. In addition, new variables are often added and existing spatial units (dis)aggregated. Consequently, making longitudinal (time) comparisons becomes difficult, as is discussed with reference to the Modifiable Areal Unit Problem (MAUP) below (Box 11.1), and sometimes simply impos-

---

### Box 11.1   Ecological Fallacy and the Modifiable Areal Unit Problem

While the inherent limitations of secondary datasets are cause for concern, so too is the relationship between secondary datasets and a number of spatial analytic issues. In particular, we examine secondary data and their propensity to increase the occurrence of ecological fallacy as well as their relationship to the Modifiable Areal Unit Problem (MAUP), both of which are important concerns for geographers.

#### Ecological Fallacy

Ecological fallacy refers to the assumption that all individuals in a group share the average characteristics of that group. In the case of spatial data, we should be careful to not assume that all people residing in a particular geographic area, such as a census tract or school district, have properties identical to the average for the area as a whole. The following example developed by a graduate student illustrates this problem. The objective of this student's research was to find out whether differences in the recycling behavior of New Yorkers are determined by differences in their attitudes toward and knowledge about recycling. Individuals from areas with low and high levels of recycling answered questions about their attitudes toward recycling. Their recycling behavior, however, was only assessed using the so-called "diversion rate," or percent diverted from disposal, estimated for each district in the city. Survey respondents were assumed to recycle less or more based upon the average statistic for their district rather than their actual behavior, which is a case of ecological fallacy. Avoiding it involves asking the respondents directly about their recycling behavior. Secondary databases make the occurrence of ecological fallacy more likely insofar as a wealth of data resembling the data the researcher needs, in this case recycling behavior, already exists and is readily accessible across multiple spatial units.

# Modifiable Areal Unit Problem (MAUP)

Another common analytical problem for geographers is the Modifiable Areal Unit Problem (MAUP) which refers to the effect of political boundaries on spatial data and its analysis. In particular, these boundaries are social constructs that may have little to do with the phenomenon under study. In the United States, for example, the effect of state and especially county boundaries on the diffusion of diseases, residential segregation, or migration may be very limited and yet data are frequently collected, analyzed, and mapped using such boundaries. In other words, we often identify patterns in data based upon boundaries that are unrelated to the phenomena in question.

Two other aspects of the MAUP are important to consider (Wong 2004). First, the boundaries of units for which the data are collected change with time making it difficult or impossible to compare datasets that describe the same territory but in different time periods. The dramatic changes in administrative boundaries in Moscow (discussed previously), which are the basis for organizing socio-economic data represent an extreme case of MAUP. Second, the scale at which data is presented and analyzed can affect one's results. For example, analyzing the same census data at different scales (the level of census blocks, census block groups, or census tracts) may yield different statistics and different spatial patterns (see Exercise 11.2). An awareness of the effect of choosing one or another spatial scale is vital. In some cases, choosing a single scale for analysis will precisely address the problem at hand, while in other cases analysis at multiple scales will be necessary to capture those processes that manifest themselves differently at different scales.

Thus, for example, the geographic evidence for connections between Soviet-era structures of political and economic control (economic ministries and Komsomol headquarters) and subsequent capitalist development (new private banking and financial firms) is only visible at the finest spatial resolution of a single street addresses. Only at this scale is it possible to see the concentration of new enterprises within the very locations and offices of Soviet-era structures of power, and at coarser resolutions this locational coincidence is not visible.

A study of access to open space in New York City conducted by a graduate student illustrates the necessity of a multi-scale approach. Open space ratios that measure access to open space (see below for details) and their correlations with socio-economic variables were calculated at three levels: that of the community board district (CBD), census tract, and the neighborhood (measured as open space within walking distance). While a number of socio-economic variables were significantly correlated with open space at the scale of the CBD (positive in the case of median household income and negative in the case of percent people of color), the same variables could not be used to predict access to open space at finer spatial scales. At those scales, associations were more complex. For example, at the neighborhood level both wealthy and poorer neighborhoods had access to open space but in wealthy neighborhoods there were large open spaces (urban parks), whereas poorer neighborhoods had access to only very small open spaces.

sible. For example, after the Soviet Union collapsed, the new administration of Moscow reduced the number of major districts from 33 in 1992 to only 9 in 1993. This was done in order to radically modify and break away from Soviet era power structures. As a result, it was no longer possible to directly compare the socio-economic situation in Moscow before and after the transition to capitalism because all socio-economic data was tied to one of two incomparable spatial systems. Unfortunately, very important and interesting research opportunities were lost by the change in data collection and organization (Pavlovskaya 2002).

## Full coverage or interpolated sample?

Despite their size and scale, few databases fully cover the populations they claim to represent. Most often, variables are estimated from selected **samples** and, therefore, may be subject to sampling errors and biases. As a census, the United States Bureau of Census provides full coverage every 10 years. Only 1 in 6 households, however, fill out "the long form" that solicits some of the most important socio-economic data. Similarly, the so-called micro-data from the census (Public Use Microdata Sample or PUMS) provides detailed information about housing units and people in them (as opposed to geographic areas such as census blocks, block groups, or tracts). While it enables the tabulation of information in the ways that the regular census dataset does not, the findings are valid only if the sample is adequate and includes a statistically valid number of cases (see Chapter 6).

In the case of fisheries, NMFS sea sampling of fish stocks is organized at a spatial resolution that is appropriate for a regional inventory and regional-wide regulatory mechanisms that might, for example, place limits on catch size. The resolution at which NMFS samples, however, makes the assessment of local fish stocks and habitats difficult at best. As a result, small-scale fishermen whose fishing practices are local find it difficult to relate to NMFS pronouncements of stock health/demise and find the ensuing region-wide regulations out of step with their local experiences or needs.

## "Silences" in secondary data

Secondary datasets are fundamentally partial representations. They only contain information about selected phenomena or their aspects and, therefore, always omit information about other phenomena or their aspects. The result is the effective silencing and disempowerment of processes, people, or places that are not represented. For example, only certain types of "formal" phenomena are described by socio-economic data that is regularly collected by state agencies. Such phenomena are, however, only the tip of the iceberg and other informal economic or social practices go undocumented and remain unseen within state sanctioned datasets (Gibson-Graham 2006). Similarly, while environmental processes and change over time are clearly affected by human activities, only some of these are accounted for in formal databases related to environmental management, thus hampering our estimation of the drivers of environmental change. For example, while commercial fishing activities are carefully monitored relative to fisheries management, recreational and subsistence fishing is not, even though these activities are thought to have considerable impact on particular fisheries resources. Formal activities are accounted for and measured and, therefore, can be made part of a secondary database. Formal employment, formal health care services, formal childcare, formal consumption, formal resource harvesting and the like are clearly important to document; yet they cannot capture the totality of human experience. No consistent information exists concerning the informal household, networks, subsistence use of resources, or community economies. What or who is not included, the "silences," in secondary datasets will clearly effect and limit our ability to construct explanations using secondary data.

Furthermore, datasets that do target a particular formal phenomenon may not represent it completely. The United State Census Bureau, for example, conducts an economic census of US businesses classified into industries, such as real estate or professional services, and aggregated into geographic units; states, counties, and zip codes. This vast dataset allows one to research regional and local economies but it incorporates only those establishments with hired workers (paid employees) and, therefore, excludes many small businesses and all self-employed workers and many family businesses. Consequently, this dataset provides only limited insight into local economies and services.

## What variables actually measure

While quantitative data, especially if distributed by government and professional agencies, seems objective, precise, and unambiguous, it is important to understand just how the variables are constructed and what exactly they may or may not measure. The most telling example is the concept of "race" as used in the United States census. Prior to the 2000 census people could only choose one racial category; this prevented them from identifying with more than one race. The resultant statistics on race concealed the racial diversity of many individuals and oversimplified the racial composition of the US population. In addition, the limited number of racial categories used by the census had a disciplining power insofar as they forced people to identify themselves and others in those terms – a powerful process that for centuries has worked to construct and maintain class and other hierarchies based on particular racial categories.

Analyzing health of individuals against other socio-economic or public health variables may also create problems. The state of health is often self-reported and, therefore, is a highly subjective measure that depends on how respondents understand the meaning of being in good or poor health. And yet, it is used in conjunction with other variables that are less subjective because they are not self-reported.

## Categories

The ambiguity of data categories, in addition to the ambiguity of variables, is another important consideration. The research undertaken by one of our graduate students on the diverse economies of Arab American communities in the northeastern US exemplifies this issue. To identify these communities she used census data, one of the only sources for such information. Yet, the census does not define Arab Americans directly using a single variable. Racial categories subsume Middle Easterners as "white." Arab populations can, however, be discerned in other ways using census data. For example, Arabs could be defined as those who speak the Arabic language, or those who come from a predominantly Arab country (national origin), or those who declare Arab as their ancestry. These definitions offer three overlapping but incongruent ways to count Arab Americans. Relatively recent immigrants are more likely to speak Arabic, national origin includes non-Arabic groups, and ancestry is an ambiguous category in itself. Using these definitions, the student produced maps that show three different although overlapping distributions of the Arab American population, which serve to emphasize that even such comprehensive datasets as the US census sometimes offer only a partial representation of the total population.

## Privacy

The contradiction between the need for detailed data and the need to protect the privacy of individuals sometimes demands that researchers make important decisions about how their research may or may not proceed. For example, certain datasets contain sensitive information collected at the level of individuals or households. On the one hand, such detail might be essential to analysis. On the other hand, its utilization in research might actually disclose an individual's private information. In order to avoid such a violation of privacy, the data are typically aggregated to relatively large spatial units which necessarily lead to information and accuracy loss. This is true for much health-related data as well as the Public Use Microdata Samples (PUMS) from the US population census.

Where high resolution data are available and their utilization is acceptable, they can still be problematic. For example, mapping detailed information on income, education levels, race, crime rates, etc., should be done carefully as it may lead to the stigmatization of particular people and places with implications for their economic and social well-being. In this case, researchers may decide to map their analytical results at scales which are smaller, that is using larger spatial units than the scale of the actual analysis (Cromley and McLafferty 2002).

## Errors and accuracy

Finally, all secondary datasets contain errors. Even professionally done surveys, including public opinion polls, may have unknown sampling problems and misrepresent the population in question. The United States Census Bureau, for example, consistently undercounts millions of mainly illegal immigrants as well as those at addresses not included in the census database. Some of these errors may systematically distort the population they represent thereby contributing to inaccurate policy decisions. For example, census undercounts of immigrant populations who satisfy the demand for cheap labor also politically disempower such working populations. In addition, where undercounting includes families with children, the demand for schools and other services may be underestimated.

There are also random errors that may not distort overall averages but do decrease the quality of the data and the researcher's ability to work at finer resolutions. In particular, errors at the data-entry stage, including typos in attribute information or mistakes in a spatial layer, are very common. For example, discrepancies in street addresses may result in the elimination of many records (in some cases as much as 40%) that cannot be matched to an address database via a process known as geo-coding. Similarly, the log book data from individual fishing trips collected by NMFS are riddled with errors. While many errors are the result of poor data-entry (the NMFS hired companies that used prison labor to enter data from forms where entries were hand-written by fishermen at sea) others derive from fishermen's deliberate misreporting. In addition, random errors may occur because of technological faults such as instrument calibration problems that reduce the quality of satellite imagery.

Locational errors are especially important in geo-referenced data. They can lead to the wrong conclusions concerning the spatial overlap of phenomena in question. For example, places may be erroneously identified with some negative social phenomenon, as was discussed with reference to privacy, or their exposure to industrial hazards as measured within the Toxics Release Inventory (TRI) database may be underrepresented (Scott et al. 1997). Other important problems are the ecological fallacy and the modifiable areal unit problem (see Box 11.1).

## Working with Secondary Data

While the advantages and limitations of secondary data must carefully be considered, they clearly contribute to and even expand the scope and power of standard forms of analyses such as **querying** (asking questions of the database and retrieving data that answer these questions) or statistical analysis. Such standard forms of database analysis are discussed in detail in a variety of introductory texts and we will not review them here (see Shekhar and Chawla 2003). Rather, we will briefly discuss three strategies for creatively using secondary data, the goal being to suggest that secondary data can be used in ways that complement creative and critical analyses in geography. The three examples we provide include transforming and adjusting secondary data to better correspond to one's original research questions, designing new measures and indicators, and using secondary data in a "mixed" method approach that combines quantitative spatial analysis with qualitative interview information. In addition, we will examine the opportunities offered by the emerging fields of **data mining** and **geovisualization**.

### Redesign the data to suit your research needs

As discussed above, categories and variables embedded within secondary datasets can influence and shape research strategies and findings. To avoid this, we need to critically examine the data and, if necessary, update, revise, and/or combine it with primary data collection. Our research on access to open space in New York City is a good example. For analysis with socio-economic census data, the student obtained from the Department of Parks and Recreation a database indicating the location of open spaces in the city. The categories of open space in this database included "publicly accessible facilities of regional importance" but excluded spaces that predominantly serve single communities such as "playgrounds, basketball and handball courts, and community gardens". And yet, the latter play a very important role in the daily recreation practices of New Yorkers. Without considering them, the analysis of access to open space would be incomplete.

Updating the database was a time- and -effort-intensive but necessary part of the research. The student acquired the additional datasets from several public agencies and NGOs and merged them with the original database, and the resultant map illustrates that there are noticeable differences in calculations of access to open space from the original to the updated database (see Figure 11.1). In many districts the difference exceeds 0.5 acres. This is a considerable discrepancy given city standards for defining severely underserved districts (1.5 acres per 1,000 residents or less). Updating the database also proved crucial for obtaining one of the key findings of the study: namely that access to open space varies differently in relation to income and minority status depending upon the size of the open spaces available within walking distance.

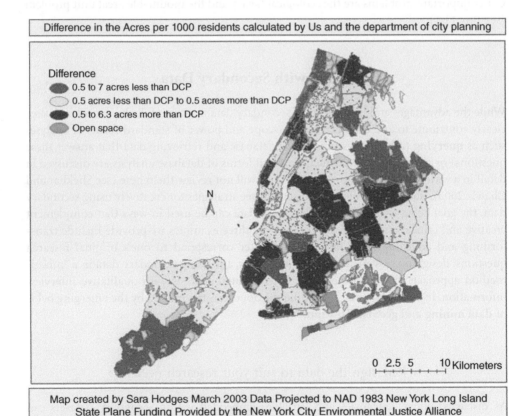

Difference in the Acres per 1000 residents calculated by Us and the department of city planning

Difference
- 0.5 to 7 acres less than DCP
- 0.5 acres less than DCP to 0.5 acres more than DCP
- 0.5 to 6.3 acres more than DCP
- Open space

N

0  2.5  5        10 Kilometers

Map created by Sara Hodges March 2003 Data Projected to NAD 1983 New York Long Island State Plane Funding Provided by the New York City Environmental Justice Alliance

**Figure 11.1**   Differences in the amount of open space in New York City (acres per 1,000 residents); calculated by researchers and the Department of City Planning
*Source*: Sara Hodges, 2004. MA thesis "Open space in New York City: A GIS-based analysis of equity of distribution and access" Hunter College, New York. Reprinted with permission. See http://www.geo.hunter.cuny.edu/~mpavlov/Articles/Ch11_Figures_H1_res.pdf for full color maps.

## Design your own analytical measures

When working with secondary data, creative thinking during the analytical stage helps to avoid limiting ourselves to standard statistical measures. Indeed, secondary datasets can be manipulated to produce novel variables and measures that lead to illuminating findings. Again our research on access to open space in New York provides a good example. Here, a new measure was designed to overcome the modifiable areal unit problem that was a result of measuring daily human patterns, such as the use of the urban parks, based upon arbitrary political spatial entities like the Community Board District (CBD). Traditionally, access is measured as a simple ratio at the level of CBDs and is expressed either as a percent open space or acres of open space per 1,000 residents in each unit. CBDs, however, are rather large entities and, for daily recreation, New Yorkers will only use open spaces within walking distance. In addition, this standard measure of access to open space is clearly tied to an arbitrary administrative boundary (the CBD) even though people disregard these boundaries and simply go to the nearest park or playground. We wanted to measure access to open space that accounted for how people use open spaces in their daily lives.

The literature suggested that children will utilize parks within a quarter-mile while adults will walk up to a half-mile. To avoid the effect of the CBD boundaries, the student converted the map of open spaces to a raster format with a cell size of 40 feet. This resolution roughly matches the size of a tax lot and, therefore, can account for even the smallest open spaces, such as community gardens. In addition, this cell size also approximates the size of a given residence from which access to open space might be measured. We then calculated for each pixel (a proxy for residential buildings) a sum in acres of open space within walking distance for quarter- and half-mile radii. Instead of a single value per large district, we constructed a surface that reflects much finer variations in access to open space (see Figures 11.2 and 11.3). These maps not only show access to open space in new terms using new measures, they also show that there are significant (but ignored) differences in the amount of open space accessible to children and adult New Yorkers.

## Mixed methods: Mapping the social landscape of fishing communities

In the above, data from a secondary database are transformed and re-worked into a new variable across a new space, the space of access to open space. In the case of fisheries we similarly produced a new measure distributed within a new space (St Martin and Hall-Arber 2008). Fisheries science and management repeatedly represents the presence of fishermen and fishing communities on the ocean as aggregate fishing effort expressed in terms of quantities of fish caught. While useful for region-wide estimations of remaining fish stock or future yields, the aggregation of effort to a single variable erases local differences and the dependencies of particular communities upon particular resources.

To express the presence of particular fishing communities and their dependence upon particular fishing grounds, NMFS biological datasets were examined to find some way to map the social landscape of fishing communities. Using the log book data, locational information by fishing trip was tied to the vessel's home port, which revealed where vessels

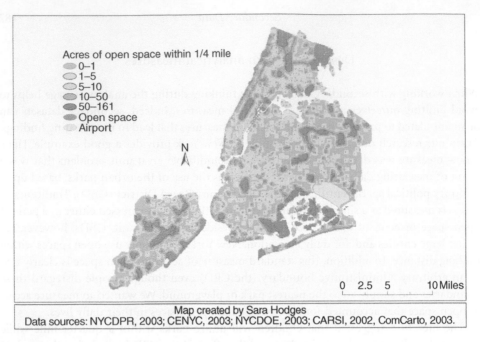

Acres of open space within 1/4 mile
- 0–1
- 1–5
- 5–10
- 10–50
- 50–161
- Open space
- Airport

N

0   2.5   5          10 Miles

Map created by Sara Hodges
Data sources: NYCDPR, 2003; CENYC, 2003; NYCDOE, 2003; CARSI, 2002, ComCarto, 2003.

**Figure 11.2**  Children's access to open space in New York City
*Source*: Sara Hodges, 2004. MA thesis "Open space in New York City: A GIS-based analysis of equity of distribution and access" Hunter College, New York. Reprinted with permission. See http://www.geo.hunter.cuny.edu/~mpavlov/Articles/Ch11_Figures_H1_res.pdf for full color maps.

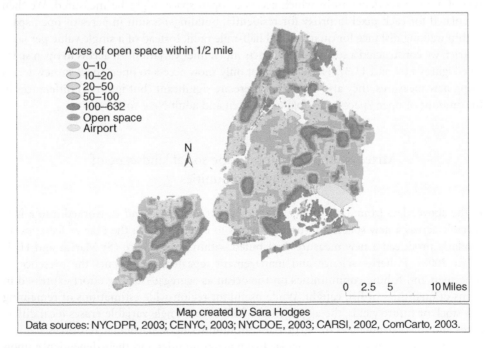

Acres of open space within 1/2 mile
- 0–10
- 10–20
- 20–50
- 50–100
- 100–632
- Open space
- Airport

N

0   2.5   5          10 Miles

Map created by Sara Hodges
Data sources: NYCDPR, 2003; CENYC, 2003; NYCDOE, 2003; CARSI, 2002, ComCarto, 2003.

**Figure 11.3**  Adults' access to open space in New York City
*Source*: Sara Hodges, 2004. MA thesis "Open space in New York City: A GIS-based analysis of equity of distribution and access" Hunter College, New York. Reprinted with permission. See http://www.geo.hunter.cuny.edu/~mpavlov/Articles/Ch11_Figures_H1_res.pdf for full color maps.

from particular communities fished, and data on the number of crew and trip length were integrated to give a measure of labor time. For each trip the number of crew on board was multiplied by the trip length to create a new variable, "fishermen days" that could be linked to particular locations or fishing grounds. This new spatial variable was used to create individual and composite maps for a variety of communities showing the areas upon which they depended (see Figure 11.4). As part of a participatory research project, fishermen from each community were then invited to correct or amend the maps, which have proved valuable as communities lobby for more localized assessments of fish stocks and for greater community input into stock management. The "mixed method" approach used in this work (statistical and GIS analysis of secondary data combined with participatory interviews and workshops) is an emerging and robust way to take advantage of secondary datasets, identify their limitations, and employ alternative methods to both address those limitations and, importantly, distribute the power of secondary data to communities and lay people generally.

## New research opportunities in a digital world

Secondary databases themselves, their attributes and the ubiquity, are making possible new forms and styles of analysis. Indeed, they are facilitating forms of knowledge production unique to secondary databases. In particular, new methods that deal with specific properties of large datasets have been employed in a number of fields, including geography and GIS. They first appeared in marketing research that demanded new techniques for the integration and analysis of the growing but disconnected and non-systematized information about consumers and their behavior. Statistically "mining" those databases promised to uncover yet unknown patterns in consumer behavior which could then be leveraged for corporate profit. What is important is that these techniques reverse the traditional approach to research; instead of testing hypotheses, the new data mining algorithms aim to detect patterns that are not yet hypothesized or observed, patterns that uniquely emerge from very large digital databases.

Today, exploration of digital data is a cutting-edge research direction in geography and GIS. In addition to statistical approaches that mine spatial data (Shekhar and Chawla 2003), new geo-visualization approaches similarly allow for the recognition of patterns in secondary data. **Spatial exploratory data analysis** involves advanced data displays that combine maps with graphs and tables that help the researcher to visually examine the data and discover new spatial patterns (MacEachren 1995; Longley et al. 2005).

## Conclusion

The quantity and magnitude of public and commercial digital datasets, and especially those with spatial information, has significantly increased and will continue to do so. Secondary data are now and will remain important to geographic research as a primary source of information to a growing number of data-intensive applications. Using these data clearly gives a researcher important advantages in terms of data coverage, quality, and costs, as well as the opportunity to analyze phenomena that otherwise would be impossible to

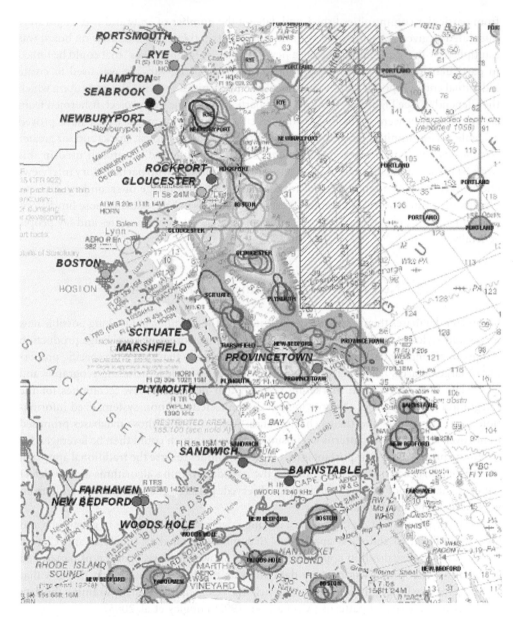

**Figure 11.4**   An extract from a map depicting the primary fishing grounds (based on labor time) of small trawl vessels from particular communities/ports in New England (color coded outlines in the original correspond to port markers, see http://fisheries.rutgers.edu for full color version). The outlines are superimposed upon a NOAA nautical chart. The map also contains a raster density surface based on the aggregate of all vessels.

analyze, such as population distribution at a national scale. And yet, the important limitations of secondary data such as the danger of data-driven research questions, incomplete representation of phenomena, ambiguity of categories, and issues of privacy should be kept in mind. In addition, geographers should clearly understand the potential ease with which secondary data can lead to ecological fallacy or MAUP.

While the advantages and limitations of secondary data are important considerations for any form of analysis, we are enthusiastic about the possibilities for new and creative analytical techniques that secondary data facilitate. In addition, we should not be confined to the original purpose of any dataset; nor should we shy from manipulating and transforming data to build new variables, measures, or maps; nor should we hesitate to combine secondary data analysis with other methods as in mixed methods research. While often associated with standard analytical techniques, secondary databases might usefully be thought of as vast territories to be explored, visualized, and understood using new critical and creative approaches.

## Acknowledgments

We used examples from the research of our graduate students Jess Bier, Sara Hodges, and Ben Mancell. Thanks to Kimberly Wolff and Valeria Treves for their research assistance on a number of projects.

## References

Cromley, E. K., and McLafferty, S. L. (2002) *GIS and Public Health*. New York and London: Guilford Press, (esp. pp. 207–9).

Elwood, S. A. (2006) Beyond cooptation or resistance: urban spatial politics, community organizations, and GIS-based spatial narratives. *Annals of the Association of American Geographers* 92: 323–41.

Gibson-Graham, J. K. (2006) *A Postcapitalist Politics*. Minneapolis: University Of Minnesota Press.

Longley, P. A., Goodchild, M. J., Maguire, D. J., and Rhind, D. W. (2005) *Geographic Information Systems and Science* (2nd edn). New York: John Wiley and Sons.

MacEachren, A. M. (1995) *How Maps Work: Representation, Visualization, and Design*. New York: The Guilford Press.

Pavlovskaya, M. E. (2002) Mapping urban change and changing GIS: other views of economic restructuring. *Gender Place and Culture: A Journal of Feminist Geography* 9: 281–9.

Scott, M., Cutter, S. L., Menzel, C., and Ji, M. (1997) Spatial accuracy of the EPA's environmental hazards databases and their use in environmental equity analysis. *Applied Geographic Studies* 1(1): 45–61.

Shekhar, S., and Chawla, S. (2003) *Spatial Databases: A Tour*. Upper Saddle River, NJ: Prentice Hall.

St Martin, K., and Hall-Arber, M. (2008) The missing layer: geo-technologies, communities, and implications for marine spatial planning. *Marine Policy* 32: 779–86.

Wong, D. W. S. (2004) The modifiable areal unit problem (MAUP). In *Worldminds: Geographical Perspectives On 100 Problems*. Janelle, D. G., Warf, B., and Hansen, K., eds. Dordrecht, Boston; London: Kluwer Academic Publishers, 571–8.

## Additional Resources

American Fact Finder http://factfinder.census.gov/home/saff/main.html?_lang=en. The US Bureau of Census online service that provides access to population, housing, economic, and geographic data. Also allows to map data interactively.

CIESIN The Center for International Earth Science Information Network http://www.ciesin.columbia.edu/ within the Earth Institute at Columbia University. Online datasets for social, natural, and information sciences. Includes PUMS from US census data.

Cromley, E. K., and McLafferty, S. L. (2002) *GIS and Public Health*. New York; London: Guilford Press. An introduction to and advanced treatment of spatial databases, mapping, and spatial analysis with a focus on environmental hazards, infectious and vector-borne diseases, and health services.

The Economic Census of the US Census Bureau http://www.census.gov/econ/census02/. Detailed portrait of the US economy once every five years from national to local level. All domestic non-farm non-government business establishments with paid employees.

GeoDa – An Introduction to Spatial Data Analysis, https://www.geoda.uiuc.edu/. Developed by the Department of Geography at the University of Illinois, Urbana-Champaign, the increasingly popular free software GeoDa provides tools for exploratory spatial data analysis.

Heywood, I., Cornelius, S., and Carver, S. (2006) *An Introduction to Geographical Information Systems* (3rd edn). New York: Prentice Hall. A well written, accessible, and comprehensive introduction to GIS, database development, management, and analysis.

ICPSR-Census 2000, University of Michigan, Institute for Social Research. http://www.icpsr.umich.edu/CENSUS2000/index.html. Access to census 2000 data files. Explains their content.

MPC Minnesota Population Center, University of Minnesota: www.ipums.umn.edu Integrated Public Use Microdata Series (IPUMS) census microdata for social and economic research. IPUMS-USA database from 1850 to 2005. UMS-International has census data from around the world.

Social Explorer http://www.socialexplorer.com/pub/home/home.aspx. Provides easy access to interactive demographic maps of the United States including historical data back to 1940.

Toxics Release Inventory (TRI) public database by Environmental Protection Agency http://www.epa.gov/tri/. Information on toxic chemical releases and other waste management activities.

## Exercise 11.1    Querying the Online TRI database

In your Internet browser, open the Toxics Release Inventory (TRI) site of the Environmental Protection Agency http://www.epa.gov/tri/. This site provides access to a public database with information on the toxic chemical releases and other waste management activities of certain industries in the US. The website provides tools that let you tabulate (summarize) these data by geographic units (for example, states), industry, and type of released chemicals. It will also let you compile a report on TRI incidents in particular neighborhoods defined by their zip codes. To find out whether any releases occurred in your neighborhood or any other neighborhood of the US, type in the corresponding zip code. For example, typing in 80524 (Fort Collins, Colorado) reveals that in this area a factory that produces malt beverages released ammonia and polycyclic aromatic compounds into the air in 2004.

The TRI website is a user-friendly interface that allows you to query, in a variety of standard ways, an enormous government database spanning many industries, at a national scale, and over many years. To build multi-attribute queries look to the bottom of the results page where there is a link to the TRI Explorer home page. Using TRI Explorer, you can search over several years (1988–2004), by type of released chemical, by geographic location, and by industry type. To use these data in more creative and geographic ways (for example, to explore the correlation between releases and poor or minority areas), you can download the type and location ($x$ and $y$ coordinates) of toxics releases and import them into a GIS.

## Exercise 11.2    Mapping with American Fact Finder

One of the great features of the American Fact Finder, the interactive database of the United States Bureau of Census, is its ability to map census and related data. In this exercise, we will map one census variable (median age) at different spatial resolutions (states and counties) and examine the effect of scale on how the data are visualized and can be interpreted (that is, the MAUP).

Open the American Fact Finder home page http://factfinder.census.gov/ in your browser and click on the link to the "Decennial Census" located in the left-hand banner under "Data Sets". When the page opens, make sure that the radio button for the "2000 Census Summary File 1 (SF 1)" is checked on. This file contains data that cover the entire US population. Click on "Thematic Maps" in the right hand portion of the screen. You now can specify the geographic scale you wish to use for displaying data. To display the whole United States by state, select "Nation" as the geographic type and "United States" as the geographic area. Click "Next". On the next screen select the theme TM-PO17 Median Age: 2000 (a specific variable from summary file 1). This variable, median age, will split the population in half. In other words, half of the population is younger and the other is older than the median age – the higher the median age, then the older the population of that area. The median age in the US in 2000 was 35.3 years. Click the "Show Result" button to load the map of median age by state. The map legend or key is on the left. It indicates what values are included into each of the five categories shown with different colors. According to these values, the median age varies significantly by state, with 10 years separating the younger populations (27.1 years in Utah) and older populations (38.9 years in West Virginia and 38.7 years in Florida). Besides Utah, the states of Texas, California, Idaho, Louisiana, Mississippi, Georgia, and Alaska have a relatively low median age (they are shown by the light yellow color). The "*i*" button activates the query function which you can use to find out the median age value for individual states. Determine which states have the oldest population.

Let us now see whether displaying the same data by county changes the distribution of older and younger populations. In the drop-down box "Display map by" above the map, choose "County" instead of "State" as the spatial unit. When the new map loads, look at the legend and note that the minimum and maximum median age values have changed. At this spatial level, the median age varies from 20 to 58.6 years for individual counties, yielding a gap of almost 40 years instead of 10 years as in the previous map. While both statistics were computed from the same data, the county data retain more variation than does averaging to the state level. Examine the map and determine whether the "younger" states (for example, Utah, Texas, California) are uniformly young? Do the "older" states have homogeneously old populations? What are the possible explanations for median age and its variation within different parts of the country? What erroneous conclusion from these data might you draw that would be an obvious ecological fallacy?

# Chapter 12

# Social Surveys, Interviews, and Focus Groups

*Anna J. Secor*

> ■ Introduction
> ■ Survey Research
> ■ Interviews and Focus Groups
> ■ Conclusion: Doing Good Research

Keywords

Coding                    Reliability
Content analysis          Response rates
Discourse analysis        Snowball technique
Narrative analysis        Survey
On site recruiting        Validity

## Introduction

When I think about the challenges of using verbal research methods such as surveys, interviews, or focus groups, I remember a conversation I had before I went into the field to do my dissertation research. I was up in the Rocky Mountains working as a field assistant for my friend and fellow graduate student, a geomorphologist. I was sitting by the bank of the stream that she was surveying for her research, feeling the sun on my back. In that quiet moment, I shared with her that I was worried about the challenges I might face when I left to do my fieldwork in Turkey. Looking up from the field journal where she was thoughtfully recording today's observations, she responded, "At least you are survey-ing people. You ask them questions, and they tell you the truth. You can never get the truth from a river." I was struck by her comment because, as a human geographer, I had naively assumed that what *she* did was based in "facts" and "truths," while what *I* did was far removed from such certainties. What my friend said didn't assuage my worries, but it did give me a broader perspective on the problems that inhere in "surveys" of all kinds.

Even though we cannot expect people, any more than rivers, to tell us the truth, we still often do rely on talking to people to answer our research questions. What can you learn, what kind of research questions can you answer, from surveys, interviews, or focus groups? And what sorts of ideas about knowledge, experience, and representation are embedded within these "talking" methods? We may need to talk to people if we are trying to learn about things that we cannot observe for ourselves, such as how a new city ordinance was experienced among residents of a particular neighborhood. Interviewing might supplement archival work if we need to learn about events and stories that haven't been recorded in newspapers or other documents. Or we might choose to talk to people to learn about their everyday lived experiences. Most often, we choose to ask people questions if we are interested in what people think, know, or feel – or more accurately, if we are interested in *how people talk about what they think, know, or feel*. Yet talking to people does not give us a clear window into their "experiences" or "perceptions." After all, for our research subjects as for us, the way that we know and talk about our lives is always creative. Experience is, as feminist scholar Joan Scott (1992) has argued, a discursive product of contingent processes.

Talking methods give us access to the ways in which people represent themselves and the world in the context of a (very particular kind of) conversation. An interview is a form of conversation between the researcher and the research participant. The conversation can be unstructured, following the rhythms of an unfolding interpersonal exchange with perhaps only a few planned questions. Alternatively, an interview can be highly structured and formal, with the interviewer sharing little of herself and guiding the conversation with a set series of questions. Of course, anything in between these two polarities is also possible, and the form of interview conversations varies greatly. Focus groups are another form of conversation: a group conversation in which the focus group moderator may play a large or a small role in guiding the exchanges of participants. Finally, a survey can also be thought of as a conversation – but a very highly scripted one. Nonetheless, like the other two methods discussed in this chapter, a survey is also a context within which information is not simply being conveyed, handed over from one person to another, but rather is being created on multiple levels, including in the process of exchange. In short, the methods discussed in this chapter, as Mike Crang (2003) put it in one of his progress reports on qualitative methods in geography, tend to produce very "wordy worlds." This isn't necessarily a bad thing, but it does mean that when we work with these methods we need to be aware that what we are working with, what we have gathered, are *words*, statements that are shaped by conversational contexts and variously situated within wider discursive formations.

This chapter proceeds by discussing surveys first and then interviews and focus groups together. Each discussion highlights the reasons for choosing each method, the selection of research participants, some of the basic logistics of using the methods, and data analysis. Of course these discussions are only introductory, and cannot substitute for in-depth engagement with the many good manuals that are available. They should, however, help you to recognize the strengths and limitations of these methods, and to identify the kinds of things you will need to address if you do choose to use one or more of these methods in your own research.

## Survey Research

Why do researchers conduct surveys? A **survey**, or a standardized set of questions administered to a number of respondents, allows researchers to gather information about a population. In statistical terminology, *a population* is the group of all individuals under examination in a particular study. A research population can be defined in many different ways, such as, "all of those who work for a women's NGO," or "all of those who live in the city," or "all of those who are the CEOs of textile companies." A survey can answer questions about the distribution of characteristics (that is, measurable attributes) within a population and across particular sites (such as different urban neighborhoods) or population subgroups (such as women, ethnic minorities, or young people). It can also answer questions about the relationships between and among these characteristics, sites, and subgroups. Unlike the other two methods discussed in this chapter, a survey does not leave room for spontaneous engagement with research participants, for supplementary explanations, or for the emergence of narratives. Because a survey-based analysis usually begins with a tightly honed questionnaire and ends with tables and boxes representing data (whether descriptive or inferential), results from survey research always reflect the categories with which the researchers began. Surveys and questionnaires are often used interchangeably, although, strictly speaking, the former is the process of assessing a sample/population and the latter the instrument through which you do that.

A survey is a good choice for two purposes: (1) to represent in summary form basic characteristics of research subjects; or (2) to present findings that are statistically valid and accurate for a population. The first use is likely to be supplementary to an interview-based study. For example, you might ask your interviewees to also fill out a short survey form that you then use to compile and present summary data. In this case, the survey does not necessarily involve a *random sample* (discussed below), and in many ways analysis does not differ from interview research. The second use is more particular to the survey, since it is often very difficult (though not impossible) to conduct enough interviews to arrive at general conclusions about a population. An example of such a general conclusion is: "[T]hose who find their jobs through personal contacts or through daily activity patterns tend to find jobs that are located closer to home than are the jobs people find through more formal means." Susan Hanson and Geraldine Pratt (1995) derived this conclusion from their statistical analysis of 309 interviews (enough to be representative of the city) in Worcester, MA. In short, surveys are useful for making certain kinds of arguments that will be more or less convincing to different audiences (Lawson 1995) (see Box 12.1).

How convincing survey results are depends on the quality of the questionnaire and the sampling procedures. *Sampling* refers to the selection of research participants from within the target population (e.g., NGO workers, urban dwellers, or CEOs). This is done for practical reasons, since it is often too time consuming or too costly to survey each member of the population. If you intend to use your survey results to make general statements about a population, then you will need to construct a *random sample* in which each member of the population has an equal chance of being included within the survey.

**Box 12.1   Should Women Count?**

Survey research is often placed on one side and interviews and focus groups on the other side of a seeming divide between quantitative and qualitative methods. Throughout the history of the discipline of geography, both quantitative and qualitative methods have at various times been central or marginal to geographic research. In the 1990s, feminist and poststructuralist approaches became associated with the renewed focus on qualitative methods in the discipline. These new approaches were critical of the masculinist notions of "science" that were embedded in the use of quantitative methods in geography. In 1995, a special section of *The Professional Geographer* (Volume 47, Issue 4: 427–66) invited a number of scholars to consider the question of whether quantitative methods had a place in the tool box of feminist geographers. While many different arguments were put forward, one of the important points to emerge concerned the destabilization of the quantitative/qualitative binary, and the importance of using statistics to represent and counter social inequality and oppression.

Appropriate sample size depends on what is being studied (e.g., population averages or proportions), how close you wish your sample's characteristics to be to those of the entire population, and how certain (or *confident*, in statistical terms) you wish to be that your results are accurate for the population. A sample can be constructed in a simple random fashion (such as using an online website to generate a random set of numbers), or it can be stratified. A *stratified random sample* still uses random selection methods, but uses them within specific subgroups or geographical areas that are the target of the research. For instance, maybe you are interested in comparing how Whites and African Americans use public spaces in a small city in the US. Your sample need not be larger than about 400 people for most analyses. And yet, what if Blacks comprise only 10 percent of the city residents? With a sample this size, you are likely to receive no more than 40 responses from this subgroup. If you are interested in how race, class, and gender intersect, you may find yourself with numbers (e.g. 10 middle-class Black women) much too small to determine whether the differences you find between groups in your sample reflect inter-group differences in the population, or are the result of chance. In other words, once you break your sample up into subgroups for analysis, your results will not be *significant* in the statistical sense. Under such circumstances, you might create a stratified sample that *over-represents* residents in Black neighborhoods in the city. The careful construction of samples (whether simple, stratified, or another design) requires a basic knowledge of the statistical principles that you can find in introductory texts (Burt and Barber 1996).

Designing the questionnaire that will yield **valid** and **reliable** results is one of the most challenging, and most important, aspects of survey research. Surveys may be conducted over the telephone, in person, or in writing (using either regular mail or email). Depending on the study, one method might yield better results than another. For example, mail surveys often have very low **response rates** (i.e. ratio of responses to people contacted). Mail surveys also exclude those who are illiterate, while telephone surveys exclude those who do not have phones. Whatever method you choose, the questionnaire should take no more than 45 minutes of your respondents' time to complete – and it is advisable that it take quite a bit less. After all, if a respondent quits in the middle, for analysis purposes you will have to toss the whole questionnaire. A survey usually begins by gathering basic information, such as income, educational level, marital status, party affiliation, or employment. In addition to collecting information about respondents' characteristics or activities, surveys also ask simple questions about attitudes and perceptions. Respondents usually choose their answers from a list of options. For example, respondents may be presented with five income categories to choose from, each represented by a range. Sometimes, respondents may be asked "open ended" questions, in which their answers are not immediately categorized, but these too are usually short answers that are coded by the researcher later on. For example, respondents may be asked to name what they consider to be the biggest problem in their neighborhood. Later, the researcher can go through and code these answers according to general categories, such as infrastructure, crime, or green space. Another common form of survey question is the question that asks respondents to place themselves, or something they are rating, on a continuum. For example, a survey question might ask respondents to rate, on a scale of one to five, where one is "never" and five is "every day," how frequently they read their local paper. Another strategy is to ask survey participants to respond to statements by registering their agreement or disagreement using a scale of one to five, where one is "strongly disagree" and five is "strongly agree." However, one must be aware that there is a tendency, especially pronounced among less self-confident respondents, towards the "agree" end of the continuum. For this reason, it is advisable either to avoid this format or to include multiple statements, some positive and some negative, addressing the same problem. Writing good survey questions takes practice, so it is a good idea to study the techniques before attempting to write your first survey (see references). Finally, it is always important that you pre-test your survey with a small number of respondents before taking it to the field. After your pre-test participants answer the survey, ask them to discuss the questions with you. You may need to revise your survey after finding that it is too long, or that some of your questions are unclear, or that the answer choices are frustrating to respondents.

Survey data lends itself to statistical analysis, whether *descriptive* or *inferential*. This is because survey data usually take a numeric form or, more often, are translated into numeric codes (e.g., female = 1, male = 2). Survey data that are gathered without using statistical sampling techniques (e.g., a short survey administered to interviewees or focus group participants for the purpose of gathering standard information from all respondents) can be used to summarize characteristics of research participants. For example, the professions of all research participants might be represented in table form, broken down by gender and race. However, in the absence of a systematic statistical sample, you would not use these data to make inferences about a whole population. With a representative sample, statistics can be used to evaluate and generalize relationships, such as those between gender (or income, race, geographical location, etc.) and views regarding the efficacy of different levels of government. While this example highlights the use of basic statistics to describe a relationship between two *variable*s, survey data are also analyzed using inferential statistics to create models that assess the contribution of multiple variables (e.g., age, gender, income, and neighborhood) to particular outcomes (e.g., voting for a particular party). These are the strengths of a well constructed survey.

## Interviews and Focus Groups

The goal of interview or focus group research is usually not to generalize to a population, but instead to answer questions about the ways in which certain events, practices, or knowledges are constructed and enacted within particular contexts. Both interviews and focus groups provide opportunities for in-depth, flexible engagement with research participants. Focus groups are sometimes called "group interviews," and indeed these methods share many characteristics, from the recruitment of participants to considerations of privacy and ethics. Focus groups are used to generate interchange and debate between respondents. Both methods can be used as part of an ethnographic study, as supplementary to survey or archival research, or as primary field methods (Hay 2000). Despite their many commonalities, interviews and focus groups differ from one another in the data produced, the analytic methods best suited to these data, and the kinds of research questions these analyses can best answer. Interviews are often used for studies in which participants are "experts" from whom you hope to learn how certain practices, experiences, knowledges, or institutions work – or at least, *how your participants talk about* these things working. Interviewees may be public figures or individuals occupying particular positions (for example, NGO leaders, politicians, urban planners, or managers and employees of a particular firm). For respondents such as these, it might be both socially awkward and logistically difficult to create focus groups. Interviews are also the best choice if you are interested in learning about the life stories of your participants, or if you anticipate long and on-going conversations (whether spending several hours interviewing or conducting multiple interviews with the same person).

What focus groups do well is to produce interchanges between groups of people. Focus groups are especially appropriate for studies of how certain issues or experiences are talked about and debated. Focus group dialogues show how ideas, positions, and representations are taken up and put into play within a conversational setting. Because participants may support or challenge one another, focus groups can provide insight into how meanings, events, or experiences are contested. Focus groups may also be a good choice for decentering the researcher and engaging in feminist research methods (see Box 12.2). This is useful in generating empathy and understanding for different people and ideas, beyond simple pre-conceptions (Montell 1999, Wilkinson 1998). It is important to keep in mind, however, that group interviews may be both uncomfortable and risky if your study involves asking sensitive questions. Further, focus groups are not the best method for learning about the individual characteristics or the life stories of participants.

The number of participants in focus group or interview research is highly variable. Interview research usually includes 10 to 30 participants, depending on whether the interview material is supplementary (to archival, focus group, survey, or other research) or at the center of the project. For a study with multiple researchers and research assistants, the number of interviews may be much higher. Focus groups are comprised of 4–12 (but optimally 8–10) participants. Most of the marketing manuals and many of the social science how-to books on focus group research recommend that participants should not be previously acquainted, both so that participants can maintain anonymity and to encourage them to explain themselves more fully than they would with people known to them (Krueger 1994). Academic researchers, however, often conduct focus groups with people

**Box 12.2**   Focus Groups as Feminist Method

Focus groups were used in marketing research for decades before social scientists became increasingly interested in their applications in the 1980s. By the 1990s, scholars in a range of fields began to think about how focus groups could contribute to feminist qualitative research. Some have argued that, by encouraging participants to find solidarities and recognize shared experiences, focus groups can be "consciousness-raising" and empowering for research subjects. Further, focus group discussions can incorporate participants in a collaborative project that decenters the role of the researcher. In focus group research, the main interaction takes place among the participants in the group rather than with the researcher. This allows the participants more opportunity to frame the terms and categories of the discussion. For this reason, it has been suggested that focus groups are a more egalitarian conversational method. Of course, focus groups are not a panacea for the complicated problems of power and ethics that accompany all of the "talking" methods.

who know each other, such as groups of friends, neighbors, or association members, and the numbers of participants may vary greatly. Each of these designs has its own strengths, and which is more appropriate depends on your study. In either case, it is recommended that groups be as homogeneous (along the lines of age, gender, ethnicity, etc.) as possible to maximize connections between people. Of course, which positionalities are important for determining the dimension of "homogeneity" will depend on the nature of the study. Also, the number of focus groups you will need to conduct will vary depending upon the range of viewpoints you wish to unearth. Multiple homogenous groups give the advantage of connectivity *and* breadth, but most researchers agree that there are quickly diminishing returns with focus groups, and that four or five groups on the same topic with the same population (e.g., discussing the decline of tobacco farming with small farmers in central Kentucky) is usually enough to learn the range of a particular discussion. Focus groups usually last two to two and a half hours, while the length of interviews varies greatly.

In interview and focus group research, participants are chosen for their relevant position or situation in relation to the research question. The selection process begins with a careful assessment of the diversity of positions occupied within the field of potential participants. In other words, if you are interested in how urban residents are talking about a recent incident of police brutality, you will need to consider the diversity of urban residents (in terms of age, race, gender, education, location, etc.) when deciding whom to interview and how many interviews you will need to conduct. To take another example, if you are interested in the effects of the work of a grassroots women's organization in a particular community, you will need to assess the different positions that community members occupy in relation to that work, such as participant, beneficiary, observer, outsider, detractor, etc. The wider the diversity and the more relevant it is with respect to your research topic, the more interviews or focus groups you will need to conduct. Incidentally, this is the same principle that underlies equations for determining sample size.

Recruiting participants for interview and focus group research can be challenging. A research project may require contacting specific individuals (e.g. personnel of specific NGOs or firms, or government workers within a particular bureau). In this case, it is a good idea to confirm that you will have access to your target interviewees (by, for example, contacting the institution in which you hope to conduct your research) before you complete your research proposal. Or you may be hoping to talk to people who fit a general category (e.g., girls at a particular high school, or migrants from Mexico living in your city). In this case, you may need to use what is called the **snowball technique**. Snowball sampling begins by finding an entry point (e.g., a friend's daughter at the high school, or a migrant association in your city) and making contact with some members of the group. These contacts are then asked to provide names of others. This can work for either interviews or focus groups, though of course this method would lead to focus groups where participants are mostly acquainted with one another. Sometimes, if your research is on a group such as "all those who are using the wireless Internet in the park," you simply must approach people on the street and ask for some of their time. This is called **on site recruiting**. You should always keep careful track of how your selection of participants evolved and be transparent about your methods when you write up your research.

While the success of a survey depends on a set of clear, ordered, relatively unambiguous questions, the success of interview and focus group research often depends on being ready to diverge from the question guide. Although the degree of structure in interview and focus group research varies, the most common choice is the "semi-structured" form. In semi-structured interviews or focus groups, the researcher enters with a *guide*, that is, a set of possible questions arranged so as to proceed in the most natural and inviting way possible. There is often a warm up period, which in focus groups includes going around the room for introductions. Interviews are usually more productive if questions are phrased in terms of "what" and "how" (e.g., "How did you become involved in the cooperative?"). These kinds of questions elicit description and place the interviewee on comfortable ground. "Why" questions are discouraged because they are seen as challenging, either of participants' knowledges (because they do not know why) or of their actions (because they feel asked to justify themselves). In focus groups, after initial introductions are complete, it is not appropriate to ask questions that must be answered by going around the room and taking turns. Instead, questions should be designed to foster discussion and debate. Questions prompting discussions of meaning (e.g., "What does it mean to be from Istanbul?") or other "brainstorming" tactics (e.g., "What are some of the words people use to describe immigrants?") work well in focus group settings. Both methods allow researchers to change or rearrange questions, and to ask follow up questions to extend participants' narratives. And in both cases, the warm, respectful curiosity of the interviewer or moderator is a key element in the success of the exchange.

Finally, because the choice of location can impact conversations, *where* and *when* interviews and focus groups take place is an important consideration. For example, if you interview housewives in their homes in the evening, what you will be told about the division of labor in the household might be quite different from what you would learn talking to these subjects in the afternoon when their husbands are at work. A less obvious scenario might be one in which you have arranged to meet with recent migrants in an unused university classroom. Such a setting can be very intimidating for participants who have not set foot in such places before, and this may have implications for the ease of the conversation that ensues – not to mention attendance. Locations should be selected to maximize the comfort and privacy of the research participants. It is also important to find locations where there is not too much background noise, since most interviews and focus groups are recorded and then transcribed – a time consuming process under the best of conditions!

The analysis of focus groups and interviews should not be thought of as an isolated stage of the research project. Analysis permeates both the design of the research and takes place throughout the encounter with research participants. During the interview or the focus group, the researcher asks questions for further clarification, reflecting back her own understandings to the participants. By the end of each encounter, the researcher has already engaged in many acts of interpretation and analysis. The act of transcribing interviews and focus groups is also one of interpretation and, often, translation. The transcripts that result are not mere reflections of the interview or focus group encounters, but new texts that have been assembled through these processes.

Interview and focus group conversations are coded and recoded throughout the process of designing and doing the research, transcribing the conversations, and conducting final analyses. **Coding** can be a systematic process, in which themes, words, phrases, and interpretations are flagged within and across focus group and interview transcripts. Qualitative analysis software may be used to help organize this process. Once the coding process (however formal or informal) is complete, there are many different approaches to the next level of analysis (Denzin and Lincoln 1994). One of the most basic approaches is to "condense" interviews by extracting themes and points and presenting them in summary form. Another systematic approach is **content analysis**, in which themes, words, and phrases are tracked and analyzed within and across transcripts. In **narrative analysis**, the researcher examines the stories told within the interview or focus group context and analyzes how they are put together, the resources they draw on, and the social work that they do. All of these methods may be used with either interviews or focus groups, though it is important to remember that in focus groups, the "unit of analysis" is not the individual quote, but rather the interchanges between group participants.

Interview and focus group conversations open up towards a horizon of possible interpretations. These interpretations will depend both on the theoretical perspective of the researcher and the openness of the conversations to interpretation. Rather than searching for true, fixed meanings, or trying to "reveal" the experiences or perceptions of the participants, it is often more productive to explore how interview and focus groups texts unfold into broader discourses (**discourse analysis**). Doing so requires attention to the silences, paradoxes, and unspoken assumptions that bound and underlay these conversations. Such an analysis returns to the spirit with which we began this chapter on "talking methods," setting aside the question of "truth" in favor of an analytic of process and the unfolding of dialogues and narratives within social contexts.

## Conclusion: Doing Good Research

For survey research, there are clear standards for significance, validity, and reliability. These standards do not produce "objective" results, but they do set internal standards

**Box 12.3** Interviewing "Others"

Whenever a researcher asks questions of others, she enters into a power relation with her research participants. After all, it is the researcher who has set the terms of the engagement and has initiated and framed the conversation. Usually, it is the researcher who has much to gain from the interaction, while the participant contributes without significant reward. Compounding this embedded imbalance, the researcher may also occupy a position of higher social status than her participants. In other cases, if the researcher is "studying up," the respondent may be of higher status. Each of these scenarios presents its own set of challenges, but in both cases the respondent should be treated as an expert whose knowledge is valued. While it may not be possible to undo the power differentials of the interview, researchers can acknowledge their own positionality and respect that of their respondents. Finally, it is always important to consider what you can give back to your research participants.

for survey research to live up to. What about interviews and focus groups? What are the measures of good interview or focus group research? While there are no equations that can assure that interview or focus group research is doing what it claims to do, there are standards of credibility. Research procedures should be transparent, the results evident, and the conclusions convincing. Convincing conclusions arise from thoughtful analyses that present a range of expressions (not just those that fit most snugly with your thesis). Further, in order to be convincing, your analyses must connect to broader arguments and debates. Through transparency and argumentation, all of the "talking methods" can lend powerful insights into processes, events, and discourses that are of broad significance.

Good research takes research ethics seriously. When research involves talking to people, it may present some risks for the participants. For this reason, it is often very important to assure the confidentiality of your conversations. This may be difficult to do if your research participants are easily identifiable figures within a particular community. The level of protection required can be negotiated with your research subjects and depends also on the kinds of questions you will be asking. Importantly, it is your responsibility to represent your research subjects respectfully, even if your study is ultimately critical of their practices. Some researchers bring their analyses back to their interviewees and ask that they participate in the presentation of results. While this is one way to address the inequities of interview research (see Box 12.3), it runs the risk of either censoring critical research (for example, research critical of the operations of a particular firm or organization) or of acting as a cure-all for researcher power differentials. Finally, it is always the researcher's responsibility to consider the effects of the research project, both in terms of the encounters that take place in the field and the dissemination of the texts that result. After all, the "talking methods" are powerful tools for learning about the wordy worlds that we inhabit. Their effects should not be underestimated.

# References

Burt, J. E., and Barber, G. M. (1996) *Elementary Statistics for Geographers*. New York: Guilford.

Crang, M. (2003) Qualitative methods: touchy, feely, look-see? *Progress in Human Geography* 27(4): 494–504.

Denzin, N. K., and Lincoln, Y. S. eds. (1994) *Handbook of Qualitative Research*. Thousand Oaks: Sage.

Hanson, S., and Pratt, G. (1995) *Gender, Work, and Space*. London: Routledge.

Hay, I. ed. (2000) *Qualitative Research Methods in Human Geography*. Oxford: Oxford University Press.

Krueger, R. A. (1994) *Focus Groups: A Practical Guide for Applied Research*. Thousand Oaks: Sage.

Lawson, V. (1995) The politics of difference: examining the quantitative/qualitative dualism in post-structuralist feminist research. *Profession Geographer* 47(4): 449–57.

Montell, F. (1999) Focus group interviews: a new feminist method. *National Women's Studies Association* 11(1): 44–71.

Scott, J. W. (1992) Experience. In *Feminists Theorize the Political*, Butler, J. and Scott, J. W. eds. London: Routledge, 22–40.

Wilkinson, S. (1998) Focus groups in feminist research: power, interaction, and the co-construction of meaning. *Women Studies International Forum* 24(1): 111–25.

# Additional Resources

Czaja, R., and Blair, J. (1996) *Designing Surveys*. Thousand Oaks: Pine Forge Press. This book is a complete guide to designing survey research. It includes chapters on the stages of survey research, the advantages and disadvantages of mail, telephone, and face-to-face surveys, questionnaire design, pre-testing, sampling, reducing sources of error in sampling and data collection, and how to write up a report on your methods.

Fowler, F. J. (1995) *Improving Survey Questions*. Thousand Oaks: Sage. The entire focus of this book is on how to write survey questions. It is very clear, thorough, and helpful. Fowler works through the ways in which survey questions sometimes *don't* work, and then shows how to write clear and relatively unambiguous questions. It includes an appendix that reviews commonly used measurement dimensions and another one that addresses open-ended questions.

Kvale, S. (1996) *InterViews*. Thousand Oaks: Sage. This book is an excellent introduction to the theoretical and methodological challenges of interview research. It begins with discussions about conceptualizing interview research and problems of knowledge and science. It then moves into questions of design and the stages of interview research. It provides in depth discussions of ethics, interpretation, and methods of analysis.

Limb, M., and Dwyer, C. eds. (2001) *Qualitative Methodologies for Geographers*. London: Arnold. This book is not a how-to manual, but instead collects the perspectives of a range of scholars on the challenges of qualitative research. The chapters work together to make qualitative research more transparent. It includes sections on research design, interviewing, group discussions (focus groups), participant observation and ethnography, interpretive strategies and writing. It concludes with a series of vignettes in which geographers share moments from their own research.

Morgan, D. (1996) *Focus Groups as Qualitative Research*. Thousand Oaks: Sage. This is one of the leading texts on focus group research in the social sciences. It begins with a review discussion of how focus groups are being used within the social sciences today. Morgan also offers a discussion of the strengths and weaknesses of focus groups compared to individual interviews. Finally, the book offers a guide to the design and execution of focus group research.

## Exercise 12.1    Giving Back

Think through what you as a researcher might be able to "give back" to the people that you want to work and research with, and the relative merits and drawbacks this process might entail. You might think in terms of policy recommendations, photographic essays, interview transcripts, or entire theses. In particular, pay specific attention to what exactly the results of your surveys, interviews, and focus groups mean for the people that you extracted the information and data from.

# Chapter 13

# Ethnography and Participant Observation

*Debbie Allsop, Hannah Allen, Helen Clare, Ian Cook,*
*Hayley Raxter, Christina Upton, and Alice Williams*

Keywords

| | |
|---|---|
| Coding | Participant observation |
| Ethnography | Research ethics |
| Overt/covert | Situated knowledge |

## Introduction

This chapter is written by six former undergraduate students (Debbie, Hannah, Helen, Hayley, Christina and Alice) and their lecturer (Ian). It's an introduction to ethnographic research and its core method of **participant observation**. Ian introduced us to this in a second year research methods module. We'd never heard of it before. He told us it involved "being there," participating in, but also observing, what happens in other people's lives. We all chose to use it in the (undergraduate) "dissertation" research that we did during the summer vacation before our final year. So we became experts on its use in a variety of projects. Debbie studied auditing culture in an international accountancy firm, and how this was affected by some new legislation. Hannah wanted to make sense of the moral panic in the UK over childhood obesity. Helen traveled to Cambodia to try to meet factory workers who had made one of her t-shirts. Hayley wanted to find out what had been lost and gained in the Hollywood remake of a Japanese horror film. Christina wanted to know how migrant factory laborers in her home town fitted in with local people. And Alice tried to find out why young girls were dressing in what were, for many, worryingly provocative "tweenager" fashions.

We're writing this chapter together because our combined "expertise" might best help novice researchers appreciate what this method can involve first time round. Rather than providing the kind of literature review plus illustration chapters that established academics often write, we're going to talk through our own experiences of encountering and doing **ethnography** and participant observation. Most of this chapter is based on a two hour discussion we had about this with Ian. We want to give a sense of the hard work, creativity and nervous excitement that made it so worthwhile for us, as students. We'd like this chapter to allow you to step into our shoes, to imagine the kind of research that you might do using this method, and the kinds of problems that you might have to plan for and deal with. You'll have to read a lot more than this chapter to appreciate what it's all about. This is just a start.

## Alice Takes Some Unusual Lecture Notes

We begin with an extract from Alice's coursework for that second year module. She used participant observation to describe what it was like to attend a geography lecture at the university. She had to assume that her reader had never been to a university, let alone a lecture. The extract below sets the scene and describes the first eight minutes (compared with five to ten weeks of action in our dissertation research!). A good participant observation account is supposed to allow a reader to vividly imagine themselves in the author's shoes. See how this works for you. It is customary for published participant observation accounts to be altered to hide the identities of the people under the microscope. Here, therefore, all proper names, dates and other identifying information have been altered. This is part of the ethics of participant observation research.

Today's lecture, on February 11, 2005, 2–3pm, was for a second year geography module. The lecture theater used is located in the Mechanical Engineering building at the university. This building has many lecture theaters of this size and is used by several departments for large scale lectures. The building is located on one edge of the campus and is built of brick, but not in the grand manner of those in the centre of campus. This building has a reputation as being uncomfortable, old, "a bit smelly" and generally that it should only be used by "those boring people that do Mechanical Engineering"! This year group of geography undergraduates have been lectured to in this theater since their first year, and are familiar with its location and atmosphere.

The theater has approximately 30 rows of benches and desks located in three sections, with two two-meter wide pathways of steps rising from floor level to the top of this "amphitheater" style room. The benches and desks are made of heavy, "orangey" wood, with the desks being heavily graffitied and scratched from years of use. The wood is slightly sticky to touch, in the way old, varnished wood is after too many coats. The benches have burgundy plastic cushions on them in two meter sections. These are attached to the bench by old, gold-looking pins. The cushions do not meet and have 15 cm gaps between them, making students sit in approximately groups of three to a cushion. Some benches are missing cushions, so instead the students sit on the hard, uncomfortable high-backed benches. At the floor of the benches and steps is a walkway into the theater and a large, wide four meter long desk made of the same heavy "orangey" wood. On the right hand end of the desk (from where I am sitting) near to the door, is a computer, keyboard and mouse linked to the projector hanging from the ceiling in the centre of the room. The projector transfers the "PowerPoint" images from the computer onto the six by four meter screen on the wall in front of me. Below this screen, behind the desk are large sliding green boards for chalk use. But these are rarely used. The left hand wall of the theater has windows stretching from the ceiling to approximately half way up the wall. The windows are always covered in thick, heavy green curtains that block the light. Small gaps appear where the curtains meet and it is possible to see one of the walkways out of campus. The curtains and windows meet the middle row of benches and form a panel that extends to the front of the room.

I arrived early for this lecture and waited outside, following the text messages from friends seeing if I was going and whether they should meet me there. I replied with the usual, "Yes, I am going. Yes I shall meet you outside." This week, however, would be slightly different. Jo, a friend I usually sit with, was not coming. So, I was waiting for another friend, Gareth. I had told Jo I was going to do my participant observation practical on this lecture. I would be looking at the actions, reactions and processes that occurred in this usually very dull lecture – in which often many students leave in the break in the middle – and hoped it wasn't because of this that she was not attending. Gareth and another friend Steph arrived at about 1:55 and immediately remarked, "God Al, you're here early! A bit keen aren't we?" My initial reaction was one of shock, "Have they found out what I'm doing already?" I passed off the comment with a quick "Oh, shut up" and we walked in. The lecturer was setting up at the front. He was fiddling with the computer, so I imagined he was sorting out the PowerPoint presentation. The theater was already quite full. This was a surprise due to the mass "walk out" experience last week. It was quite noisy, with people chatting, walking up the steps, sitting down and removing coats, taking out paper and pens, etc. I led the three of us to our usual spot, three rows from the back on the left hand side. As I walked up the steps, I found myself avoiding eye contact. I almost felt that by looking at people they would know what I was up to. I sat next to a friend, Ben. Gareth then sat next to me and Steph next to him. I felt slightly uncomfortable already, a little hot and flustered from the walk here, and from the fact that the room was really hot. I got out my paper and pencil case. Gareth asked to borrow some paper and a pen as the lecturer began to speak. It was now 2:02pm. At this point, I started to note some things. I was concerned. What should I write? Where do I belong? Should I be describing what I feel or what I see? And do I need to ignore the lecturer or pay attention in case what he says affects the students?

My attention was drawn to the lecturer. He explained how this lecture was going to be an hour long and put the first slide on the screen. "Another fun topic," Gareth stated sarcastically as he leant over. Muffled voices around the theater, I imagined, pointed out the same fact, along with the slight excitement at it only being an hour long. Coughing and whispers followed as it appeared that the lecturer's microphone was not working. He didn't appear to notice this and simply continued, explaining the first few slides. Gareth's next comment was, "Oh great. We can't even hear him." He leant over again and, this time, looked at my notes. He immediately realized that I was not taking lecture notes, and must have seen his name on my pad. He quickly grabbed my pen and read the notes I had taken. "What are you doing?" "Nothing." "Yes you are. This is your participant observation thing, isn't it?" "Nope." I tried to battle on. "Yes it is! Oh, go on, I'll help you. It'll be well funny." I gave in and quickly explained to him what I was doing and that he must shut up and pretend he does not know. Feeling really annoyed that he found out, and even more that I had had to reveal myself, I tried to regain my concentration. The microphone problems still continuing, I looked around at the students, all furiously scribbling down the slides. I seemed to be drawn to those I knew, though: various friends and associates were sat in different places around the room. As I did this, I unintentionally caught some of their eyes and they mouthed "hi." This made me feel even more vulnerable after Gareth's discovery. Feeling slightly worried that I was only focusing on those I knew, my attention was drawn to the door. It was my friend Si arriving late. It was now about 2:08pm. The room stirred as Si, at six feet two, broad, with a shaved head, stormed in with a large sports bag. I knew he had been at badminton. People started to chat. I could hear little bits of laughter and movement. Gareth's word "legend!" seemed to sum up how many people felt about Si's continued lateness to this lecture. This caused me to laugh slightly. My uncertainty was then drawn to whether I was only observing or if I was now in fact participating in this situation.

## Ian Starts his Lecture on "Ethnography and Participant Observation"

Alice didn't write this "live" in that lecture theater. Participant observation accounts usually start off as "scratch notes": bits of information, key quotations written down word for word, and reminders of things to elaborate upon later. These are the building blocks of detailed accounts that are written somewhere else later in the day, while the experience is still fresh. Participant observation's data are pieces of creative writing based on what you see, hear, and feel under specific and often unpredictable circumstances. This data isn't "collected." It takes hours to write. And it's tricky. That's what Alice was finding out. Ian set that practical in a (boring) lecture. It was our introduction to participant observation, and how to do it "properly." You might find this useful.

> I usually put questions up on the screen, then answer them, ask some more, and off we go. For this lecture, the first slide asked *"What's ethnography"?* I replied, "While 'geography' means 'earth-writing,' 'ethnography' means 'people-writing.' This has been done by anthropologists and geographers for at least a century. And its central methodology is 'participant observation.'" *What's participant observation?* "It's research that involves 'being there' and 'stepping into others' shoes,' as much as this is possible. It involves participating in and observing social life, and conveying this to others mainly through writing." *What's this writing like?* Cloke et al. (2004: 200–4) suggest that, to give an outsider a vivid sense of being somewhere, you need to include six "layers of description" in your participant observation notes. Let's briefly go through them.
>
> *Layer one: locate your ethnographic setting.* "Where in the world did you do your research? Which country, region, town, neighborhood, street, building and how can they be characterized?" *Layer two: describe the physical space of that setting.* "How was that space you worked in set out? What were its dimensions? What was it made up and out of? What was it's atmosphere?" *Layer three: describe other people's interactions in that setting.* "Who was there, and where, in the space that you've described? What were they doing? How did their interactions unfold?" *Layer four: describe your participation in that setting.* "What were you doing in that

setting? What was your role in those interactions? To what extent were you also participating in what happened?" *Layer five: describe your reflections on the research process.* "Participant observation is quite unpredictable. You'll never feel in full control of how your research takes and changes shape. So here you have to write notes to keep track of this process, to regain and/or change your focus and direction." *Layer six: describe your self-reflections.* "Most researchers find participant observation quite stressful, and write many pages of notes (along with letters and emails to friends and family) to let off steam."

## The First Participant Observation We Read

Alice followed these instructions quite closely in her coursework. Layer one is followed by layer two. But, after that, the others were mixed up. This isn't unusual. Participant observation isn't the kind of method that you can carry out exactly as planned. Of course, it has to be planned and you have to read about it as a *methodology*. But we wouldn't advise you to do this first. We started by reading some *results* of participant observation research. Ian gave us a list of past undergraduate dissertations to read. A year later, after we'd handed in our own, we talked about how they'd inspired us.

> DEBBIE:   You get one out and you think, "Oh God. This is an academic dissertation. It's going to take forever to read." But I read the one about children's understandings of Eminem lyrics and it was a page turner" (Griffiths 2002, 2003).
>
> CHRISTINA:   I read one on migrant labor in British agriculture (Crook 2005). It was so readable and she'd written it so you could actually step into the shoes of those farm workers. You could *really* do that. And I don't think it was until I read this that I appreciated how participant observation could be done.
>
> HELEN:   The one I read was about people in Sri Lanka and England whose lives were connected through the tea trade. It was amazing (Wrathmell 2003). It was so real, reading about the people she had gone and visited. You could imagine being there. It scared me quite a lot, actually. I thought, "Could I do this?" "Could I actually go somewhere and do this?"

## "You Just Can't Beat Experience ..."

After reading these dissertations, we wanted to do the same thing ourselves, with topics that mattered to us, in settings that were relatively accessible. Alice and Hannah went back to their old schools. Debbie had already got an internship with that accountancy firm, and Christina had worked in that factory before. Helen was already going to Cambodia to do voluntary work. And Hayley had a computer and broadband. We'd tried out other research methods in that second year module and – even though it wasn't everyone's cup of tea – this was the one that we liked the most.

> CHRISTINA:   I don't think you can beat experience in what you're talking about. I could have interviewed migrant and other factory workers. I could have said, "Okay, I live in the area and so-and-so said that, and said that, and around the area people think this." And I could have concluded that, "This is what the population of the village think about it." But, it

wouldn't have portrayed the *actual* situation. It's the people, and the experiences, that matter most.

ALICE:  If we'd done a survey or a questionnaire, then we wouldn't feel anything for those people. But you're right by their side in their space in their bedrooms, in their school canteen, in their factory, wherever you are. You're part of their lives. So you're seeing why they're picking that top from Top Shop, or why they don't want to wear that because they look fat in it. Because you're with them, you understand.

DEBBIE:  You're connecting with the people. You create friendships. The people I worked with were very passionate about that legislation and how it had dramatically changed their working experiences. That created a passion in me.

IAN:  Did you get that passion too, Hayley, doing your online ethnography?

HAYLEY:  Not so much. But that's what the community is based on anyway. All the people who are talking to each other are having the same experience. They're not right next to the other people they're talking to. They're from maybe the other side of the world. That's just the nature of online communities.

## "So, Your Participant Observation Was Like an Ice-Breaker"

Ian was a bit shocked at how strongly some of us felt about our research. But the reading he gave us said that participant observation involved developing relations of trust with strangers, trying to appreciate the issue studied from their perspectives, sometimes finding out quite private things about their lives, empathizing with them, and communicating that empathetic understanding to others. So we were bound to feel things personally. To us, that's "real life." But, our research was *ethnographic*. Participant observation was *one* of the methods we used. It wasn't responsible for everything. None of us used it on its own.

HANNAH:  I think that point about participant observation being just part of an ethnographic research process is really crucial. That's why I chose it in the first place. I couldn't just walk into a school and talk to a load of kids because they'd be thinking, "Who are you? You're a teacher." So doing participant observation as a classroom assistant allowed me to get to know them and for them to get to know me, and for them to be comfortable with talking to me and for me to be a bit more comfortable with talking to children because it's been years since I was their age and knew how they relate to things. It was a key part of the research design, but it was also just a step toward my interviews.

IAN:  So your participant observation was like an icebreaker.

HANNAH:  Yeah.

CHRISTINA:  I did the same. If I had walked into the factory without being part of the workforce and said to migrants, "Can I come round your house please to talk about why you're in England working?," they would have been suspicious. They were there legally, but were still worried that someone's going to throw them out. So I got my job, worked alongside everyone, and waited until I got to know them. Then I could tell them what I was doing and why, and try to get across that I wasn't trying to pry into their lives. I just wanted to access something that I didn't think had ever been done before. They didn't understand what a dissertation was. So I said I was writing this massive book or doing this tiny little school project. There were a lot of issues with team leaders, and management as well, being incredibly racist towards the workforce. And that was something that they wouldn't have told me about otherwise. We had to have conversations outside work because they were so worried that they'd say something and then get fired. So, you have to build up relationships of trust. And I don't think I would have been able to do that without starting off with participant observation.

HAYLEY:   For me it was different. I started off observing, and then participated in the online discussions of my horror film. I started by going through loads of threads trying to pick out relevant things about the remake. So, it wasn't like a complete participant observation study. But it could have been.

IAN:   But you needed this observation..?

HAYLEY:   … to get what I wanted to be able to ask questions.

## "I'm Not Setting It Out Like They Say I Should, But It Is Okay"

We seemed to spend the first few days and weeks of our participant observation research noticing how much we didn't know, how wrong or dated our expectations were, and research findings came to us outside our planned field setting. This wasn't surprising. Ian had told us that research always takes place in an "expanded field." And, you have to (prepare to) be flexible. You often feel like your research is falling apart. That's scary. But a bit of experimentation, and using those layer five notes to try to keep track of changes and to regain and/or change focus, can get you through (see, for example, Crang and Cook 2007, and Cook 2001).

CHRISTINA:   If you read all the textbooks that tell you "How to do participant observation," they can give you confidence in how you're doing it. During my research, I had one and would refer to it. I'd ask myself, "Is what I'm doing academic? Is this the way I'm supposed to be doing it?" Going back to that book every now and again gave me the boost I needed. I'd say to myself, "Yeah, actually, you are doing it right." Or, "Okay. Something just happened that I wasn't expecting. But it's okay because this book says that it might."

HELEN:   That's a big part of this kind of research. You have to keep re-arguing your point and reason for doing it.

CHRISTINA:   You have to keep going and going. A methods textbook, and maybe your research proposal, can give you the confidence to keep plowing on, feeling that your research will end up somewhere interesting in the end.

DEBBIE:   You need a basic framework to follow.

CHRISTINA:   But everyone's is different.

ALICE:   So, you adapt it. That "how to" writing just gives you a framework. I tried to write those six layers separately, but you can't do it. You just can't have your paper laid out that neatly.

CHRISTINA:   A lot depends on what you can do in the place where you're working. I was in a food factory. I couldn't have a piece of paper with me. I couldn't say every five minutes, "Can I go to the toilet to write down secretly what you've just said?" So I had to write things down at break times, and at lunch times. And, often, I didn't use a notepad because I didn't always remember to bring it. So, I would be writing on a napkin that I found in the kitchen. Even when I wasn't at work, I saw my work people. We went to the same pub. And, sometimes, someone would say something and I'd immediately think to myself, "Ooh! That's going in the dissertation." So, I'd be writing things down on beer mats.

DEBBIE:   I wrote a message on my phone and saved it, pretending I was texting someone. But really I was just writing down exactly what they'd just said. Then I put the phone back in my pocket!

CHRISTINA:   And the point is, you won't find any of this "field note" advice in a textbook. They don't tell you whether you're allowed to do things this way. You have to say to yourself, "I'm not setting it out like they say I should, but it's okay because I'm achieving the same result. I've just got to adapt this advice to my situation."

IAN:   So you're saying that, in the beginning of your participant observation, you have to be well-behaved, "good" students. Then, after a while, you say, "Oh sod it. I don't need to do that any more."

ALICE:   Yes!

CHRISTINA:   Sometimes I'd be thinking, "I shouldn't be doing this." But I couldn't see any other way. So I thought it's better to go with it because …

DEBBIE:   … *you* know what's needed, don't you?


## "And You Start to Put these Tiny, Tiny Things Together …"

We didn't always feel this confident. It often took a long time to get to the point where we felt we were actually finding something out, creating data that we'd actually be able to use. A methodology that, at least at the start, is based on writing down more or less everything that happens and everything that you think about it is not going to give you a sense of achievement straight away. But that does eventually come … , if you just keep at it (Bennett and Shurmer-Smith (2001) describe this dilemma and how to get through it).

HANNAH:   When I got to my research field, there weren't any obese children or "fat" children or people with a poor diet there. I thought, "Where are my fat teens?! I need them for my research!" I panicked and thought I had to change my research drastically. But I stuck with the methodology and it led me to a whole different view on things. I had planned to get children to do food diaries, and then to talk to them about the diaries, go into their homes, watch them at lunchtimes. I wanted to see how accurately they'd written down what they ate, where and who with. Sticking to that led me to concentrate on diet and what children eat *in general*, and the different spaces and people that influence this. It took me a long time to see that something else would come out of it.

ALICE:   My story's similar. You go in hoping to find this big conclusion within the first week, and it's not there at all. And you think, "Oh no! This is going to go wrong." But it isn't going wrong. It's real life and things are slow. Sometimes you feel you're not going any-where. But you keep writing these things down that seem irrelevant. Then, sometimes every week, the same thing comes up, or it comes up in a different area or a different space or in a different way. And you start to put all these little *tiny*, tiny things together that might happen in a second and they start to form this personality of someone or the way they are and it starts to make more sense. Sometimes, it just takes someone to say or do one thing. There was this girl who spilled paint down her jumper and she just completely fell apart. She didn't want to go and put her PE top on because she'd look different from everyone else. But, if I hadn't been there to see that, I wouldn't have known how she would have reacted at the time. I think it's the mundaneness of participant observation that's important. You just plod through and think you're getting nowhere. But in the end, it all somehow becomes clear. And you find things out after you've finished, too, especially when you cut things up and you **code**.

HANNAH:   I know what you mean. The main conclusion of my dissertation was something I didn't notice while I was doing it. But, if you do your analysis properly, you can see pat-terns emerging in all those damn notes and transcriptions!

## "I Wanted Them to Think I Wasn't Some Freak"

We weren't, of course, just observers looking for paint spills and patterns. Our presence on these "stages" and the ways we could "act" with different people, helped to create the "dramas" that we wrote about. What people thought about us affected where we were allowed to go with whom, what they showed us, and what they told us. We sometimes tried to influence this though dressing or talking differently. But there were some things – like our gender, age, and skin color – that weren't so easily changed. Our identities as researchers were never separate from our data. We made a difference and had to think that through.

> DEBBIE:     There were two groups of people in the place I worked: managers and auditors. The managers were the elite and didn't accompany audit teams. It was really difficult to interview them. So I tried to engage with them by saying, [in a high pitched voice] "Hi, I'm a little innocent auditor and I'd like to know what's happening." And that seemed to work. They started to think, "Oh, I know, I can educate this girl. I can tell her all about what I know. Oh! The Sarbanes Oxley Act. It's amazing. It does this, this, and this." But I couldn't say to an auditor, "Educate me," because they didn't care. They were on my level. We'd go for a drink, and they'd say, "I can't believe you're doing your research on this. It's the most boring legislation ever." It took a completely different angle to understand their perspective. So, it was all very up and down. I spent a lot of time finding out how to be, and evolving my "character" as well as doing the research. I just had to go with the flow really. You need to be quite adaptable. And use common sense.
>
> HANNAH:     When I spoke to the parents of the children I worked with, I wanted to make the best impression. I wanted them to think that I wasn't some freak. I was just a student doing my dissertation. And I'd gone to the school that their children were at. So I felt I had a sort of reputation to live up to. But with the children, I wanted to be really relaxed and casual with them so that they wouldn't treat me the same way that they treated teachers and adults. Lots of them called me "Miss." And I kept saying, "No, no, no. Call me Hannah. It's fine." But I could see them thinking, "Ooh [i.e. not sure], we're at school and you don't do that."
>
> HELEN:     In Cambodia, I was treated as the rich western white-skinned girl. People thought I was too young to be travelling on my own. People would say, "Your parents are okay with you coming out here? Are you from a good family?" When we visited a compound where some of the factory workers lived, it attracted almost the whole village. And there was this one lady who asked, "Could I come back to England with you? Could I be your servant?" And I was like, "Oh my God! We don't have servants. I do my own washing. I clean my own house." And she just couldn't understand it at all. It was seriously bizarre.
>
> HAYLEY:     It's so different online! You can be whoever you want to be. You never know if someone is telling the truth because you can't see them, can't look in their eyes. And, because you have time to think about what you're saying, you can premeditate your answers completely.

## "You're Not Neutral At All, Inside"

Most of us act differently with different people in "real life." We may be quite good at it. But, as a researcher, you don't want that difference to be too controversial. If, you're trying to appreciate what it's like to live other people's lives, you don't want to jeopardize this by correcting or disagreeing with them, or breaking confidences. It could cause them embarrassment or harm – to most, that's "unethical" – and ruin your research. So you usually have to agree with everyone, [almost] regardless of what they say. This is hard to deal with, especially when your research is changing *you*.

CHRISTINA: When I interviewed the British workers in the factory, some of them were incredibly racist about the migrants, and my initial reaction was [sharp intake of breath], "You can't say that!" I'd feel like defending the migrants and the relationships that I'd built with them.

ALICE: It's difficult to put your morals aside, sometimes. There was a time when I was thinking "Maybe I can tell these parents what I think of them." But you can't. You can't rat people out to their parents, or vice versa, because you've said you're not going to do that. But, if I was shopping with the girls and they'd say, "Oh no, I look fat in that," I would want to say, "No you won't." And I did because I became their friend. These issues can relate straight back to your childhood or the way you felt about yourself at that age or the way your friends said things to you. And you feel like, in this stupid way, you want to make a difference to them and say, "Don't worry about this stuff. It's completely irrelevant." But you can't.

HANNAH: I had to stop myself from being judgmental. When a parent would say, "I don't give them bad treats. But we do go to McDonald's on special occasions," I'd be thinking, "Duh, that's giving kids a treat that's bad food. They're going to associate bad food with good things, with reward." But you can't say that. I had to be totally neutral on everything.

ALICE: But, you're not neutral at all inside.

IAN: So, sometimes you'd be saying one thing out loud but, in your head, screaming something else?

HELEN: Yeah. But sometimes I couldn't keep it in. I was so close to tears that it was obvious to everyone there. There was this guy who lived in this tiny house with his wife and daughter. His wife also worked in a garment factory. He came across and wanted to interrogate me about what I'd found out. Like, "How much do people get paid here?" I did my best to answer his questions. We spoke through a translator and used very basic English. He started telling me about how he worked in his factory's ironing department and sometimes had to work from seven in the morning until three the next afternoon. He told me that they got a little package of rice to keep them going through the night. Then he said that sometimes he'd be so tired he'd iron over his hand. And he was smiling when he said it. His daughter was running around outside. She was peering in at me and my friend. She was so intrigued by us. And then he told me that he can't afford to send her to school. It was just *so* real. The initial aim of my research was to identify the people in Cambodia who had made my t-shirt. But it turned out more that I was identifying *with* them. That turned out to be a big theme in my dissertation.

ALICE: But that takes it right back to, "Why do participant observation?" If I'd read that, I'd feel just as awful as I do now. You couldn't have done that with a survey or a questionnaire. You *have* to be there, seeing that man and seeing his daughter.

## "You Can't Hide Behind Posh Words, Can You?"

We wanted to write up our ethnographic research so that it could have the same kind of effect that those dissertations had had on us a year before. But we were used to writing only in the third person. This made things tricky again, because only the first two layers of our notes were like this. The rest were so personal. So Ian told us to write in the first person. Set out *your* perspectives and try to pass on and compare the perspectives of those with whom you've done your research. Make this writing vivid, academic, and situated. Write so that your readers can step into your shoes. Help them to see and feel what you saw and felt. That took some doing.

CHRISTINA:   I wrote in my dissertation that I wanted it to be accessible to geographers, to non-geographers, to the migrants, and to the other people that worked in the factory. I wanted them to be able to pick it up and read it. They're not going to completely get the literature review and the methodology. But the majority of it, the analysis definitely, is me talking about my experiences in that factory. Everything is in my voice or their voice. So they could see where it's going and they can follow it. My mum read it and said to me, "I really understand where you're coming from. And I can step into their shoes. I can step into your shoes."

IAN:   Does that mean it's not as clever as that other stuff, though? Someone might say, "It's just description, it's not scholarly."

CHRISTINA:   It is though! There's no other way you can access the things you want to access. So how can it not be academic?

HAYLEY:   But also, it's about linking your own lived experiences to academic ideas. You can show that you understand it more when you can link it to your own life.

DEBBIE:   I think some people would find that very hard to do.

HELEN:   To start with, it *is* hard.

CHRISTINA:   I think you have to get over a sort of mental block.

ALICE:   It's about feeling all right to put what's *there* in the "real world," *there* in your notes and then in your dissertation. Throughout school, and at university to an extent, you're not allowed to give your opinion as if it was all right. You had to give Joe Bloggs' or some other academic's opinion. They wrote in a difficult way to understand. And you wrote like that too. When you're doing your participant observation, you have to explain what a classroom looks like because you know it's important. But, at the back of your mind, you have this person saying, "What are you doing? This is ridiculous. I don't need to know this."

DEBBIE:   That's why, when I wrote my draft, I thought my interview material was more solid. They'd actually told me that. But I didn't use my participant observation notes. I just thought, "This is just me sitting on the train writing *everything* that happened throughout my day. That's irrelevant."

ALICE:   Before, no one ever said that *you* writing something down serves as evidence or "data." But, you come to realize that that's *just* as important. It's the way *you* saw it. That's why *you* were doing the research and not someone else. But you don't realize that until you've read a bit more and your supervisor says it's alright to write this way.

DEBBIE:   But it's hard because you think it's going to sound awful.

HELEN:   But, then, when you do get into it, it's really easy. Because it's just you speaking, you writing things down.

ALICE:   Once you've done it, you think, "Oh, I can do this." I can read it and it sounds all right. Somebody who doesn't do geography can read it and understand it.

HELEN:   And enjoy it, too.

ALICE:   And you say to yourself, "Oh, awesome. Just keep writing like this."

IAN:   But did that writing make you feel exposed?

ALICE:   Yeah. Definitely. Because it's you [laughs].

DEBBIE:  You can't hide behind posh words, can you?

ALICE:  You're not hiding behind Joe Bloggs. You're saying "It's me."

DEBBIE:  You have to be brave to give your own opinions, link them to your own experiences and to what you've read. It's difficult to get over that barrier. But, once you're there, it just makes much better writing.

IAN:  But some might say that writing is "biased."

DEBBIE:  Yeah. But you can't give a universal opinion that everyone is going to agree with.

ALICE:  And someone can argue with us, if they want. They're our opinions. So you don't say, "This is right." You say, "This is what *I* found!"

CHRISTINA:  "I"m fully aware that this is *my* interpretation." I accepted that and was open about it.

HELEN:  Mine was totally "biased." But then, hopefully, the people marking it can understand that "bias" because I was there, experiencing those things. What I said was backed up by everything I saw. That's why I think what I think. It's situated knowledge. You don't have to like it. Just read it (Box 13.1).

## "They Might Be Highly Offended If They Read My Dissertation"

It was the end of our discussion and, it seemed, Ian didn't want to let us off without one last question. He said that "some people" think that entering into trusting relationships with people in order to write about them because it's good for their degree is a bit "unethical" and exploitative. It's a kind of deliberate betrayal (Stacey 1988). There are dodgy power relations here. Participant observation seems so straightforward when you first come across it. But it's a minefield, when you really get into it. He expected us to refer to standard ethical guidelines (and would have expected us to refer to Institutional Review Board processes if we had had them at the time (see Marshall 2003, Plattner 2003)) and then to get a bit stuck. But we'd thought a lot about this.

DEBBIE:  If your participants *know* what you're doing, surely that's "ethical?"

ALICE:  Isn't it more "unethical" to show someone as a statistic, on a chart, on a graph, and not let their voice or opinion appear in academic work "about" their lives?

CHRISTINA:  That's even worse. My migrants and Helen's garment factory workers would probably have never had people like us asking them about their lives before. And, in a sense, being subjects of the research has given them a chance to talk about their situations and talk about actually what's going on. It's not big people who have never met them before telling everyone, "This is what their situation is." It's someone small like me saying, "Yeah, okay, I'll have this voice, you can have this voice." Surely that's more "ethical?"

DEBBIE:  But I was giving managers a voice they *wouldn't* have wanted me to give them. They knew what I was doing. But they didn't know what I found out. They might be *highly* offended if they read my dissertation. Their job was to make the Sarbanes Oxley Act look worthwhile. And a lot of people say that it's not worthwhile, including their own auditors.

IAN:  So you sort of ended up playing people and perspectives against one other when you wrote this up.

DEBBIE:  Yeah [laughs nervously]. And that's probably "unethical." But how else could I do it? I'd wanted to get a rounded view of the organization and the Sarbanes Oxley Act. A one-sided account would have been bad research.

**Box 13.1    Just Read This!**

This brief passage from Helen's dissertation (Clare 2006: 44–6) gives a flavor of the evocative writing that ethnographic research needs to produce. Here, she describes arriving in a garment manufacturing district of Phnom Pehn with her friend Louise and their guide Tola (who she met through the Cambodian Labour Organization) to meet some garment workers in the compound where they lived.

"It's early afternoon as we hurtle down National Road #4, the rain pouring down. It's a particularly wet day, monsoon season. We're squashed onto the moto, in ponchos. Tola points out the factories, huge white warehouse-type buildings, usually with the name of the factory emblazoned across the side, in Khmer, Chinese or English. He points out people driving motos carrying *huge* bundles of clothing on the back. The factories have thrown them away and people root through the bins, to then make into doormats and sell on. I am still desperate to find out if the people we are going to visit produce for H&M. Tola doesn't think so. Feeling quite disappointed, we turn into the road leading to the compound.

We drive through the gate and dismount outside one of the doors. Tola greets the lady like a long lost friend. We take off our flip-flops, and leave them outside in the courtyard. I now have very muddy feet! Tola introduces us to Sarom, the owner of the room, a younger teenage girl (her niece), and another young woman, Sreymom. We greet them with the *sompiah*, pressing our hands together in a prayer like position and bow, saying "chim-rip-sur" – "hello" in Khmer. Each of the rooms in this compound, of which there are about eight around the courtyard, is no more than six by four metres, with three camping-style beds. It is small and dark. There are no windows, and an open door. The walls were covered in magazine pictures, which I find out are of famous soap actors and actresses.

We sit cross-legged on the floor. A man then enters the room – Seythung. Tola asks if I have the t-shirt with me. Yes I do. "Show them," he says. I take it out of my bag. Everyone wants to have a look. They pass it round. Tola says that they may not know who they are producing for a lot of the time. They don't recognize the H&M label, though. Sethung leaves the room and comes back with his daughter who is about two years old. She had been running around in the courtyard and was interested by these two white girls with funny yellow hair! Sethung also had a shirt. He pointed at the label. "Puritan." "Do I know it? Is it sold in England?" He asks me if customers in England are forcing the prices of the products to fall. I try to explain that it's not the same as in Cambodia. We don't barter, but that we are forcing prices to fall by demanding cheaper products."

HAYLEY:     And, you didn't use their names in your dissertation, did you?

DEBBIE:     No, it was all anonymous.

CHRISTINA:     And you didn't misrepresent them. Surely it's more "unethical" to portray something that's not going on.

ALICE:     Yeah, but some things that are important to your argument *have* to be left out. When I was writing my dissertation, I didn't include the way I wanted to rant at the parents. I didn't think it was my place to cause a big mother–daughter rift that's going to wreck their lives! I didn't think that was fair.

## Conclusion

We hope that this chapter has given you an appreciation of our journeys through participant observation research, and that you can see what we mean when we say it's like, and about, "real life." We hope this might encourage some of you to have a go yourselves. If so, we've provided a list of additional resources below, and suggest that you get into them next. These are mainly the kind of literature review plus illustration chapters we mentioned at the start. They contain the kind of detailed academic discussions that are "between the lines" of this chapter. Each contains accounts of participant observation research, plus other methods that could be combined in an ethnography. They also contain theoretical debates and discussions of situated knowledge, **research ethics** (including questions of **overt** or **covert** research, participant anonymity and (not) being neutral), data analysis/coding and writing that have to be seriously considered for any such study. If you like what we've described in this chapter, you'll *have* to get into these debates and discussions before doing your own participant observation research. You might also like to read this chapter again, when your research is underway. It might give you confidence when you need it. Good luck!

## References

Bennett, K., and Shurmer-Smith, P. (2001) Writing conversation. In *Qualitative Methodologies for Geographers*, Limb, M. and Dwyer, C. eds. London: Arnold, 139–49.

Clare, H. (2006) *Made in Cambodia*. Unpublished BA Geography dissertation, School of Geography, Earth & Environmental Sciences, University of Birmingham, UK.

Cloke, P., Cook, I., Crang, P., Goodwin, M., Painter, J., and Philo, C. (2004) *Practising Human Geography*. London: Sage.

Cook, I. (2001) "You want to be careful you don't end up like Ian. He's all over the place": autobiography in/of an expanded field. In *Placing Autobiography in Geography*, Moss, P. ed. Syracuse, NY: Syracuse University Press, 99–120.

Crang, M. A., and Cook, I. (2007) *Doing Ethnographies*. London: Sage.

Crook, J. (2005) *Seasonal Migrant Working in the Horticultural Industries of Hesketh Bank and Tarleton – Implications for People and Place*. Unpublished undergraduate dissertation, School of Geography, Earth and Environmental Science, University of Birmingham.

Griffiths, H. (2002) *Children's Consumption of Popular Song Lyrics: The Eminem Phenomenon*. Unpublished undergraduate dissertation, School of Geography, Earth and Environmental Science, University of Birmingham.

Griffiths, H. (2003) A tale of research. Children's consumption of music lyrics: the Eminem phenomenon. In *Cultural Geography in Practice*, Blunt, A., Gruffudd, P., May, J., Ogborn, M., and Pinder, D., eds. London: Arnold, 269–71.

Marshall, P. (2003) Human subjects' protections, institutional review boards and cultural anthropological research. *Anthropological Quarterly* 76(2): 269–85.

Plattner, S. (2003) Human subjects' protection and cultural anthropology. *Anthropological Quarterly* 76(2): 287–97.

Stacey, J. (1988) Can there be a feminist ethnography? *Women's Studies International Forum* 11: 21–7.

Wrathmell, S. (2003) *The Connectivitea of Britain and Sri Lanka*. Unpublished undergraduate dissertation, School of Geography, Earth and Environmental Science, University of Birmingham.

## Additional Resources

Blunt, A., Gruffudd, P., May, J., Ogborn, M., and Pinder, D. eds. (2003). London: Arnold. Gail Davies' chapter on studying the networks of natural history film-making would be interesting for anyone, like Helen, who is thinking of doing a "following" ethnography.

Clifford, N., and Valentine, G. (eds) (2003) *Key Methods in Geography*. London: Sage. Eric Laurier's "participant observation" chapter is a vivid introduction to this method, and is nicely illustrated with his hand written notes, etc.

Crang, M., and Cook, I. (2007) *Doing Ethnographies*. London: Sage. A combination of a detailed literature review and the stories of their student ethnographies (undergraduate, masters and PhD) from start to finish.

Flowerdew, R., and Martin, D. eds. (2005) *Methods in Human Geography: A Guide For Students Doing Research Projects* (2nd edn). Harlow: Prentice Hall. A book aiming to introduce research methods to undergraduate students. Ian Cook's participant observation chapter is illustrated with extracts from undergraduate dissertations. Those students convinced us that we could do this kind of research too.

Hoggart, K., Lees, L., and Davies, A. (2002) *Researching Human Geography*. London: Arnold. An extremely thorough and wide-ranging account of research (including on-line research) methods. This seems to be aimed at upper level undergraduates and graduate students. So, it's not a good place to start your reading, but you should end up finding it essential reading.

Limb, M., and Dwyer, C. eds. (2001) *Qualitative Methodologies for Geographers*. London: Arnold. There's a whole section here in which different geographers give accounts of their ethnographic research, "warts and all." These stories are full of the excitement, dilemmas and "real life" that we liked so much in those dissertations we read.

Shurmer-Smith, P. ed. (2002) *Doing Cultural Geography*. London: Sage. You don't have to be a cultural geographer to do an ethnography, but this book is a great introduction to the ways in which theoretical, ethical, and practical issues have to be considered and combined in any research project. Katie Bennett's chapter on participant observation is a fascinating read.

## Exercise 13.1   Participant Observation in Action

Alice's first attempt at participant observation was to describe what it was like to attend a geography lecture at her university. She used the "layers of description" that Ian had outlined in his lecture to provide the kind of detail that is essential to provide her reader (you?) with a vivid sense of "being there." If you want to experiment with this approach, you could start with just 30 minutes of your day – any 30 minutes – and try to describe it so vividly that a stranger could imagine "being there" with you. You could choose an ordinary or an unusual experience. You could choose a familiar or an unfamiliar place. You could choose to do this with people who know you, with strangers, or some combination of the two. You could choose a meal, a party, a journey, playing or watching sport, visiting family, reading a book, joining a demonstration, working in a book store, anything! Whatever experience you choose will raise all kinds of questions about the difficulties and rewards of this kind of research. Write some scratch notes when you're there, then use them later to write a detailed account. Have someone in mind who might read this account, a reader who wasn't there with you and may be unfamiliar with the kind of place or experience you have chosen. This should encourage you to take nothing for granted, and to develop an appropriate eye for detail.

# Chapter 14

# Cultural Landscapes

*Richard H. Schein*

- Introduction
- Cultural Landscapes
- Data and Archives
- Sites of Information
- Conclusions

Keywords

Archive
Authorship
Discourse
Epistemology

Palimpsest
Serendipity
Vernacular
Whiggish or teleological

## Introduction

Cultural landscapes in this chapter are defined firstly and primarily as the tangible, visible impress of human activity on the surface of the earth – the everyday "stuff" of the material world we have created over time. Landscape scholars often speak of the **vernacular** land-scape to emphasize that cultural landscapes are not only about "landscaping" (or con-sciously designed lawns and gardens) but are seen more generally as comprising common houses and fences and public buildings and parks and backyards and fast food restaurants and light poles and streets and public squares and so on. The cultural landscape is a geo-graphical **palimpsest**, or an accumulation of "geo-graphy" – literally human "earth writing." It is a material record of our activity, and as such we can gather information on its creation and meaning, through many sources, but especially through historical records collected in **archives**. This chapter introduces you to some starting points for collecting data about cultural landscapes, first through a brief introduction to the idea of landscape, second in a discussion of data and archives, and finally by describing some key sites for gathering information about cultural landscapes.

## Cultural Landscapes

While the cultural landscape is a material "thing" or set of things, it also is simultaneously a way of visually and spatially ordering and organizing the world around us. As a way of seeing and knowing, the cultural landscape is, in effect, an epistemology that has long been central to human geographical traditions of observation, interpretation, and analysis (Cosgrove 1998). Commentators on the cultural landscape have called it vague, duplicitous, and ambiguous, and have devoted much time and paper to discussing just how landscapes have come to be, how we live in them, and how they might work in reproducing the everyday world around us (see Boxes 14.1 and 14.2; Daniels 1989; Groth 1997).

Let's take a seemingly simple, nineteenth-century bird's eye view of an American city (Figure 14.1). This 1852 image of Syracuse New York is reasonably accurate in depicting the city's material fabric at that time. That is, the buildings and roads and industry and railroads you can see really were there. This view could then be used as *data* or *evidence* of the landscape's material evolution, of what was on the ground at one time, of what has disappeared, and, when compared to the contemporary scene, of what persists as remnant in the landscape's always-palimpsest appearance. But that bird's eye view also has a

---

**Box 14.1  Landscape Across Disciplines**

Paul Groth and Chris Wilson (2003) have referred to the "polyphony" of landscape study to illustrate that it is a "many voiced endeavor." It is true that scholars from many disciplines (e.g., geography, anthropology, art history), practitioners (e.g., planners, historic preservationists, landscape architects), and others (e.g., commercial developers) all have legitimate claims to the term (see for example Bender 1995; Herman 2005; Mitchell 1994; Potteiger and Purinton 1998). As you explore different literatures as well as different sources of information about cultural landscapes, you may find yourself reminded of the apocryphal observation about the United States and Great Britain: that we are "two nations divided by a common language." Landscape enthusiasts of many stripes may use the same words, all with slightly different meanings, and this often can be confusing, if not downright exasperating. The trick when coming up against this potential communicative obstacle is not to withdraw into the parochialism of your own discipline, or your own pedantic definitions of landscapes or interpretive and analytical preferences, but to search for "common ground," in realizing that we all are interested in some aspect of the tangible, visible scene. Once past the immediate problems of terminology, there is much to be said for the cross- and inter-disciplinary practice of cultural landscape study.

**Box 14.2** Landscape in Geography

Geography's cultural landscape tradition generally is traced to the work of Carl Sauer (1925), whose famous landscape maxim posited the physical landscape as the medium, "culture" as the agent, and the cultural landscape as the result. Although Sauer did not consistently subscribe to that conception (or its theoretical implications) in his own work, the basic idea persisted abroad for generations, first as a guiding principle for an empiricist tradition in the United States, and later as a conceptual foil for geographers interested in re-thinking the importance of landscape in line with post-empiricist moves in human geography across the Anglophone world. "Cultural landscape" was one of the spoils in the so-called "Civil War" that erupted around debates in cultural geography over geographical traditions, guiding concepts, theoretical foundations, and empirical focus in the last 30 years. The (re)theorizing that took place through those debates has enlivened cultural landscape study to the present day. It has mandated attention to the imbrications of class, race, and gender in and through the landscape, to the place of landscape in power relations and questions of identity at a variety of scales, and to other broadly socio-spatial concerns of human geographers including, most recently, a renewed phenomenological interest in our everyday experience with landscapes. As a result, our interest in the landscape as the tangible, visible impress of human action has been extended to asking questions about the place of cultural landscapes in constituting the world – through their symbolic qualities and material presence, through their normative qualities, through their capacity to mask social process, through their role as a site of action and intervention into the everyday world.

particular *point of view* beyond the simple location of the (imagined) viewer on a hill overlooking the city from a distance; and that point of view helps to frame or constitute the cultural landscape itself – the tangible, visible scene that seems so realistically represented in the view. The image adheres to an American art genre of romanticism, and this bird's eye view also expresses an opinion about the burgeoning of American industrial cities and, by implication, the changes being wrought in American society at the time, an opinion that hoped also to shape the future of American urbanism (Schein 1993).

Cultural landscapes are indeed complicated and fascinating. We can, at once, study cultural landscapes as material artifacts, with traceable and documentable empirical histories and geographies, and simultaneously use cultural landscapes to understand and question ideas about and ideals of everyday life. Ultimately, cultural landscape study is important to critical human geographies if we see the landscape as **discourse** materialized, the tangible and visible scene serving to normalize or naturalize social and cultural practice, to reproduce it, to provide a means to challenge it (Schein 1997). In that latter capacity, then, the cultural landscape is subject to any number of methodologies, if we understand methodology to create the link between **epistemology** and method or technique in interrogating the empirical scene before us. Gillian Rose (2003) has provided one example of the methodological possibilities for landscape study in a book focused upon interpreting visual

**Figure 14.1** Bird's eye view of Syracuse New York, circa 1852. The image provides reasonably accurate historical information on the built environment even as it presents an opinion about the city and burgeoning urbanism in the mid nineteenth century.

images, especially as they are implicated in questions of cultural meaning and power. In what she calls a critical visual methodology, Rose offers a broad analytical framework for understanding cultural landscapes through a range of methodologies, including content analysis, semiology, psychoanalysis, and various forms of discourse analysis which take seriously ideas of textuality, intertextuality, and context. Ultimately, she points to the importance of understanding, at least, the production of an image or landscape, the image or landscape itself, and finally what she calls the "audiencing" of the image or landscape, or the manner in which it is received and circulates as an image.

Rose's work highlights the central empiricist problematic of assuming that landscapes simply "speak for themselves" and reminds us that cultural landscapes are themselves representations embedded in, and that embed meaning in, everyday life. As such they require the same sorts of methodological considerations we would apply to any social science problem, and in gathering information about cultural landscapes we can employ many of the research methods described elsewhere in this volume (see Box 14.3). This chapter, however, assumes that most, if not all, good cultural landscape study begins with the material thing or set of things that we identify as the landscape, and draws a fine, if heuristic, line between finding out empirical information about a landscape and asking questions about what the landscape means or how it works. First we must be able to describe the landscape and its particular history, documenting when and where the landscape was created, by whom, why, how has it been altered, and so on.

**Box 14.3** Approaching Landscape Study

This chapter's primary focus on landscape data collection leaves largely unanswered the question of why and how we study cultural landscapes at all. The polyphony of landscape study suggests that there are many reasons and approaches. It is possible, however, to inductively parse the landscape literature into the following descriptive categories. Many studies will fit more than one category, and this framework is really a heuristic device for making sense of the field (Schein 2009). Although there will always be studies that do not quite fit any typology (its purpose is to help describe, not to force things into categories), we can see landscape studies as being variously about:

- *Landscape history:* Empirically documenting when/where the landscape was created, by whom, why; how has it been altered, and so on.
- *What the landscape means:* To identity (individual and collective, of people who live in and through the landscape) and interpretively (the landscape as our "unwitting autobiography," to quote Peirce Lewis (1979)).
- *The landscape as facilitator/mediator:* of political, social, economic, and cultural intentions, and debates.
- *Landscapes as discourse materialized:* Asking questions about how landscapes "work" to normalize/naturalize social and cultural practice, to reproduce it, to provide a means to challenge it; seeing the landscape as the material through which these relations flow.

## Data and Archives

All cultural landscapes are local, which is to say that all landscapes exist somewhere *in particular*. It is the sites of particular knowledge that concern this chapter. That does not mean that we should be parochial or antiquarian when interrogating cultural landscapes, especially when we remember that our ultimate goal is to understand the place of landscape in larger discourses about material culture and social life. A house might be of a type (that has diffused, contagiously or hierarchically), a city park may have been designed by Frederick Law Olmsted, a debate over a local historical plaque commemorating slavery may be a racialized landscape that mediates a general American ambivalence about the structural imperatives of racism. But we must start somewhere when looking at cultural landscapes, and that somewhere can be thought of as obtaining data about particular landscapes in particular places; about *a* house, *a* park, or *an* historical plaque. Historians often think in terms of primary and secondary data; the former being "raw" information, supposedly unmediated by interpretation, the latter being data that has been organized according to interpretive rationale. While this fundamentally empiricist distinction does not really hold up on inspection (even raw data has been collected according to some organizational rationale), it helps to get us thinking about where there might be basic information about a specific cultural landscape that has to do, firstly, with its creation, order, organization, and appearance. There is a good chance that, secondly, there already exists a secondary literature that can help put your cultural landscape in context or in perspective, that can help you see and interpret the landscape you are interested in as one of a type, or one linked to other places and landscapes transcending the immediately local scale of your initial inquiry. To follow the examples raised, there are scholarly literatures on house types; on the origins, forms, and functions of American public parks and Frederick Law Olmsted's role in their evolution; or on slavery and public memory and historical commemoration that will push you to see your particular landscape as linked to other spaces, other places, other landscapes, other ideas. In this chapter, you will get some ideas about how to find the particulars that are at the heart of those literatures – information about particular landscapes.

There is an enormous amount of information about cultural landscapes out there, especially when we realize that we (as students of the landscape) are not the only ones who are interested in landscape. At some level, the landscape also is an engine of capitalist accumulation, a focus of humanist aesthetic desire, a utilitarian means to social, economic, and political ends. The trick is to figure out who else is interested in "the cultural landscape" even as they might not think of cultural landscapes per se. For example, Peirce Lewis (1979) wrote some years ago about the value of trade and professional journals, which often contain technical information for the people who design and build and maintain our vernacular landscapes. Highway engineering bulletins, the in-house journals of fast-food franchises, meeting minutes of the local planning commission, and architectural and landscape architectural journals are full of information about the making of American landscapes. These kinds of periodicals are especially useful for gaining the view of those who actually make or produce specific landscapes. While getting at landscape consumption is a bit more difficult, there are available popular fashion and taste magazines – *Home Beautiful, House Beautiful, House and Garden, Southern Living, Sunset, Architectural Digest* in the US for instance – that contain clues about the manner in which we have adopted certain preferences for particular landscapes, especially the more personalized landscapes of home and garden that are the most easily manipulable by the average person.

With some diligence and ingenuity, we often can find information about cultural landscapes in unexpected places, or through "proxy" means. Slave narratives, for example, comprise an American literary genre and many were recorded through the Federal Writers Project in the first part of the twentieth century. They can contain rich descriptions of everyday southern landscapes and of the plantation life lived through them. City directories, described below, do not provide specific information about, say, the urban landscape of 1900 Cincinnati or Bridgeport or Seattle. But if those same directories include a simple notation about telephone ownership, you can extrapolate a picture of at least one visible aspect of urban infrastructure at that time, telephone poles and wires, which in turn allows speculations on urban class structure or wealth distinctions made visible in the landscape. Institutional records of all kinds, especially those having to do with surveillance – like police records or the records of the Freedman's Bureau after the Civil War – similarly can be used as proxy indicators of particular landscape features if not actually providing detailed descriptions of a whole scene.

There are many ways of gathering information about cultural landscapes, both historical and contemporary. Approaching the landscape as a palimpsest – albeit a messy and often seemingly illegible one – of accumulated human occupance on the earth's surface allows us to collapse the distinction between a contemporary and historical landscape, realizing that those designations – historical and contemporary – are really just two idealized snap-shots or moments of a cultural landscape continuum. The closer we are to examining a cultural landscape of the present, the more options there are for gathering data on it. Whether we are interested in a landscape's production or consumption, its **authorship** or its reader reception, we can employ methods such as participant observation, interviewing, or survey techniques to gather information or to generate data from the testimony of living people. Other types of information about cultural landscapes comes from written and visual records, generally kept for specific purposes (other than landscape interpretation), and identifiable as a particular source material (such as fire insurance maps, city directories, or deed records). These records often are kept in repositories specifically created for them (such as the records of real and personal property – deeds and wills – in every county clerk's office in the United States), or are found in special collections, in historical societies, in special libraries, or are in private possession. In short, there are particular kinds of records that lend themselves to cultural landscape description and interpretation, and there are particular places, storehouses of that information, that are more likely to bear fruit. The latter we generally call archives, and before the final section of this chapter enumerates some of the records and archives that are especially fruitful in attempting a local landscape study, it is worth making a few observations about the place of the archive in cultural landscape research, in broad conceptual terms as well as for the practicalities involved in actually getting at the source material you want.

The term archive has strict professional definitions – there are, after all, professional archivists, the people who create and look after archives, and they have a scholarly literature that looks at everything from basic technical information (such as how to conserve records in acid free binders) to deep philosophical problems (such as the role of the archive in guiding our normative epistemological understandings of the world). See, for example, the journals *Archivaria* (especially number 61, Spring 2006) and *Archival Science* (especially volume 2, numbers 1–4, March and September 2002) and *Historical Geography* (especially volume 29, 2001). We can leave aside those more nuanced contemplations to think of an archive as simply a place that stores visual and written information about, in our case, cultural landscapes. The term archive used to conjure up damp, musty basement storage rooms and old books with loose bindings and pages falling out, carefully guarded by an archivist or librarian whose sole purpose, it seemed, was to deny a young student access to the precious and rapidly deteriorating material. These days, however, an archive might be entirely digital, and be widely and easily accessible in an on-line format. Archives can be public (and thus legally accessible to all citizens), such as the libraries of land grant universities in the United States, or county clerk's offices (as long as a state has enforced sunshine laws). They can be private, such as historical societies or private museums or clubs (and private archives might include corporate or business records of a specific company or other private enterprise). Gaining access to an archive is half the battle; the other half is finding what you want or need there. There always are rules for accessing archival collections, just as there are rules for using the public library or for how you behave in class. If the archive is a busy public space, such as a local county clerk's office, you may have to figure out the generally unwritten rules as you go along. If the archive is an estab-lished scholarly venue (such as a special collections wing of a university library) there will be clearly delineated rules for using the material. If it is private you may be subject to certain requirements regarding the manner in which you use the information. In all cases, of course, professional and scholarly ethics pertain as to the manner in which you present yourself and represent your position.

Once inside an archive, you may be asked to put all your belongings except a pencil and paper in a storage locker. You will likely be asked to use only pencil (not ink) in taking notes. You may or may not be able to take your computer or to xerox fragile and unique materials. You may have to pay a hefty charge in order to do so. And so on. It is important to respect the rules of the archive, no matter how supercilious or silly they may seem to you. After all, these rules are what stand between you and the information you are after, and at their best they are meant to protect archival materials for future use. Finding what you want in an archive might be as simple as asking for the fire insurance maps you came for. At other times you will rely upon the archive's cataloging system, which is almost certainly not oriented to your desire to research a particular cultural landscape. Librarians and archivists are invaluable aids to finding information, but you must be prepared with a brief and succinct explanation of your project so that they may understand what kinds of heretofore undiscovered materials deep in their collections might be useful to your project. Once you get access to useful material, it often is very easy to get misled by the archive, especially if you come across documents or other material you had not expected and which contain a rich variety of information (I have whiled away many an hour in a library reading room looking at historical materials that were fascinating in their own right, but ultimately bore little relevance to my original quest).

More insidiously, the archive may hijack your project; that is you may find yourself redirecting your inquiries based on the material you find in an archive. If this is only taking advantage of carefully managed **serendipity**, then that is not a problem. You want to be careful, however, not to lose your critical edge, to or to lose sight of your original focus so that the manner in which the records are kept starts to dictate your research questions and direction. After all, records are kept for a specific reason (they are not value neutral), and that reason is seldom the same as the one that brought you to the records in the first place. You will more often than not be using records for reasons other than those that were originally intended. This can be a good thing, as your perspective may facilitate not only basic data collection, but might illuminate the social consequences of record keeping itself.

You should try to be aware, as well, of information that is missing from the archives. It is a truism that the winners of history write the stories, and that goes especially for historical geographical records. You are more likely to find records of the rich and powerful than you are of the poor and marginalized, or of men than of women, in most cases, or of white people rather than people of color. This is especially true when we realize how many records in the United States, at least, are related to the demands of free-hold property systems and capital accumulation. This should not stymie your project, as there are ways to read absences in the archive, even as we realize that there also may be questions we might ask of our landscapes that have no archival record at all. More prosaically, we should always remember that archives are fallible – they are, after all, records kept by humans for particular reasons, and sometimes they present outright lies, sometimes they are full of errors, and, when dealing with historical records of some age, you might find that particular kinds of penmanship and cursive presentation have changed significantly over the past few hundred years and can make reading old documents a challenge. Similarly, the meanings of words, terms, occupations, and landscape features can change over time, and so you must be careful to understand archival records in their own terms before you use them for your own ends, whether simply to account for and describe particular landscapes or landscape artifacts, or in order to achieve a more critical take on the place of a landscape in social formations. Finally, do not be constrained in your search for a particular (i.e., local) landscape to only local archives. While it always is easier to start locally, there may be significant records on your particular site housed elsewhere in the country. To go back to an earlier example, you might find information on that seemingly unremarkable neighborhood park you are researching in local archives, perhaps in a photographic collection in the local history room of your public library, or in the records of the planning commission at city hall. But if that park was designed by one of the several permutations of the Olmsted landscape architectural firm, there may be records in the Library of Congress in Washington, or in the Olmsted papers in Massachusetts. And if the traveling Chautauqua visited there in the nineteenth century, you might find records of the park in a national repository covering that national phenomenon. Of course, on-line searching has these days made the task of making those connections much easier. The final onus is upon you, the researcher, to have a well defined sense of what you would like from your project and an open mind to the serendipitous possibilities of archival research even as you try not to get hijacked by what will undoubtedly be a plethora of information.

## Sites of Information

While there may be a surfeit of information about cultural landscapes out there, limited only by your creative imagination, there are a number if sources that are "tried and true" and are useful starting points for getting the most for your time and effort when beginning research on a particular cultural landscape. Some of these are listed and briefly described in this section. This is not by any means an exhaustive list, and can be read as an indicator of the myriad possibilities of data collection on a cultural landscape as much as it is a blueprint for starting your study. Both particular kinds of information (or records) as well as general kinds of archives are listed below, without distinction, for you will soon realize that maintaining the distinction is difficult (and some kinds of records may be kept or found in several different places; and some archives hold several useful data sources). Remember that cultural landscapes are, at heart, about spatial as well as visual order, and so included in this section is a number of sources for maps. In no particular order, then, you might find the following sources of data, the following general archival descriptions, useful when beginning to prosecute your own study of a particular landscape, when you are after "primary" data about *a* portion of the tangible, visible world. The examples in this section are primarily about US sources, but many of the kinds of sources enumerated also are available across the world, and most especially in Anglo-derived or dominant settler societies created in the 500-year wash of European imperialism and colonialism and its attendant transformations to urban, industrial, and post-industrial life.

### Bird's eye views

These have been produced for centuries but, like so many historical urban records, became especially popular in the nineteenth century when so many cities across the industrializing world were rapidly expanding and lithography became cheap and popular. Over 5,000 views of 2,500 US cities and towns, for example, were drawn and printed in the century after 1825. Bird's eye views depict the city from an imaginary point high above the ground, and while there are changing aesthetic conventions regarding the manner in which the city is represented (ranging from abstract cartography to romantic portrayals), they still are useful for the wealth of empirical detail they contain. What is printed in the views generally really was "out there" in the cultural landscape. With the caveat of an attention to artistic license (or the framing intent of the view maker), you can use these views as an historical record of the physical urban fabric. Bird's eye views also are housed in the Library of Congress (and many are available on-line at http://www.loc.gov/index.html), and many of them have been commercially reproduced over the years and can be inexpensively purchased. Most local historical repositories will have one on file for any city of even the most modest size in the late nineteenth or early twentieth centuries.

## Fire insurance maps

These became popular in England in the late eighteenth century and soon were being produced in rapidly industrializing American cities as well (Figure 14.2). These large scale (often 50 feet = one inch), generally urban maps are organized according to the practical information needed by insurance underwriters (building materials, placement of water and gas lines, presence of steam boilers, road access, distance to fire department, etc.), who used them to issue policies on a building sometimes without having to physically inspect the property. They were at first hand-colored, according to building material, and reproduce detailed pictures of over 13,000 American towns and cities for the past century and a half. Although they are available in microform, microfiche, and, increasingly, digital format, older, paper versions often reproduce the cultural landscape of fire insurance underwriting as a three dimensional palimpsest, as periodic updates by subscription provided the map owner with "paste on" additions that simply went on top of the existing map. A number of different companies produced these maps in the United States (including Perris and Browne and the Rascher Map Company) and Great Britain (such as Charles E. Goad Ltd), but the Sanborn Map Company (founded in 1867) eventually monopolized the business in the US, Mexico and Canada, and for that reason fire insurance maps, no matter who produced them, are often simply referred to as "Sanborns." Many local history collections keep paper copies of these maps. The Library of Congress holds almost three quarters of a million of them. Even though they were created for a specific purpose, they contain so much historical information about American urban and small town landscapes that fire insurance maps are among the most satisfying archival documents for recovering and describing the cultural landscape. Like all maps, of course, there is much missing (properties not deemed "insurable" for instance, which might mean African American or other marginalized parts of the nineteenth century city), but as long as you remember not to rely upon them as the "absolute truth," fire insurance maps are invaluable.

## City directories

These might be thought of as phone books before there were phones! Like "Sanborns" they have been around for a couple of centuries, but generally are associated with burgeoning urban industrial societies (including Canada, Britain, Australia, and the United States) in

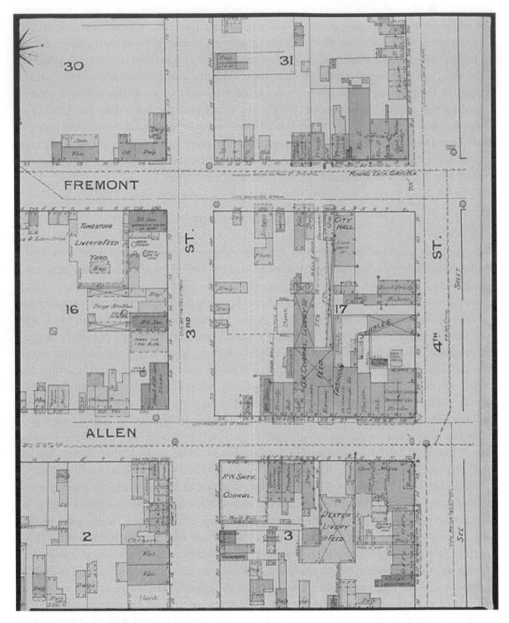

**Figure 14.2**   Sanborn fire insurance map. These maps provide a wealth of large-scale information on individual material attributes of a cultural landscape. They can also serve as clues to a landscape's social and cultural use and meaning.

the late nineteenth and early twentieth centuries (even though the Polk Company, synonymous with city directories, is still in business). City directories were and are commercial ventures, and included advertising even as they provided information for commercial travelers (salesmen) and other people interested in the commercial and industrial opportunities in any one of thousands of cities across the continent. They generally had an opening section with advertisements and general commercial and civic information about a place (including maps of trolley and bus lines; how many churches there were and their locations, the kind of city government, etc.). The bulk of the city directories were given over to listings of a city's residents, and the best directories (for our purposes) did this in two ways. First, they presented a geographical, street-by-street listing of addresses and residents (grouped by household). Second, they presented an alphabetical list of a city or town's residents, which often included personal information such as occupation or racial designation (presumably assigned by the directory enumerator) as well as household information, such as whether a house had a telephone in the late nineteenth century for instance. By cross referencing these two lists and by working with several in a series of directories (which were usually reproduced yearly) you can learn much about the social composition of a city on a street-by-street basis, as well as get a sense of how neighborhoods developed and changed over time.

## County histories, atlases, and gazetteers

These were especially popular in the United States in the nineteenth century, although plenty of more recent examples exist, as do reproductions of earlier publications (especially after the time of the US bicentennial in 1976). County histories and atlases were especially popular in rural counties, and often were produced after the first few generations of (primarily) Euroamerican settlers on an American frontier had transformed the landscape into a prosperous patchwork of farmsteads that they viewed with a sense of accomplishment and pride. These works generally are celebratory and tell a **whiggish** or **teleological** history, but are nevertheless invaluable sources of information, as long as you remember that they often were produced commercially by subscription, and so the representation of particular landscapes may have been swayed by either the ego or the pocketbook (or both) of the landowner whose property is richly illustrated. They are excellent "snapshots" of settlement evolution and often include illustrations of important landscape elements as well as maps of settlements and the features then deemed important to residents. Gazetteers were especially popular early in the nineteenth century, and often include both narrative and census information on localities in a particular state or region. Land ownership maps are still produced, and appear as more utilitarian documents that record the basic spatial order of county-wide landscapes as a support for free-hold land tenure across, especially, the US Midwest and far west. There are other kinds of local histories – especially for larger urban areas, and these often include collections of local or regional photographs that may or may not be accompanied by narrative description or analysis (and you might do well to trace the archival source of the photographs used in any one of these particular collections, to see if there are others that did not make the editing cut but which might still be of some use to your project).

## Topographic maps and aerial photographs

These provide spatial information about cultural landscapes at several scales or resolutions (the so-called 7.5 minute map at a scale of 1:24,000 is probably the most useful large-scale map providing the kind of detail useful to intensive landscape study). In the United States topographic maps were produced primarily by the US Geological Survey for more than one hundred years, and they generally are available in local repositories in paper copies (which are especially useful if you are comparing time sequences and want several temporal permutations or scale representations of the same coverage area) or on-line, these days through either the USGS website (http://topomaps.usgs.gov/), which is now superseded by the national map (http://nationalmap.gov/), which coordinates spatial data from multiple sources. Similar topographic mapping programs exist in other countries, such as the Ordnance Survey in Britain. Aerial photographs in the United States were especially used by the US Department of Agriculture (Farm Service Agency) and its local representatives beginning in the 1930s. They now are associated with the National Aerial Photography Program (NAPP), and are available in a number of places, including on-line (http://www. apfo.usda.gov/). Like so many potentially surveillant technologies (such as GPS and GIS these days), the development of aerial photography was linked to military needs, in this case World War I and World War II.

## Survey notes, land records, and travelers' accounts

These are ubiquitous wherever Europeans took up native or aboriginal lands as their own, and then marveled at the transformations they had wrought (Figure 14.3). The process of "transferring ownership" (by whatever means) necessitated field survey and record keeping, in the United States by both the individual states and, especially after the 1780s, by the federal US government under the township and range survey system and as prelude to establishing transcontinental railroad routes. Field surveyors often archived detailed notes (including sketches) while marking boundaries, and these are especially good as one of the few sources for noting both "remnant" indigenous or native landscape features as well as ecological and environmental conditions at the time of survey (at the turn of the nineteenth century for instance, tree cover was deemed an indicator of soil fertility, and so was duly recorded in some detail to aid settlers and speculators alike on American frontiers). Travelers' accounts, such as those by Alexis de Tocqueville or Frances Trollope, were popular throughout colonial periods and, in the US especially in the first half of the nineteenth century. Europeans often traveled the US to marvel or scoff at the newness and crudity of the American landscape, especially as compared to (generally deemed) superior European counterparts.

## Historic preservation

This has provided an impetus to landscape data collection that you can tap into if a particular neighborhood, rural district, building, or set of buildings have been subject to preservationists' attention. Listing and preserving a building requires documentation, and while much historic preservation documentation draws upon the same kinds of resources outlined in this chapter, often, in the US for example, a historic preservation nomination form will include bibliographic information directing you to particular sources for your

**Figure 14.3** Map from the first state survey of central New York, circa 1788. This map helps document natural drainage systems, survey lines (which became property boundaries and road networks), remnant native American presence, and a public reserve for producing salt on the frontier (with permission; from the personal collection of D. W. Meinig).

site. Historic preservation in the US operates at several different scales and through several entities, including the Federal government, which operates a registry of historic structures and landscapes, often through the activities of state historic preservation offices (SHPOs) as well as local private societies (often organized as preservation trusts) and local government offices of historic preservation, which usually exist if preservation efforts have been translated into zoning overlay codes designating historic districts (the only time in the US when historic preservation has any legal "teeth"). Local preservation trusts, the historic preservation officer in your local planning office, or your SHPO are places to start.

## Local newspaper records

These are an excellent place to start if you are interested in a particular landscape or site, especially if it bears some local, "newsworthy" significance (such as an urban redevelopment project, or a public park, or a site of contest or political protest). Often local history rooms in public libraries or the local history society will keep "clippings" files on such sites. Your local newspaper has likely been microfilmed and indexed (and is probably available at your local library), or, increasingly, has been put on-line (although the on-line option can get pricey as you will either pay for each news item downloaded, or be charged a significant subscription fee for unlimited access).

## Other records

These are readily available and those that provide a good return for the time invested include deed records, census information, paintings and murals, and New Deal make-work literary projects. While deeds (in the US) can refer to any number of legal transactions, the most useful as landscape data sources are property deeds that record sale transactions and detailed information for individual properties or buildings (lot lines, buildings present, sale price, adjacent properties, prior sales, etc.). They are most easily traced back in time from an existing address, although property subdivision or unusual transfer circumstances can sometimes muddy a trace. Census information is available in both published form (on a decennial basis in the US since 1790, and available at a variety of scales, from the block to the nation) and in manuscript form (comprising the enumerators' actual field notes, available after a 70 year waiting period, so that the 1930 manuscript census is just now coming "on line"). There are also periodic special censuses, such as agriculture or manufacturing. Landscape painting is a longstanding western art genre, and in addition to nationally known works, you are very likely to find local or regional artists over the years who have committed local scenes to canvas or even to the walls of public buildings during the New Deal, when artists recorded local scenes in murals across the United States. The New Deal also employed scholars and artists and writers to document the American scene during the Depression. The most famous of these efforts is the American Guide series, all of which have been republished and which provided detailed regional descriptions of land and life across the nation

## Thinking archivally

This can take you beyond specific records to types of archives containing a potential bounty of data on cultural landscapes. It is impossible to comprehensively list such archives, but in order to get you thinking about the possibilities, a list might include: county courthouses (for county-level public records such as deeds, wills, probate, and auction records); local, county, and state historical societies and museums (which were mandated by law in some states, such as New York, and which vary in their collections, coverage, and strategies); local and state government planning offices (these are the professionals who are responsible for thinking about the future order of the American landscape); historic preservation offices at local, state, and federal levels; special collections in libraries across the world, including university libraries (for instance, my university library has a collection of papers donated by members of the famous Kentucky Clay family, who operated farms and planta- tions and kept detailed records and diaries about the operation and appearance of their landholdings); local history rooms at public libraries; and of course, any good search engine these days will grant you access to an ever-burgeoning list of on-line sites devoted to written and visual records about the cultural landscape.

## Conclusions

Collecting data on cultural landscapes is not hard. It simply requires a basic sense of what you are after, where you might get started, and a little diligence and perseverance. In fact, like any research project, it is likely that you will require more time than you initially budgeted to track down and collect useful information. The harder part is figuring out what to do with the information you collect once you have it, for the most important lesson to remember when collecting "data" about a landscape is that the data and the landscape do not speak for themselves. Even simple landscape histories and descriptions require a point of view, and the best landscape descriptions and interpretations and analyses require a basic understanding of the empirical particulars of a cultural landscape in order to ask questions about what that landscape does, about why it is important, about how people live in and through that particular landscape and to what consequence. These, of course, are harder questions, and there is a very large literature extant to help you ask and answer those questions, once you have got started documenting the cultural landscape as an object of study – you might start with Cosgrove (2000) or Wylie (2009). And in answering those questions you are likely to employ any number of social science or humanities methodolo- gies, many of which are discussed further in this volume. For a long time in the Anglo- speaking world, at least, cultural landscape study was insulated from the methodological controversies of either the human geography-as-a-science paradigm or its post-empiricist multiple challenges. One of the more exciting developments in landscape study over the past generation is the extent to which the cultural landscape is seen as an important com- ponent of everyday life, a central aspect of human existence that is subject to the same sorts of methodological considerations and possibilities as any other human geography problem, issue, hypothesis, or description.

## References

Bender, B., ed. (1995) *Landscape: Politics and Perspectives*. Oxford: Berg.

Cosgrove, D. (1998) *Social Formation and Symbolic Landscape*. Madison: University of Wisconsin Press.

Cosgrove, D. (2000) Cultural landscape. In *The Dictionary of Human Geography* (4th edn), Johnston R. J., Gregory, D., Pratt, G., and Watts, M., eds. Oxford: Blackwell, 138–41.

Daniels, S. (1989) Marxism, culture, and the duplicity of landscape. In *New Models in Geography, Vol II*, Peet, R. and Thrift, N., eds. London: Unwin Hyman, 196–220.

Groth, P. (1997) Frameworks for cultural landscape study. In *Understanding Ordinary Landscapes*, Groth, P. and Bressi, T. W., eds. New Haven: Yale University Press, 1–24.

Groth, P., and Wilson, C. (2003) The polyphony of cultural landscape study: an introduction. In *Everyday America: Cultural Landscape Studies after J. B. Jackson*, Wilson, C. and Groth, P., eds. Berkeley: University of California Press, 1–22.

Herman, B. L. (2005) *Townhouse*. Chapel Hill: University of North Carolina Press.

Lewis, P. F. (1979) Axioms for reading the landscape. In *The Interpretation of Ordinary Landscapes*, Meinig, D. W., ed. New Haven: Yale University Press, 11–32.

Mitchell, W. J. T., ed. (1994) *Landscape and Power*. Chicago: University of Chicago Press.

Potteiger, M., and Purinton, J. (1998) *Landscape Narratives: Design Practices for Telling Stories*. New York: John Wiley & Sons.

Rose, G. (2003) *Visual Methodologies*. London: Sage.

Sauer, C. O. (1925) The morphology of landscape. *University of California Publications in Geography* 2: 19–54.

Schein, R. H. (1993) Representing urban America: nineteenth century views of landscape, space, and power. *Environment and Planning D: Society and Space* 11: 7–21.

Schein, R. H. (1997) The place of landscape: a conceptual framework for interpreting an American scene. *Annals of the Association of American Geographers* 87: 660–80.

Schein, R. H. (2009) A methodological framework for interpreting ordinary landscapes: Lexington, Kentucky's courthouse square. *The Geographical Review* 99: 377–402.

Wylie, John (2009) Landscape. In *The Dictionary of Human Geography* (5th edn) Gregory, D., Johnston, R. J., Pratt, G., Watts, M., and Whatmore, S., eds. Oxford: Wiley-Blackwell, 409–11.

## Additional Resources

## Data sources

These references will help you to identify other sources of data for cultural landscape study, get you thinking about the various methodologies that are central to landscape study (including field work), and give you a sense of what kinds of secondary literatures might help you to contextualize your landscape "data."

Carter, T., and Cromley, E. C. (2005) *Invitation to Vernacular Architecture: A Guide to the Study of Ordinary Buildings and Landscapes*. Knoxville: University of Tennessee Press.

Conzen, M. P., Rumney, T. A., and Wynn, G. (1993) *A Scholar's Guide to Geographical Writing on the American and Canadian Past*. Chicago: University of Chicago Press.

Delyser, D. and Starrs, P., eds. (2001) *Doing Fieldwork* (special issue of *The Geographical Review*, Vol. 91, Nos. 1 and 2). See especially articles by Cole Harris (Archival Fieldwork, pp. 328–34) and Karl Raitz (Field Observation, Archives, and Explanation, pp. 121–31).

## Methodology and theory

There are many essays and books that present methodological and/or theoretical contexts for interpreting the polyphony of everyday landscapes. Those listed here will help to get you thinking about how and why we might interpret ordinary landscapes in the first place.

Meinig, D. W. ed. (1979) *The Interpretation of Ordinary Landscapes*. New York: Oxford University Press. Even though it is over 30 years old, this edited collection of essays is still an interdisciplinary standard for cultural landscape study. You might find particularly relevant chapters by Peirce Lewis (Axioms for reading the landscape: some guides to the American scene, pp. 11–32) and Donald Meinig (The beholding eye: ten versions of the same scene, pp. 33–50).

Mitchell, D. (2008) New axioms for reading the landscape: paying attention to political economy and social justice. In *Political Economies of Landscape Change: Places of Integrative Power*, Westcoat, J. L. and Johnston D. M., eds. Dordrecht: Springer, 29–50. Mitchell plays off Lewis's "Axioms" for reading the landscape in that 1979 essay from a political economy perspective, and with the hindsight of an additional 30 years of landscape literature.

Other good beginning points include:

Johnson, M. (2007) *Ideas of Landscape*. Oxford: Blackwell.

Robinson, I., and Richards, P., eds. (2003). *Studying Cultural Landscapes*. London: Arnold.

Wylie, J. (2007) *Landscape*. New York: Routledge.

Wilson, C., and Groth, P. eds. (2003) *Everyday America: Cultural Landscape Study after J. B. Jackson*. Berkeley: University of California Press.

## Case studies

These books are extended landscape studies which put into practice the interpretation of ordinary landscapes based on detailed data collection and analysis. They provide an illustration of what kinds of cultural landscape work are possible, and a close read of the footnotes and citation sections of these books will reveal the empirical foundations of the research.

Conforti, J. (2001) *Imagining New England*. Chapel Hill: University of North Carolina Press.

Domosh, M. (1996) *Invented Cities: the Creation of Landscape in Nineteenth-Century New York and Boston*. New Haven: Yale University Press.

Duncan, J. S., and Duncan, N. G. (2004) *Landscapes of Privilege*. New York: Routledge.

Jackson, J. B. (1997) *Landscape in Sight: Looking at America*. New Haven: Yale University Press.

Matless, D. (1998) *Landscape and Englishness*. London: Reaktion.

Mitchell, D. (1996) *The Lie of the Land*. Minneapolis: University of Minnesota Press.

Olwig, K. W. (2002) *Landscape, Nature, and the Body Politic*. Madison: University of Wisconsin Press.

Price, P. (2004) *Dry Place*. Minneapolis: University of Minnesota Press.

Schein, R. ed. (2006) *Landscape and Race in the United States*. New York: Routledge.

## Exercise 14.1   Beginning a Cultural Landscape Study

These following are intended to have you start looking at the cultural landscape in your own backyard, or at least in your own neighborhood or city by seeking out the data sources described in this chapter.

### Comparative Map Analysis

If you are in the United States, visit the US Geological Survey's National Map website (http://nationalmap.gov/). Find two topographic maps of "your" city (where you live;

where you grew up; where you now attend university – make it a place you know well) from different times (they may be at different scales; what kinds of possibilities/problems does that pose?). Compare the maps to draw some conclusions about landscape change. Now, visit the website of the US Library of Congress (http://memory.loc.gov/ammem/pmhtml/panhome.html) and find a bird's eye view of the same place. Compare the bird's eye view with the topographic maps; compare both representations to your own contemporary knowledge of the city (go out and walk around a part of the landscape that is depicted in both scenes – take notes and photographs). What narrative might you write about your city based on these three "data sources" (bird's eye view; topographic map; personal observation). What did you find? What questions did you raise? If you could talk to anyone about questions raised by your narrative, who would you talk to? What would you ask them? Can you think about how you might include attention to issues of race, of class, and of gender when interpreting your landscape, or when talking to people about it?

## Historical Reconstruction

Choose a house or building where you know the occupants (it may be your own house; or a building in which you rent an apartment while at university). Try to trace the ownership of that building *and* the land upon which it stands, back in time, from local records that may or may not be on-line, or housed in a local government office. What can you find about the history of ownership of that property? What kinds of problems are there in undertaking this kind of "backwards" tracing? Now find historical insurance maps and city directories that will include the house or property you have just traced. What speculations or hypotheses can you draw about the neighborhood or immediate surroundings of the property you previously traced through the deed record?

## Collecting Local Resources

Imagine you wanted to collect data for a landscape study in your town, or the place where you are living. What local archives exist for your city? Government offices? Libraries (local and university)? Other organizations? See if you can put together your own list for starting a data collection project on local landscapes, complete with locations and opening hours, and other information relevant to a researcher who might want to use these archives.

# Chapter 15

# Human-Environment Field Study

## *Paul F. Robbins*

- Introduction
- Three Human-Environment Problems
- Measuring Flows and Connections
- Assessing Boundaries and Categories
- Gauging Human Impacts
- Conclusion

Keywords

Cultural ecology
Desertification
Environmental perception
Human dimensions of global change
Human-environment geography

Livelihood
Local knowledge
Political ecology
Rapid ecological assessment

## Introduction

This chapter takes as its charge the question of field methods, and in particular the methods appropriate for human and environmental analysis. It is quite possible to argue that **human-environment geography** and its many differing thrusts and flavors – which include risks and hazards, **environmental perception**, **cultural ecology**, **political ecology**, **human dimensions of global change**, health geography, and energy research, among many others

– has no exclusive suite of field methods unique to its practice. In fact, a majority of research methods practiced by human-environment researchers are those used in human and physical geography, albeit applied in pursuit of somewhat different questions. For example, a valuable approach to human-environment research involves time series analysis of remotely sensed data (air photography or satellite imagery) for areas where mutual influences of people and certain land covers (e.g., forests) may be occurring. The field component for this exercise might require nothing unique, involving the rigorous "ground truthing" of the image to confirm and evaluate the accuracy of classifications and inter-pretations of imagery (see Chapter 10).

In another example, a survey of people's perception of organic foods need not differ dramatically from survey instruments from some kinds of human geography. A survey might be created to evaluate the demographic and economic influences on people's percep-tion and consumption of differing foods. This might involve a carefully crafted question-naire, sampling strategies appropriate for a representative sample, and interpretation using simple descriptive or inferential statistics.

Insofar as the forms of evidence required to explore a specific nature/society problem are not very different than that of other fields or areas, one might conclude, therefore, that human-environment research is simply the application of physical geography methods to humanized questions (e.g., explaining tree cover changes in frontier forest settlements in Brazil) or human geography research techniques applied to problems related to non-human questions (e.g., studying political agencies that manage rivers in Romania). In essence, the field demands the match of human methods to non-human problems and systems, and vice versa.

On closer examination, however, the human-environment interface presents several unusual problems, each of which requires methodological strategies that, while they may use or combine methods from other areas, do so in configurations that make necessary certain kinds of distinctive field practice. The problems presented by the encounters between people and the environment specifically involve the relationship between humans and non-humans. For field-based research, moreover (as opposed to remote sensing, for example), the research problems are more clearly related to understanding immediate and intimate processes in which human beings use, alter, or relate to non-humans and vice versa. To measure those relationships, describe them in meaningful detail, and empirically evaluate their form and status (either materially or conceptually), an enormous range of techniques are at the disposal of the researcher, but decisions made about the form of collected evidence pursued have enormous implications for what can and cannot be explored and explained.

What this chapter seeks to do, therefore, is lay out a handful of the more common methodological approaches to dealing with human-environment research problems, describe their basic forms, and reflect on their strengths and weaknesses. What the chapter cannot do, however, is consider the difficult epistemological positions and positionings that occur prior to the selection of method and that set the terms for what counts as *evi-dence* (see Chapter 2). In this sense, the most important decisions come in linking the research question, through its theoretical framework, to a decision about what new infor-mation needs to be learned. To answer the fundamental question, to learn something new, test a hypothesis, or evaluate a claim, what does the researcher need evidence of, and what would count as credible evidence?

# Three Human-Environment Problems

Answering this question is further troubled by the dual character of "nature" and the environment itself, which is simultaneously understood and experienced as all of those things "outside" of the human self, while also always fundamentally a part of being human "inside" and frequently a product and outcome of human action (Williams 1976). This curious condition makes research a task of trying to pin down an unstable object and gather evidence of a fluid process. As a result, typically in human-environment research, the empirical search is one for evidence of: connections and mutual flows between the human and the non-human; varying conceptual and categorical arrangements that make the non-human world conceivable by human beings; and impacts, alterations, and changes of state, especially of non-human systems. Each of these presents somewhat different problems.

*   *The problem of flows and connections:* Objects, forces, and resources travel between human and non-human systems and actors, making each such system or actor different as they do, meaning that field study often depends on tracking, quantifying, or qualifying the movement of things and inventorying how and why objects (e.g., fuelwood, coffee beans, dung) wind up where they do and to what political, economic, and ecological effect.
*   *The problem of boundaries and categories:* The distinction between objects (e.g., forests, species, diseases) is often the product of social or cultural consensus (or "construction"), meaning that field study often must address the knowledges and categories through which people and communities order the natural world.
*   *The problem of impacts and influences:* Determining whether and how human actions influence natural systems becomes a difficult question of defining "change" in systems that are often highly dynamic, meaning that field study requires isolating key drivers of change, establishing baselines, and understanding inherent environmental variability.

To solve each of these problems, differing schools of research, both inside and outside academic geography have developed methodological tools relevant to their specific needs and research designs. Figure 15.1 summarizes a few of these, though it is by no means

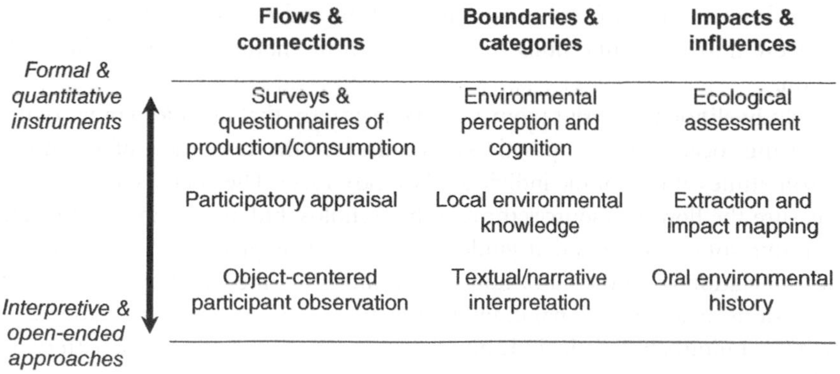

| | Flows & connections | Boundaries & categories | Impacts & influences |
|---|---|---|---|
| *Formal & quantitative instruments* | Surveys & questionnaires of production/consumption | Environmental perception and cognition | Ecological assessment |
| | Participatory appraisal | Local environmental knowledge | Extraction and impact mapping |
| *Interpretive & open-ended approaches* | Object-centered participant observation | Textual/narrative interpretation | Oral environmental history |

**Figure 15.1**   Typical approaches to human/environment research problems

comprehensive. It is tempting, in considering a list of options like the one laid out here, to imagine that mixing and matching is a simple question of ordering off a menu a la carte. As we shall see, however, the underlying assumptions and conceptions behind some of these methods make some more or less compatible with others, with more formal quantitative techniques contrasting somewhat with more interpretive tools.

## Measuring Flows and Connections

Flows between human and non-human systems and actors are ubiquitous and come in many forms, from wood entering homes in a rural village to be burned for fuel, to wastewater sewage exiting suburban toilets and flowing from subdivisions into streams during times of overflow caused by heavy rainfall. Secondary data sources sometimes provide only a crude picture of such flows, often aggregated to where processes are invisible and rendered in extremely unreliable statistics. Field study is often essential to make sense of these flows and produce reliable accounts of how they work and to what effect.

Quantities of garbage produced from households flow into landfills or informal dumpsites, for example. Regional, national, and international statistics on such waste are highly dubious and do not allow us to ask some basic questions about the processes that drive waste generation. What accounts for the variability of its production, how is it linked to the economic systems and strategies of households, and how does its flow impact or depend upon different members of the household, by age and gender? To get any kind of answer to these questions typically means individual, firm, or household scale research, directly asking about or measuring quantities of waste, the role of waste in the household or business economy, and the position of waste in people's daily lives. Tracking flows is an essential field task for many research projects.

### Survey approaches

Survey approaches allow a detailed and often quantitative analysis of individual, household or firm level practices and behaviors. By directly querying a small company's energy usage, for example, understanding the logic of its appropriations and practices, and measuring the rate of flows and role of resources in production and reproduction, research allows a detailed view into the connections that objects from "nature" make in the context of production.

Similarly, **livelihood** analysis (a somewhat generic term that includes a range of methods for measuring local economic practices), is aimed at the resolution of the household, though sometimes at that of the individual (see Box 15.1). The goal of this approach is to both measure the flow of resources through households, but also to analyze the decision-making framework and logics that guide these flows. It is predicated on the assumption that the association of human and non-human actors is mediated by constrained individual decisions, made by rational individuals, whose goals may vary by context.

Specific techniques usually include questionnaire-based surveys with a battery of questions related to aspects of the economy: how much woodfuel is used, how many animals are kept, how much water is required for differing crops? Other formal techniques include

## Box 15.1    Groups or Individuals?

For many years, the household was the accepted unit of analysis for understanding social systems, especially in rural areas. This was based on a somewhat patriarchal assumption that since production and decision-making were based on the joint labor and decision-making of the family as a whole, a survey of aggregated household behaviors (e.g., fuelwood collection, farming practices, pesticide usage) were sufficient to understand human–environment interactions. In the last 20 years, however, feminist analysis has demonstrated that not everything is harmonious in households, that the division of labor in the home often produces differing interests, strategies, and decisions, and that different members of households know very different things (Rocheleau et al. 1996). Similarly, choosing coarser resolution units (cities, firms, regions) opens a window for larger scale comparative analysis, but inevitably implies hypotheses about the scale or level of causation. Choosing whether to survey groups or individuals is a crucial question, and will depend heavily on the questions being asked. Nevertheless, it is essential not to mistake one for the other in nature/society research.

time diary-keeping, where respondents record in detail their actions related to specific activities, like forest product collection or driving a vehicle. These instruments are commonly supplemented with more intensive observations to elicit understandings, in the researcher's terms, of why certain decisions are made. Why use a woodfuel rather than a gas stove, for example; is the decision constrained by availability of resources, costs, or labor demands? Why drive rather than take public transit; is the decision conditioned by distance, time, or pay?

As a result, survey analyses tend to produce careful censuses of material practice and emphasize that non-human resources both enable and constrict decision-making. These censuses are analytically powerful since they are often measured in quantitative metrics that allow some inferential analyses. Are households with higher incomes more or less likely to exploit timber from adjacent conservation areas? Do firms with recycling policies actually increase their paper consumption? Cross-tabulation, regression, and analysis of variance are all available for data collected in this fashion (see Chapters 17 and 18). Similarly, these sorts of data can be easily translated into other metrics, including energy, capital value, and labor time, allowing comparison between sites and over time. Expanded somewhat from its largely underdeveloped rural context, such an approach also pertains to other forms of economy. Household water use in Phoenix Arizona, for example, might be understood in terms of time constraints, in-house technology, demands from household economics, and overall water prices; the relationship between which can be assessed through survey techniques.

While this approach provides one-of-a-kind forms of specific, high-resolution data, it is not appropriate for many questions. Firstly, the categories and variables in such a project are typically determined *a priori* by the investigator, rather than being derived locally. So too, the *meaning* of these objects and flows to individuals and households is mostly inaccessible through surveys.

## Participatory appraisal approaches

Conversely, one of the longest-standing and most well-codified approaches for assessing the conditions and role of the non-human world *on people's own terms*, is that established in over three decades of development research. This approach takes varying forms and comes under different titles, including participatory action research, participatory appraisal, and "rapid" assessment (the last so-called to distinguish it from long-term residence and participant observation). These approaches encompass an enormous range of techniques specifically designed to elicit from people the key elements of their daily lives in terms of the time, labor, and the geography of their use of resources, recorded in ways that are locally intuitive. More importantly, the work is generally predicated on the assumption that even (and especially) the most poor and marginalized communities are capable of analyzing their own realties and articulating them in ways that might make sense to outsiders. More normatively, this research is predicated on the understanding that, if it is executed in any meaningful way, people are actually empowered through the process of articulating these realities and that outsiders (e.g., researchers) should act in a facilitating, as opposed to dominating, fashion.

As such, this work tends to produce a mix of qualitative data (including lists, abstracts from discussions, and interview-derived impressions) along with graphic data products (like maps, timelines, and pictures). Space-related techniques include sketch mapping of community resources and hazards, graphic inventories of land uses, and transects of natural conditions. Time-related techniques include daily labor schedules, seasonal diagrams of activities, and annual calendars of changing opportunities and conditions (Kumar 2002). These techniques were designed for work in Third World and rural conditions, but most certainly are amenable to other contexts, from suburban homes to ministries and other agencies.

Like more formal surveys, these can be used to inventory the use of resources and the role of important plants, animals, and ecosystems in people's daily lives. Unlike such instruments, these responses are generally derived directly from people's categories and priorities. While they are more difficult to translate into universal and quantitative metrics, they adhere far more closely to local experience and reveal local process with greater clarity. Even so, the roots of rapid assessment techniques in action-oriented research, typically related to improving material provisioning in development, makes many of the complex contextual and meaning-laden details of human/non-human interaction difficult to elicit.

## Object-centered participant observation approaches

More intensive exploration of the flows between people and the environment typically take the form of "participant observation." Catalogued and explained in greater detail elsewhere in this volume (Chapter 13), participant observation aims at an intimate understanding of group and individual practices and meanings through immersion. Whether a researcher is assessing why city governments are increasingly pursuing global change policy, or instead trying to determine the impact of animal feedlot runoff on rural communities, participant observation opportunities abound, either in meetings of political staffers or around a breakfast table with ranchers. It should be noted that participant observation can vary greatly in the level of formalism and even quantification involved. Some studies may involve counting, timing, and recording precise activities by groups or individuals, while others may eschew these metrics for more impressionistic and interpretive activities (Jorgensen 1993).

In either case, however, engaging in participant observation in order to understand human-environment interactions usually involves a directed effort to explore relationships between people and specific objects. As such, the *participant* portion of participant observation often becomes more crucial. Interaction with the environment, after all, requires a great deal of daily practice, embodied knowledge, and know-how. Determining the impact of tubewells on rural social structure, for example, is greatly aided by understanding how these mechanical objects actually function, how they break, and the techniques through which they can be maintained. To participate can and should mean not only immersion into a social setting, therefore, but also into an environmental one, with the material labor and learning that this may require.

The product of participant observation data collection in a human-environment context, therefore, should have an "object centered" skew to it, detailing social/environmental facts about how crops, technology, or environmental ideas are used, deployed, handled, treated, stored, and manipulated. In the reverse direction too, participant observation can and should reveal the influences, impacts, pressures, limitations, opportunities, and contexts that the object, species, or environmental condition imposes on people (see Box 15.2).

Such "deep" detail is always context-specific, of course, and normal caveats concerning qualitative data apply: these data are hard to generalize from, difficult to translate into universal metrics, and may or may not represent larger populations. Even so, such data are also unique to participant observation and provide insights into the material and ideational flows between people and other parts of nature that are unavailable through other approaches.

---

### Box 15.2  Actor Network Theory and Human Environment Research

Actor-Network-Theory (ANT) is a loosely assembled methodological and theoretical approach to explanation that seeks to trace the connections between differing players (or actants) including objects, people, animals, plants, and microbes. Essential to such a form of explanation is the acknowledgment of the active role of all of the participants in making the network and transforming one another in the process. In his own formulation, Latour states simply: "A good ANT account is a narrative or a description or a proposition where all of the actors do something and don't just sit there" (Latour 2005: 128). The approach is not without critics, however. Some have argued that it fails to account for the structured differences in force or power of different players and that it provides more of a description than an explanation. Nevertheless, it is one of the most prominent recent methodological efforts to make the environment matter more in human-environment research.

## Assessing Boundaries and Categories

Many problems and questions in human-environment research hinge on how individuals, groups, cultures, and political bodies understand, define, and divide the non-human world. Explaining the successes and failures of conservation policy in Africa, for example, inevitably involves considering and investigating how disparate actors conceive of wilderness, nature, and ecological process. Pesticide application by American homeowners, in another example, is in part tolerated because middle class consumers conceive of the high-input turfgrass as an important environmental asset and a collective community good; the social and the natural join through the category work of the lawn.

Where do humans leave off and non-humans begin? What distinctions are there between differing species? What causes environmental conditions and outcomes, like soil quality, forest growth, or even the weather? Different people and communities answer these questions with striking variability, which holds implications for pressing human-environment questions, including biodiversity decline, water supply, and global warming.

## Environmental perception and cognition

The most formal and well-established methodological approach to these questions comes from the field of environmental perception. In geography, this field developed through the "behavioral" turn in the 1970s, which centered on understanding the cognitive processes through which an "outside" world is apprehended by individuals, and then translated into actions that, in turn, effect that "real world." The thorny epistemological issues raised by these distinctions (outside and in – real and perceived) are far beyond the scope of this survey, but are essential to remember in considering environmental perception techniques in research (Bunting and Guelke 1979).

Generally these techniques involve formal instruments to record, as accurately as possible, the specific constructs people use to organize the world around them. Interview techniques in this field range widely, but may involve questionnaires asking people to list or rank environmental categories or concepts, mental mapping exercises to evaluate what and where are important environmental elements, and graphic tests where pictures are shown to respondents who are asked to relate their impressions or feelings (Whyte 1977).

A related branch of psychometric research, risk perception analysis, seeks to measure the varying ways different groups and individuals perceive the hazards of the world around them, and the more universal tendencies humans share in their estimation of risk, including our tendency to underestimate the risk associated with things we directly control (e.g., driving a car) relative to those we don't (e.g., riding in an airplane). Like environmental perception, the techniques and methods in this field are fairly well established and linked to fairly specific research questions in the field.

Together, these approaches provide formalized methods that are relatively transportable and comparable from individual-to-individual and group-to-group. Though these are effective tools for eliciting people's understanding of the natural world, they do rely upon measurable responses and on the predetermined categories of concern to the researcher.

## Local environmental knowledge

Techniques that seek to elicit contextual knowledge – variously identified as local, indigenous, or traditional – are equally well-established, and come from deep roots, especially in the field of environmental anthropology. Despite the wealth of terms to describe the topic, **local knowledge** research generally encompasses the same broad set of objects, what people think and know about the world. This includes declarative knowledge, which encompasses names, traits, and ordering of entities, plants, animals, temperature, precipitation, as well as procedural knowledge, which includes routines and practices like planting or operating equipment along with explanations of processes like soil degradation, climate change, etc. Sources of data for knowledge studies are numerous. Intensive interviews and observation can reveal people's understandings and categorizations, but so too can textual materials ranging from stories, personal narratives, and scientific and policy documents (Berkes 1999).

In these terms, it is crucial to recognize that the "local" in local knowledge is by no means exclusively restricted to sites like villages, but instead extends to all contextual sites where knowledge is produced and reproduced. Government staffers in their office have local knowledge. So too, scientists in their labs can be understood to possess contextual or local knowledge, with unique and culturally specific categories for the world and shared views on processes that account for and explain the world. This quality of local knowledge study, distinct from environmental perception approaches, is far more ambivalent on the question of whether the "outside" world has ontological status as a "real" universal object or condition. This ambivalence enables the symmetrical study of the knowledges of experts and non-experts alike, since the researcher privileges no knowledge as more or less "accurate" (Figure 15.2)

**Figure 15.2** Even "objective" and quantitative field tasks can be fraught with interpretation. Choosing categories for levels and types of impact involves multiple parties to agree on what they see in the landscape

## Deconstructing nature

This ambivalent perspective can be further extended into methods and analytics that embrace a deconstructive technique for reading and interpreting claims about the natural world (see Chapter 23). Following the tradition of deconstruction, established and widely-held environmental claims can be brought under scrutiny and examined in terms of the social work they do, the political conditions of their assertion, and the metaphoric associations and borrowings at their core.

Such research can reveal a number of surprising and important associations. Consider research into environmental conflicts over forests in British Columbia (Braun 2002). Such work, conducted on texts and narratives of opposed groups, unsurprisingly reveals that environmentalists and loggers have different views of the forest, with environmentalists claiming a pristine and primordial natural world and logger championing the productivity of the forest and economic bounty. Far more interesting, however, the texts of apparently opposed groups have disturbing congruities, as both erase native presence and interests in the forest. Thus deconstruction does not simply assert the obvious claim that people have different views or assumptions, rather it is a technique that seeks to reveal deeply buried common assumptions and their concomitant political and social roots.

Deconstruction represents a kind of methodological relativism, a critical approach to examining knowledge claims, especially those from scientific sources. The approach, rather than assessing the truth or falsehood of an environmental claim (e.g., whether or not deforestation is actually ongoing in North Africa), seeks instead to explain why such an assertion is made, specifically disallowing its accepted "truth" as a reason. Methodologically, it entails careful study not only of the claim itself, but also its genealogy: the social and political history of its assertion (e.g., careful tracing of the colonial iterations of the idea of North African Desertification; see Davis 2005).

As such, deconstruction is an explicitly empirical project related to local knowledge and perception research in that it attempts to derive the categorical underpinnings of the way the non-human world is imagined, viewed, and understood. Like these other approaches, it may draw from interviews or from the examination of documents. More radically than perception and local knowledge study, however, it adopts an ontological ambivalence about the "real world" that makes it more difficult, though by no means impossible, to combine with methods that are more "objective" and measurement based, like ecological assessment (see below).

## Gauging Human Impacts

A great proportion of field-based work in human-environment geography is geared towards measuring and evaluating the level of human impact and influence on the environment. Some impacts may be subtle, including selection of plant or animal species, or large scale, like denudation of forest cover through fire or land clearance. Alternately, researchers may be seeking to disprove claims of environmental change. In either case, a number of tools and records from physical geography are available to researchers, including remotely sensed data (for land cover change), soil horizon data (for soil erosion), stream-gauge data (for water use and flows), among a huge range of sources. So too, secondary historical accounts may exist in archives describing previous environmental conditions and the drivers of change. In many cases, however, no such data exist and/or the questions being asked can only be answered with higher resolution assessments using original field-based measurement.

## Ecological assessment

There are few substitutions for direct ecological measurement. While full-scale, long-term observation and assessment are ideal, satisfactory assessments can often be conducted under temporal and fiscal constraints. For these purposes, perhaps the most accessible and useful guide is *Nature in Focus*, which reviews **Rapid Ecological Assessment**, a combination of techniques for surveying the condition of flora and fauna developed for the Nature Conservancy over the last two decades specifically to survey rare endemic species, but which serves as a useful step-by-step guide to sampling, surveying, recording, and analyzing the condition of any plant and animal community (Sayre et al. 2000).

No technique is universally applicable, however, since the specific impact analysis will vary immensely depending on the project. The methodological problem for much of this work therefore comes prior to any field visit, and rests in selecting the appropriate index of change or impact. A project concerned with declines in diversity of species in a forest is considerably different than one concerned solely with biomass or productivity, for example.

Similarly, there may be several ways to sample or measure impact, but with varying levels of efficacy and confidence offset by difficulty and time constraints. Consider for example, the range of assessment techniques available for assessing the impact of white-water rafting recreational activities on salmon reproduction (not a trivial issue in the North American Pacific Northwest). Besides the remote observation of salmon runs using air photos and satellite images over time, which closely examine the egg-laying areas of the fish, more intensive, reliable, and time-consuming field methods are also available. These might include videotaping of salmon runs before and after the passage of a boat to assess the movement and return of fish.

Demonstrating impact typically requires demonstrating change, and this presents serious sampling problems for researchers. An assertion of impact may require time-series analysis to demonstrate that a landscape or system is different before and after an action or event (such as the introduction or increased presence of people). In other cases, researchers may collect data on different areas, using space as a temporal proxy that enables a comparison of places with different levels of, for example, human presence (See Box 15.3).

Several further crucial caveats also apply. Even if impact can be demonstrated, this does not necessarily mean that the system is irreparably transformed, nor on the other hand that if undisturbed it will recover to its original state, especially under changing conditions (e.g., variable precipitation). So too, if change is demonstrated over a period, this may reflect only a portion of longer-term trends, making it difficult to determine whether the anthropogenic variability is entirely outside the range of the system's "natural" variability. Even so, serious human-environment work can benefit from direct assessment of ecosystem conditions, whether in terms of a general inventory or more intensive research design to show the impacts – or absence of impacts – from human beings.

**Box 15.3    Finding Control Sites in a Humanized Environment**

Proper ecological assessment generally demands some kind of control site for comparison. It is difficult to assert, for example, that the conditions of a grassland are maintained by grazing if there are no ungrazed areas, essentially identical in all other ways (soils, rainfall, etc.) for direct comparison. For *ex situ* experimental research, this is not a difficult matter. Exclosures, for example, can be built which keep certain impacts out of small areas. The effect can be compared with adjacent sites where these influences have occurred. In many human-environment contexts, the luxury of such controls is commonly not afforded. In high density areas there are few places that do not experience some kinds of impacts. Even many conservation areas have human uses and experience some kinds of extractive pressure. Where this is the case, careful methodological design can still allow exploration of these relationships. Graded impact can be examined, at random sites or along transects, to assess whether proximity to human population correlates with differing levels of diversity, productivity, or species invasion, for example, using distance as a proxy (Karanth et al. 2006).

## Extraction and impact mapping

More directed study of anthropogenic impacts can be accomplished by following, measuring, and mapping human use. A range of techniques are possible here, from attaching GPS units to the horns of grazing cattle, to tracing the routes of fishing trawlers with on-board equipment, or conducting focus group meetings with women wood collectors around a sketch map. The result of such analysis in any case should be quantitative and qualitative measures of human modification of the landscape, whether direct or indirect.

Caution should be used, however, in the kinds of claims such data can support. It is possible, for example, to map the areas of highest human impact in a conservation area and to list the species extracted from these areas, but without an ecological sample in controlled areas, these data do not necessarily support a claim that the resulting landscapes are structurally different from areas free of human impact. Spatializing extraction and activities in a detailed way, however, does make it possible to link human use information to high resolution land cover data and ecological assessment results.

## Oral environmental history

For many places, however, there is a total absence of meaningful baseline data from which to assess impact. So too, where changes have occurred over long time horizons, records of both earlier environmental conditions and of the possible drivers of change are both typically scarce.

One of the powerful and ubiquitous resources for reconstructing and understanding the human/environment past in such cases is oral environmental history. People carry with them an enormous storehouse of often highly detailed information about the condition of past landscapes, the forces that drive change, and the historically relative condition of contemporary ecological systems. On many occasions the single most useful source for reconstructing and understanding change may be a handful of older community residents, witnesses to long periods of ecological change and upheaval.

Memory is, however, a prismatic, personal, and uneven thing, and research into memories of landscape brings with it inquiries into normative concepts of what landscapes ought to look like, and how moral and ethical concerns are grafted into the pictures people paint of the environment for themselves and others. This is not just to say that memory may be "incorrect," in the sense that it may not correspond to some actually existing previous conditions, but more forcefully that an environmental memory is threaded to a host of other cultural and political experiences.

As such, oral history has enormous advantages, in that it allows an integrative picture of the linkages of human experience and non-human conditions, feelings and actions, landscapes and personal change. On the other hand, it is a form of evidence that may be highly incompatible with questions or claims in certain kinds of research, since it will inevitably produce contradictory accounts, gaps over time and space, and murky relationships rather than crisp causes and effects. Likewise, the collection of oral histories can produce uneven power relationships, and the researcher has an ethical responsibility to her or his informants (see Chapter 24). These strengths and weaknesses are shared by all interpretive approaches, of course, but are perhaps most prominent in the problem of memory and the natural world.

## Conclusion

In sum, taking human-environment relations seriously in geography means accounting for non-humans and their dynamic status, influence over, and constitution by human beings, with difficult implications for measuring and collecting information about those relationships. It is of course impossible to provide a comprehensive survey of all the methods used in human-environment research, especially since an enormous range of combinations of physical and human geography methods are typically used in pursuit of answers in this field. Even so, this brief overview has pointed to a few epistemological tensions and practical questions facing researchers in the field.

First among these is the problem of mixing methods in a way that makes epistemological and theoretical sense in the terms set by the project and the audiences to which the work might be presented. To be sure, combining approaches seems essential. A historical analysis of discourses surrounding **desertification** may be far more compelling when *accompanied* by rigorous paleoecological data drawing such narratives into question. Measuring the benefits of organic agriculture may necessarily require an analysis *both* of the local political economy of production and the comparative quality of soils that result from its practice. Triangulation across diverse approaches often is useful as well. Imagine joining oral histories with pollen or tree ring data analysis to reconstruct regional forest history. Where such records converge and diverge, a great deal can be learned not only about environmental change, but about the character of that change, its experience, and its drivers.

That does not make the mixing and matching of techniques an unproblematic smorgasbord, however. Gathering oral histories to "correct" and "ground truth" an older air photo for example, may be met with skepticism by some more positivistic practitioners. Using a survey instrument to elicit people's "experiences" might be treated as facile by researchers from psychoanalytic and interpretive traditions. There is no essential barrier to mixing ecological assessment with participant observation or oral history with quantitative survey, but the claims made in the interpretation of evidence hinge around difficult prior epistemological decisions about what differing forms of data represent, especially for diverse populations with differing experience (Rocheleau 1995). While this is true of all forms of research, it is all the more pronounced in work that takes trees as seriously as it does logging economies, acknowledges bacteria as well as sewage provision budgets, or recognizes virus mutation as well as stigmatization in understanding HIV/AIDS. The place where these things come together, in the field, is therefore as exciting as it is confusing, but requires the formation of a careful explanatory strategy, one whose methods are carefully fitted to an epistemological framework.

## References

Berkes, F. (1999) *Sacred Ecology: Traditional Ecological Knowledge and Resource Management*. Philadelphia: Taylor and Francis.

Braun, B. (2002). *The Intemperate Rainforest: Nature, Culture, and Power on Canada's West Coast*. Minneapolis: University of Minnesota Press.

Bunting, T. E., and Guelke, L. (1979) Behavioral and perception geography: a critical appraisal. *Annals of the Association of American Geographers* 69(3): 448–62.

Davis, D. K. (2005) Potential forests: degradation narratives, science, and environmental policy in protectorate Morocco, 1912–1956. *Environmental History* 10(2): 211–38.

Jorgensen, D. L. (1993) *Participant Observation: A Methodology for Human Studies* (Applied Social Research Methods, Vol. 15). Thousand Oaks, CA: Sage.

Karanth, K. K., Curran, L. M., and Reuning-Scherer, J. D. (2006) Village size and forest disturbance in Bhadra Wildlife Sanctuary, western Ghats, India. *India Biological Conservation* 128: 147–57.

Kumar, S. (2002) *Methods for Community Participation: A Complete Guide for Practitioners*. London: ITDG Publishing.

Latour, B. (2005) *Reassembling the Social: An Introduction to Actor-Network-Theory*. Oxford: Oxford University Press.

Rocheleau, D. (1995) Maps, numbers, text, and context: mixing methods in feminist political ecology. *The Professional Geographer* 47(4): 458–66.

Rocheleau, D., Thomas-Slayter, B., and Wangari, E. eds. (1996). *Feminist Political Ecology: Global Issues and Local Experiences*. New York: Routledge.

Sayre, R., Roca, E., Sedaghatkish, G., Young, B., Keel, S., Roca, R., and Sheppard, S. eds. (2000) *Nature in Focus: Rapid Ecological Assessment*. Washington, DC: Island Press.

Whyte, A. (1977) *Guidelines for Field Studies in Environmental Perception*. Paris: UNESCO.

Williams, R. (1976) *Keywords*. Oxford: Oxford University Press.

## Additional Resources

Carney, J. (2001) *Black Rice: The African Origins of Rice Cultivation in the Americas*. Cambridge, MA: Harvard University Press. A study of the introduction of African agriculture to North America and a paradigmatic example of mixed empirical methods, including intensive fieldwork and history.

Fairhead, J., and Leach, M. (1996) *Misreading the African Landscape: Society and Ecology in a Forest-Savanna Mosaic*. Cambridge: Cambridge University Press. Uses remote sensing, textual analysis, and ethnography to show how land cover changes, but also how and why expert authorities frequently get it wrong.

Gold, A. G., and Gujar, B. R. (2002) *In the Time of Trees and Sorrows: Nature, Power, and Memory in Rajasthan*. Durham: Duke University Press. Extremely rich and highly reflexive ethnographic oral environmental history in action.

Kepe, T., and Scoones, I. (1999) Creating grasslands: social institutions and environmental change in Mkambati area, South Africa. *Human Ecology* 27(1): 29–54. Demonstrates an empirical technique for assessing the multi-directional land cover change effects of social institutions.

Neumann, R. P. (1998) *Imposing Wilderness: Struggles over Livelihood and Nature Preservation in Africa*. Berkeley: University of California Press. A deconstructive but highly empirical environmental history of the conservation movement in Africa.

Swngedouw, E. (2004) *Social Power and the Urbanization of Water: Flows of Power*. Oxford: Oxford University Press. Economic and field-based analysis of urban environmental problems.

Turner, B. L., Villar, S. C., Foster, D., Geoghegan, J., Keys, E., Klepeis, P., Lawrence, D., Mendoza, P. M., Manson, S., Ogneva-Himmelberger, Y., Plotkin, A. B., Salicrup, D. P., Chowdhury, R. R., Savitsky, B., Schneider, L., Schmook, B., and Vance, C. (2001) Deforestation in the Southern Yucatan peninsular region: an integrative approach. *Forest Ecology and Management* 154(3): 353–70. Demonstrates seamless integration both of multi-disciplinary perspectives and field-based observations with other data sources.

Turner, M. D., and Hiernaux, P. (2002) The use of herders' accounts to map livestock activities across agropastoral landscapes in Semi-Arid Africa. *Landscape Ecology* 17(5): 367–85. Demonstrates a technique for combining local information into quantitative assessment of human impact (and its absence) on semi-arid lands.

## Exercise 15.1   Analyzing a Human-Environment Study

This exercise is based on a close reading of one scientific article, and is intended to help you clarify aspects of your research design and methodology in human-environment studies. Your task: conduct a literature search on a topic related to human-environmental

**Table 15.1**   Exercise questions

| Research problem and objectives | Preparation for going to the field | Once in the field | Data analysis |
| --- | --- | --- | --- |
| Does the work fit within an established area of research in human-environmental study? Is the literature review adequate in framing the problem at hand? | Was special language or cultural study necessary for this research? | How long did the field research last? Do you think the study would have been improved with additional time in the field? | What methods were employed to analyze the data (qualitative, quantitative, or a mixture of both)? |

**Table 15.1** *Continued*

| Research problem and objectives | Preparation for going to the field | Once in the field | Data analysis |
|---|---|---|---|
| To what degree does the approach used stress "formal" and quantitative measures versus "interpretive" ones? | What "instruments" (questionnaires, checklists, category sets) might have been prepared for differing elements of the study and how might they differ? | How is the collection of "formal" data different from quantitative "interpretive" data? Which seems to have taken more time or higher priority? | Are all of the findings in the study drawn from "formal" approaches congruent or easy to reconcile with those from "interpretive" analysis? |
| Which of the problems discussed in this chapter (i.e., flows and connections; boundaries and categories; or impacts and influences) seem most relevant to the study? | Did the researcher receive special funding for her or his time in the field? | Does the article indicate whether or not the researcher collaborated closely with local people, or with government or non-governmental officials? | Do the data analyses seem adequate and appropriate? What other approaches to the problem might the researcher have taken? |
| Do the data seem to have been collected in a way that is appropriate to the research design? Are there other ways to have proceeded with this research? | What scientific background (social or environmental) do you think the author(s) had prior to going to the field? Do you feel that their training equipped them for the study's research objectives? | What sort of social and environmental data were collected in the field? What methods were used in collecting data, and what sorts of equipment were required? | What are major conclusions drawn from the study? What implications do the findings have for human-environment study and for further research? |

study in geography. Good journals for this type of research include *Annals of the Association of American Geographers, Geoforum, Geographical Review, Progress in Human Geography,* and *Progress in Physical Geography.* Select an article that interests you; read it carefully, and then answer the questions in the (Table 15.1) . You can use bullets or paragraph style.

# Part III
# Representing and Analyzing

# Chapter 16

# Maps and Diagrams

*Stephen P. Hanna*

- Introduction
- Approaches to Map Making and Use
- Elements in Map Design
- The Power of Maps

Keywords

Cartogram
Cartography
Digital elevation model
Equal area map projection
Generalization

Geographic visualization
Map
Map projection
Map scale
Symbolization

## Introduction

**Maps** and spatial diagrams are powerful tools used to visualize, explore, store, and communicate geographic information. Thus, the skills of making and using these visual representations of the worlds around us are very important within the discipline of geography. Traditionally, we learn these skills within the subdiscipline of **cartography**, but they are also integral to the related fields of Geographic Information Science and **GeoVisualization**. In addition, we produce maps, like the places, identities, landscapes, and spaces they represent, within particular social and cultural contexts. Thus, they at least reflect – and many would argue help reproduce – dominant cultural values and power relations present in the society within which the maps are made and used.

Maps and spatial diagrams come in a variety of forms and are produced and consumed through a variety of media. Whether printed on the two-dimensional pages of an atlas, molded out of plaster or plastic to make a three-dimensional model, or animated on a computer screen, all maps and diagrams are composed of more or less abstract symbols that represent different aspects of our cultural and physical environment – what most people consider to be the "real" world. For example, we may use ♟ on road maps and highway signs to stand for a library. Making and using maps successfully involves the creation and interpretation of these symbols.

The International Cartographic Association (ICA) (2008) defines a map as "a symbolized image of geographical reality, representing selected features or characteristics, resulting from the creative effort of its author's execution of choices, and is designed for use when spatial relationships are of primary relevance" (See Box 16.1). This very broad definition encompasses the examples found in Figure 16.1. It also includes maps produced through a variety of media ranging from the traditional paper map to the whole world of digital maps and terrain simulations. Such spatial representations are traditionally grouped into two categories. *General purpose maps* are designed for a variety of uses while *thematic maps* focus on more specific topics (Dent 1999). While these categories can be useful, they are problematic as well. After all, a map's purpose or use is not solely defined by its creator. Any map could be used for purposes not even imagined by the cartographer. For this reason, general purpose maps are better described as multivariate maps.

Traditionally, the term "map" is reserved for graphics that use an absolute coordinate system, such as latitude and longitude, to "realistically" locate and represent attributes of specific places and spaces. More recently, however, many geographers have argued that we do not experience and know the places and spaces around us as mere sets of coordinates. **Cartograms** and other spatial diagrams can be employed to capture these other, often subjective and culturally specific, ways of knowing the world. For the remainder of this chapter, therefore, I will use the word "map" to refer to any spatial representation.

Most people assume that the "author" referenced in the ICA definition is a professional cartographer who makes maps for public consumption. Yet, more and more people are creating their own maps for their own private and public purposes using desktop mapping

---

### Box 16.1   Defining Maps

The ICA's definition of a map is only one of many. Most textbooks devoted to cartography or map use and analysis offer their own definition. For example, in the Preface to Volume One of *The History of Cartography*, J. B. Harley and David Woodward provide the following definition:

> Maps are graphic representations that facilitate a spatial understanding of things, concepts, conditions, processes, or events in the human world.

Is this definition more or less restrictive than the ICA's definition? Why might authors focusing on the history of cartography need a different definition?

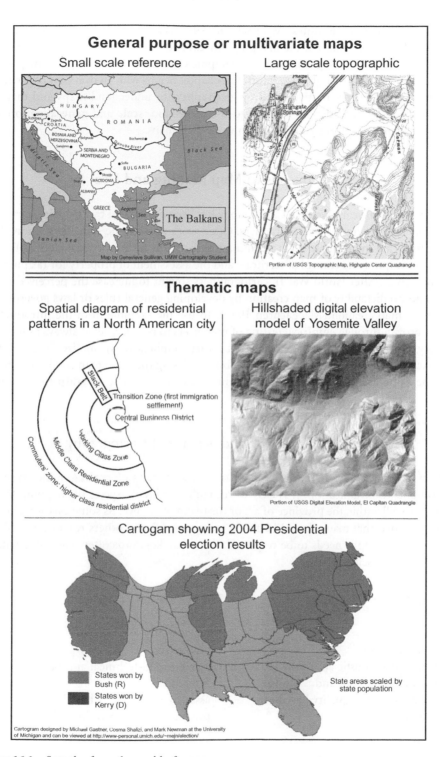

**Figure 16.1**    Samples from the world of maps

software or interactive websites (MacEachren 1995). While this challenges cartographers' traditional understandings of the mapping process and the map itself, it only underscores the importance of learning some key cartographic concepts that can help any map maker or user harness the power of maps. Thus, this chapter begins with an overview of the major Western theoretical perspectives used by geographers to understand maps and mapping and concludes with guidelines on how to use projection, **scale** and **generalization**, and **symbolization** to improve your ability to work with maps.

## Approaches to Map Making and Use

Over the past 500 years, Western Cartography has been dominated by explorers, surveyors, and academic geographers/cartographers. Often working with the state, these individuals strove to create increasingly accurate and scientific maps for purposes including navigation, the conquest and control of territory, and the assessment of property for tax purposes (see Box 16.2). After World War II, cartographers sought to increase the perceived objectivity and predictability of map creation by developing general rules or laws governing all aspects of map creation (MacEachren 1995). To use the example from the introduction, if ♣ is proven through empirical testing of map users to be interpreted always as meaning "library," then it should always be used by cartographers to symbolize "library." One approach to achieving this goal is standardization. Using the same map symbols to represent the same earth objects should result in all users being able to grasp the meanings

---

### Box 16.2   The Art and Science of Cartography

Map making is often defined as both an art and science. As a form of graphic expression and communication, it is easy to recognize different aspects of mapping that seem to fall within the province of art or the realm of science. As some cartographers have argued that maps should be made more scientifically, others respond that the art in cartography needs to be respected. Below are keywords often associated with art and with science.

**The Art in Cartography Stresses**
- creativity
- subjectivity
- intuitive-holistic approach
- emotional input and response
- assessment of quality based on aesthetics

**The Science in Cartography Stresses**
- repeatability
- objectivity
- standardized-technical approach
- functionality
- assessment of quality based on best practices sestablished through empirical testing

Are these lists mutually exclusive?

contained in any map more easily. Single map-making institutions, such as the National Geographic Society or the United States Geologic Survey (USGS) have created "house" standards for type styles and sizes, symbols, **map projections**, and uses of color. Because of the difficulty of standardizing methods and symbols across institutional and cultural contexts, however, most cartographers have pursued another path based on theories and methods developed by behavioral psychologists.

Prior to the late 1960s, behavioral psychologists attempted to define laws concerning how people respond to external stimuli apprehended through vision, hearing, and the other senses. In this method of explaining human behavior, there is no attempt to understand how people process the sights, sounds, and smells to which they are exposed. Instead, researchers seek to predict that a particular stimulus will provoke a particular response. Cartographers have adopted this approach and many researchers focus on how map readers react to various symbols, colors, and lettering sizes and styles.

Most cartographers who borrow from behavioral psychology conceive of **cartography** as a communication science (Figure 16.2). In this model, the cartographer's role is to communicate some geographic information objectively and accurately to an audience of map readers. The cartographer conceives of the map's purpose, gathers and processes the necessary data, and transforms it into a map that meets that purpose. The map, therefore, both stores the data until it is used and serves as a medium for communication. The map reader, most likely removed from the cartographer in both time and space, is responsible for transforming the map's symbols back into the geographic information they represent in order to understand the cartographer's intended purpose.

In order for this communication to be effective, the cartographer must know that the map she/he designs will stimulate the correct response in the map reader. A great deal of cartographic research in the 1960s and 1970s consisted of exposing study groups to particular symbols, type styles and sizes, and other map elements and then measuring their

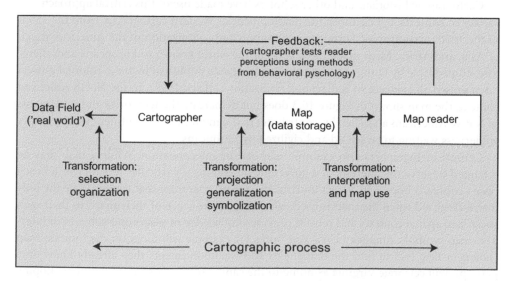

**Figure 16.2** Cartography as Communication Science (adapted from Dent, B. 1996. *Cartography: Thematic Map Design*. Boston: WCB McGraw-Hill)

responses (MacEachren 1995). A key goal was to determine the "least discernable difference." If a map designer knows the smallest difference between two type sizes, symbol
shapes, or colors that a reader can discern, then she or he can employ that knowledge to
ensure that the map reader will be able to perceive differences and similarities among the
symbols that make up the map. Within this model of cartography, such feedback is
intended to ensure that a map can be predictably interpreted into an accurate and objective
representation of some place and/or aspect of the world.

Since the 1980s, critics within and beyond cartography have noted the limitations of
the communication model of cartography. Some theorize that cartography and maps need
to be understood within their broader social and cultural contexts and argue that since
cartographers operate within such contexts, their maps cannot be viewed as merely objective communication devices (Harley 2001, Wood 1992). Others, noting that psychologists
have largely abandoned their own behavioral models in favor of a cognitive approach,
argue that simply noting map readers' responses to visual stimuli does not actually explain
human perception and interpretation of map symbols (MacEachren 1995). There is no
recognition of the variations in knowledge, interest, and purpose map readers bring to a
map in the behavioral approach. Finally, since the communication model focuses on the
means of communicating the map's content, and not on the content itself, little attention
is paid to the actual meanings of the maps themselves. These critiques have lead to two
major developments in our understanding of how maps work.

One approach was pioneered by J. B. Harley (2001) and Denis Wood (1992), among
others. In their calls for more critical understandings of cartography, they argue that map
makers, maps, and map users have always operated within particular social and institutional contexts. These ensure that all maps, both thematic and general purpose, are made
to serve particular interests, usually those of a society's elites. Individual maps and the
entire discipline of cartography, therefore, can be studied to reveal how they both represent
and reproduce the social and spatial status quo.

Cartographic historians and other scholars have made use of this critical approach in a
variety of contexts. For example, the development of Western cartography from the 1500s
to the 1800s accompanied European exploration and colonization of the Americas, much
of Asia, and Africa. Maps made these parts of the world known to Europeans and, therefore, controllable by Europeans. Later historical atlases published by these colonial powers
often erased all evidence of indigenous civilizations (Black 1997). In the North American
context, the map shown in Figure 16.3 does not demarcate the territories and settlements
of Native Americans and, thereby, contributes to the myth that the continent was an empty
wilderness waiting to be settled and civilized by Europeans.

Critical cartographers also examine map use or interpretation. While a map may be
intended to serve the interests of a government claiming a particular territory or a beach
resort company hoping to attract tourists, there is no guarantee that those will be the only
interests served when the map is used. Map users make sense of their maps in their own
social and spatial contexts and rely on previous knowledge of places and subjects to interpret map symbols. American Civil War buffs, for example will focus on the tourism map
shown in Box 16.3 to find the battlefields and war monuments they already know and,
perhaps, will celebrate Virginia as home to the capital of the Confederacy. Others hoping
to learn about the history of African American slavery and resistance will be disappointed
that this map hides any evidence of their struggles for freedom and equal rights. Thus maps

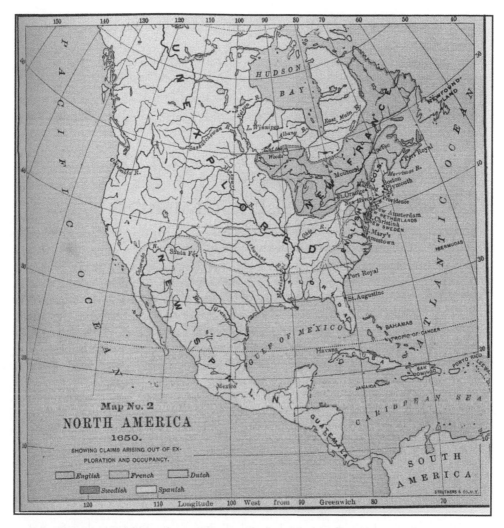

**Figure 16.3** "North America 1650" from Hart (1891). *Epoch Maps Illustrating American History.* New York

are *intertextual*; for any user they refer, both explicitly and implicitly, to books, photographs, web pages, and other texts and experiences concerning that place and/or topic (Del Casino and Hanna 2000). These texts lend their own meanings to the places, peoples, and/or events represented on the map.

Cartographers more interested in how to improve maps' functionality suggest that these critical approaches do not help anyone make better maps. Responding to computer and internet mapping technologies that have exponentially increased people's access to both maps and mapmaking, Alan MacEachren (1995) and others have reconceived the relationships among map makers, the map, and map users (Figure 16.4). Recognizing that not all maps are made to communicate a cartographer's message to a public audience, MacEachren's model of cartography leaves space for maps that help us visualize previously unknown

## Box 16.3    A Critical Approach to Map Interpretation

This placemat can be found in restaurants throughout Virginia. An example of popular cartography, it was created to entertain its readers and advertize tourism places to consumers.

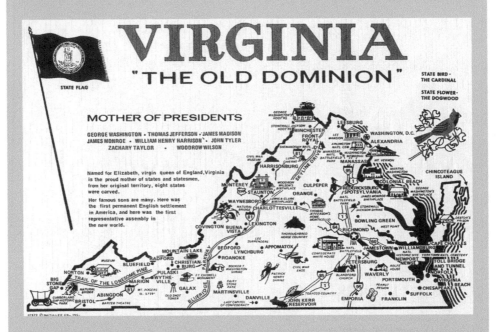

A critical analysis of this or any map involves research into the sites referenced on the map. For both the mapmaker and all map users, the symbols refer not just to places, but to social meanings people attach to these sites. Therefore, it is important to investigate the social and historical contexts in which a map is produced and consumed. While a map's meanings change through time and vary according to a map users' background, some meanings dominate others. They are reinforced by what we learn in school, in other texts, and simply by living within a society.

On this map, Thomas Jefferson's home, Monticello, may have many different meanings, but the dominant ones focus on his contributions to an American nationalism that celebrates freedom and individualism as well as its European heritage. Furthermore, most tourism marketers try to attach positive meanings to the places they sell to tourists. Thus, Monticello's presence on this map more than likely reproduces this patriotic national identity rather than drawing attention to Jefferson as a slaveholder.

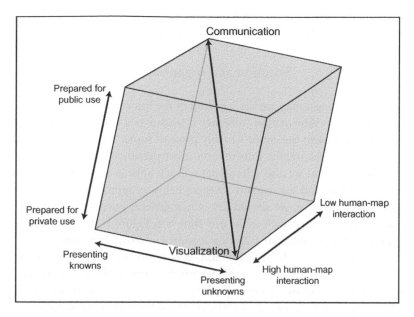

**Figure 16.4**  MacEachren's model of cartography

spatial patterns in private. It also captures the possibility that, as map users, we do not simply and passively receive the meanings encoded on a map by cartographers. Rather, we interact with maps actively by bringing our own knowledge and perspectives on the places and themes represented on a map. We may also alter a map to make it suit their purpose or actually make the map ourselves.

Understanding how maps are made and used is necessary to improve how maps help us visualize, explore, and communicate about the world around us. To do this, MacEachren argues, we need to understand the biological, psychological, and social factors that explain how people perceive and interpret the symbols comprising all maps and diagrams. This means knowing: (1) how an individual's vision and cognition work to make sense of visual stimuli; (2) how cartographers create and choose symbols; and (3) how symbols acquire multiple levels of meaning as maps are made and used in particular social contexts. While all three sets of factors are needed to gain a full understanding of how maps work, I would argue that we can *begin* to make and use maps more effectively if we focus on the last two.

## Elements in Map Design

In order to make and use maps to visualize, explore, and communicate any aspect of our cultural or physical environment, we need to unpack how the geographic information we collect is transformed into symbols that can be positioned on a map. This involves the processes of selection and organization, projection, scale and generalization, and symbolization (Dent 1999, Monmonier 1996). Each of these operations involves transforming our richly detailed, three dimensional, ever-changing, and contested world into a simplified, time-specific representation with a limited perspective. It is vitally important for both map users and mapmakers to remember that a single map is always only one of a potentially infinite number of maps that could be made to represent the same information.

## Selection and organization

The first steps in creating a map are the selection of the information to be represented and the organization of this information. Whether intended as a general purpose or thematic map, no map can represent the entirety of any place or topic. To make the map useful, we begin by selecting the geographic information most pertinent to the map's purpose. By definition, this means that we also select what will *not* appear on the map. Remembering that the maps we make and use contain absences helps us approach these representations more critically. For example, we should ask why a general purpose or multivariate map – perhaps a topographic map such as the one shown in Figure 16.1 – represents churches, but does not note that a particular factory is a source of toxic emissions.

When organizing the selected information, it is useful to categorize it into layers or themes (Figure 16.5). For a thematic map, we might think of the map's subject, perhaps population, layered on top of a base map comprised of the locations for which we have collected data. A general purpose map may be composed of several thematic layers, perhaps topography, drainage, and buildings. This is an exercise in abstraction and reductionism. We tend to use only one or two measurable attributes of complex entities, such as forest regions or ethnic enclaves, to represent these spaces on our maps. Variations within regions and other aspects of the spaces we map are intentionally and necessarily disregarded. Once organized into thematic layers, we usually attach each attribute to specific locations using the map's coordinate system.

## Map projections

To describe locations on the surface of the earth we use a spherical coordinate system in which latitude measures distance north or south of the Equator and longitude determines distance east or west of the Prime Meridian. A map projection is a systematic method of transforming this spherical coordinate system, and the locations on the Earth's surface it describes, into a flat or planar coordinate system suitable for use on a piece of paper or a computer screen. This process inevitably produces distortions in shape, area, distance, and/ or direction (Figure 16.6). These distortions are especially noticeable in small-scale maps – maps that show large portions of the Earth's surface, but with few details.

Over the past 500 years, cartographers have developed hundreds of map projections designed to control these distortions and/or to serve specific purposes. When choosing a projection, we must decide whether it is more important to measure distances accurately or to preserve the angular relationships that make up the shapes of the territories and water bodies shown on the map. For small-scale thematic maps, most cartographers agree that areas should be preserved accurately. This is in part because most people associate an increase in size with an increase in importance. Imagine using the Mercator projection in Figure 16.6 to make a map of percent of population living in poverty by country. Then imagine making the same map using the Molleweide **equal area projection**. Given that most of the world's wealthiest countries are clustered in the northern hemisphere while

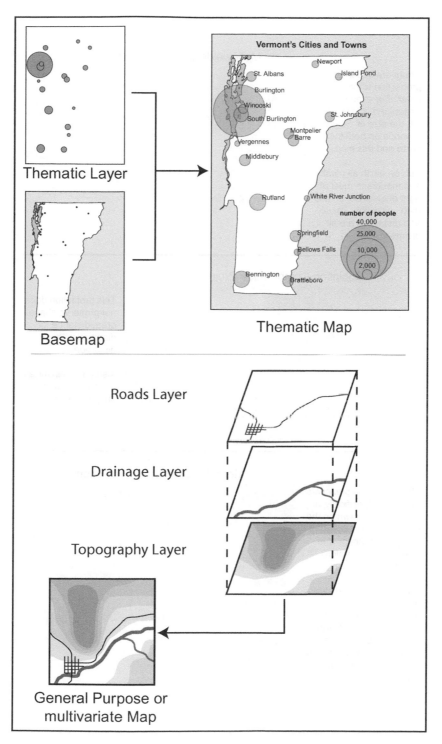

**Figure 16.5** Organizing information on thematic and general purpose maps

## Molleweide Equal Area Projection

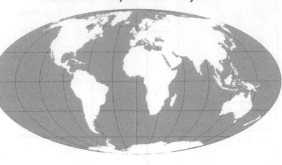

Areas are in correct proportion to each other. For example, Greenland is less than 1/8 the size of South America on both the earth and this map.

But, on earth all parallels and meridians meet at right angles. So, on this map, direction, shape, and distances are all distorted.

## Robinson Compromise Projection

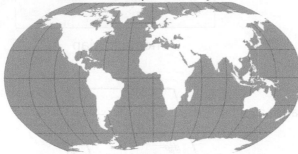

This projection does not preserve shape, area, distance, or direction without distortion.

Instead, it minimizes distortion in both shapes and areas.

## Mercator Conformal Projection

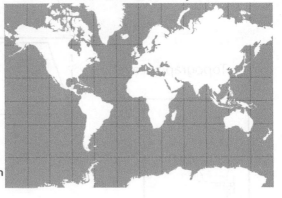

This projection preserves direction. Specifically, any straight line drawn on the map is a line of constant compass bearing. It is most useful, therefore, for navigation.

This map greatly distorts areas, especially as distance to the equator increases. Thus, North America, Europe, and northern Asia appear much larger in relationship to Africa and South America.

**Figure 16.6**   Map projections

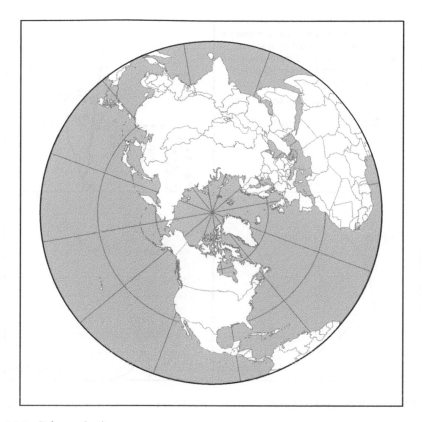

**Figure 16.7**   Polar projection

the world's poorer countries are concentrated nearer the equator, you can imagine how the two projections represent the same economic geography very differently.

   This is just one example of how projections used to map whole hemispheres or the entire globe can influence people's knowledge and understanding of our world. For example, the widespread use of the Mercator projection for wall maps in US classrooms has been criticized for helping to creating both isolationist and Eurocentric world views. On this projection, the United States appears separated from Europe and Asia by wide expanses of ocean. During World War II and in the early days of the Cold War, polar projections (Figure 16.7) were used to combat isolationism by showing how close Europe and the Soviet Union "really" were. And, as hinted at above, the tendency for people to associate size with importance means that the Mercator projection reinforces world views that rank the countries in Europe and North America as somehow "better" than those in Africa, Latin America, and Asia.

## Scale and generalization

The transformation of our three-dimensional complex and contested world into a flat map does not only involve projection. In almost all cases, we map the world at a much smaller

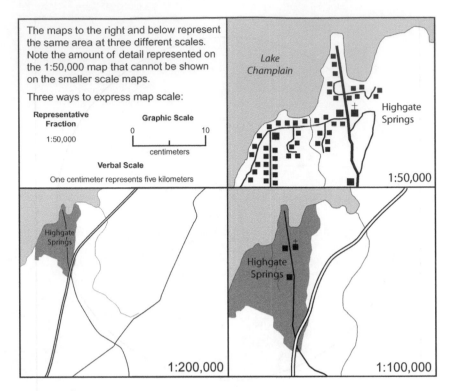

The maps to the right and below represent the same area at three different scales. Note the amount of detail represented on the 1:50,000 map that cannot be shown on the smaller scale maps.

Three ways to express map scale:

**Representative Fraction**

1:50,000

**Graphic Scale**

0          10

centimeters

**Verbal Scale**

One centimeter represents five kilometers

Lake Champlain

Highgate Springs

1:50,000

Highgate Springs

1:200,000

Highgate Springs

1:100,000

**Figure 16.8**   Scale and generalization

scale than "reality." In other words, earth distances we typically measure in kilometers or miles are represented in centimeters or inches on our maps. A map's *scale* is a mathematical measurement of how much smaller the map is than the portion of the world it represents (Figure 16.8). A scale of 1:50,000 means that a unit of distance on our map represents 50,000 of same units of distance on the Earth's surface. In other words, the linear distances shown on our map are one 50,000th the size of the distances they represent on earth.

This reduction in scale inevitably involves generalization – the process of selecting and organizing the geographic information to be represented on the map and transforming that data into map form (Dent 1999). Selection and organization of geographic information has been covered above, but generalization also involves the simplification and classification of the already selected information to suit the map's scale and purpose. Most obviously, the smaller our map's scale (the more reduction that occurs), the less detail can be shown (see Figure 16.8). As scale is reduced, coastlines and rivers are simplified and individual buildings may be agglomerated into urban regions. But generalization is not only regulated by scale; the map's purpose also plays a role. A map feature that is more central to our purpose can be shown in more detail than elements that are less important.

Another form of generalization is classification. On most thematic maps, symbols representing the same or similar phenomena located in different places are made to appear the same. For example, on a **digital elevation model**, all locations with an elevation between 200 and 210 meters are given the same color despite the actual variations in value within that range. This classification of geographic information can help us to see patterns more clearly by suppressing details that may "clutter-up" the map.

## Symbolization

A final aspect of generalization that is often treated separately in cartography is symbolization – the transformation of the selected geographic information into map symbols. Every point, line, area, icon, letter and word on a map is a relatively simple symbol or sign that represents some aspect of the more complex world we wish to visualize, explore, or communicate to others. In a very real way, these symbols make places that are distant in time and space present whenever we work with maps.

Choosing, creating, and interpreting map symbols can be approached from the scientific, artistic, and/or critical cartographic perspectives. Despite the limits of the behavioral approach, decades of testing groups of map readers has yielded useful guidelines on how people perceive and react to different symbol colors, shapes, and sizes. At the same time, map symbols continue to be evaluated on aesthetic grounds as well. Cartography textbooks continue to quote the map critic John K. Wright who, in 1944, wrote, "The quality of a map is also in part an aesthetic matter … An ugly map, with crude colors, careless line work, and … poorly arranged lettering may be intrinsically as accurate as a beautiful map, but it is less likely to inspire confidence." Finally, map symbols acquire social and cultural meanings and these can be reinforced or challenged when a map is made and used. The widespread use of green to symbolize low elevations on reference maps, for example, may lead us to believe that California's Central Valley is a naturally humid and verdant environment. Of course, while this semi-arid valley is a rich agricultural area, this is only due to massive, publicly-funded irrigation projects.

The concept of visual variables provides a very practical starting point for understanding how we create, choose, and make sense of symbols (MacEachren 1995, Monmonier 1996). In two-dimensional mapping, Earth objects are represented as points (buildings, individual trees, a city on a smaller scale map), lines (rivers, roads, railroads), or areas (countries, water bodies, parcels or properties on a larger scale map). Because not all roads carry the same volume of traffic and different city parcels are characterized by different land uses, we need to be able to vary symbol appearance to represent different attribute values. Figure 16.9 contains the visual variables we can employ to represent similarities and differences among our map symbols.

Of course, we measure differences among the places on our maps in a variety of ways. A highway's traffic volume is a quantitative attribute, while a parcel's dominant land use is a qualitative characteristic. Size and color value (the lightness or darkness of a color) are best suited to show quantitative differences and similarities on a map. Map readers tend to associate the size of a circle or the thickness of a line with absolute amounts. For example, a larger circle might represent a city with a larger population than a city symbolized as a smaller circle. Giving area symbols a range of color values suggests a difference in percents or rates. A darker area is usually associated with having a higher value than a lighter area. Shape, color hue, and patterns are best used to symbolize qualitative difference or difference in kind. Geologic maps often use different hues to symbolize different rock or soil types.

We can also think of map symbols existing on a continuum from the most pictorial to the most abstract. Pictorial symbols look more like the earth objects they represent. They attract a map reader's attention due to their visual complexity and also make the link between a symbol and what it is supposed to represent much clearer. The map used in Box 16.2 employs very pictorial symbols. Abstract symbols have little in common visually with the earth objects they represent. A dot (·) may represent a city on a small scale map or a monument on a city plan. It certainly has nothing in common visually with either a city or a monument.

Of course, such graphic map symbols seldom stand alone. They are labeled with place names and explained in map legends. Placing the name "London" next to a dot on a map ensures that almost all map users will know that the dot represents a particular city. In order to make our maps legible, however, we usually cannot label every attribute represented by every symbol on our map. Therefore, we rely on map legends and explanatory notes to help us attach meanings to, and discern differences between, map symbols. The methods of symbolizing qualitative and quantitative differences shown in Figure 16.9 are organized in typical map legend formats. You can imagine, for example, a map where each European country is shaded one of the three gray values shown in the "percent in poverty" legend. If the map title reads, "Children in Poverty", then the legend would provide map users with the information needed to attach the map's symbols to the attribute values they represent.

## The Power of Maps

Maps are deployed for an ever increasing number of purposes. We find them in our newspapers and on our newscasts. Governments use them to assign values to properties for tax purposes, to define their territories, and to inspire nationalism among their citizens. Maps make us aware of events, help us understand their contexts, and determine how such events will affect our futures. As Hurricane Katrina approached the Gulf Coast of the United States in late August, 2005, we watched animated representations of swirling clouds churn across the Gulf of Mexico. In the immediate aftermath, Jonathan Mendez and Greg Stoll (two Austin, Texas, residents with limited programming experience) created an online form that allowed people to report where family members, friends, and pets in need of help were located. The resulting locations were mapped online using a Google map mashup (see http://googlemapsmania.blogspot.com for more examples of mashups). Later maps produced by professionals placed the images of destroyed buildings and displaced people and helped us know where aid was most needed. Now, new government maps delineating flood-prone areas are determining which neighborhoods can be rebuilt and which may be transformed back into swamp land.

We use maps – both those we make and those made by others – to navigate, explore, and see more of our worlds than we can ever experience firsthand. Desktop mapping and GIS software, as well as a large number of interactive mapping sites on the World Wide Web, such as the map mashups mentioned above, permit rapid entry into the world of map making (see Box 16.4). These technologies, and the expanding digital geo-databases upon which they are based, are also making maps that are increasingly ephemeral – the data on which our maps are based change more rapidly then ever before. Such changes

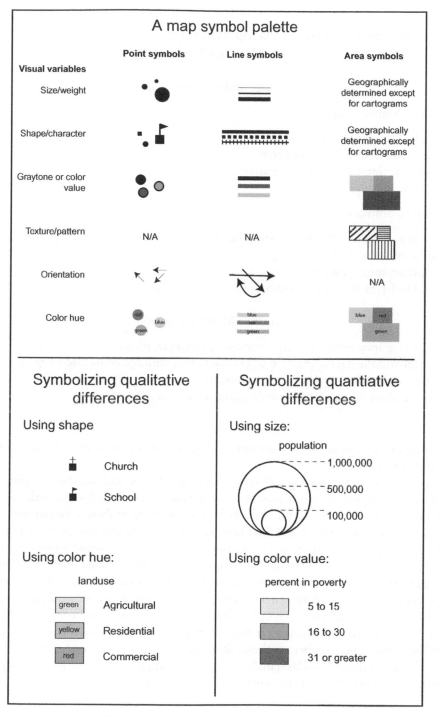

**Figure 16.9** Visual variables of cartographic symbols (adapted from Monmonier 1996. *How to Lie with Maps* Chicago: University of Chicago Press)

**Box 16.4**   Sample Desktop Mapping Software and Internet
Mapping Sites

Mapping and GIS Programs

Public Domain /Freeware

AGIS: http://www.agismap.com
ArcExplorer: http://www.esri.com
FlowMap: http://flowmap.geog.uu.nl
GRASS: http://grass.itc.it

Licensed Software

ArcView and ArcGIS: http://www.esri.com
Idrisi: http://www.clarklabs.org
MapInfo: http://www.mapinfo.com
Maptitude: http://www.caliper.com

Internet Mapping Sites

The Geography Network: http://www.geographynetwork.com/
UN Environmental Programme GEO Data Portal: http://geodata.grid.unep.ch/
TerraServer.com: http://www.terraserver.com/
Google Maps Mania: http://googlemapsmania.blogspot.com/

may increase our confidence in map currency and our ability to create our maps to serve our own interests.

Yet, there are still constraints. Desktop mapping software and internet mapping sites are based on geographic information that has already been collected, organized, and generalized into geo-databases. While we may be able to augment these with our own data, the structure of the database will limit how we can represent our worlds. In addition, to make mapping programs more accessible, most producers of such software provide symbol palettes and default design options. These may or may not be based on the suggestions and guidelines summarized in this chapter, but limit our options either way. When making maps using these tools and using maps created with such programs, we must pay attention to how they enable and constrain our ability to visualize, explore, and communicate geographically. Also, we must remember that any map we use and make is but one of many simplified, time-specific, perspective-limiting representations of our complex, ever-changing and contested worlds. Finally, it is critically important to consider the implications our maps may have for the people and places we represent.

# References

Black, J. (1997) *Maps and History: Constructing Images of the Past.* New Haven: Yale University Press.

Del Casino, V., and Hanna S. (2000) Representations and identities in tourism map spaces. *Progress in Human Geography* 24(1): 23–46

Dent, B. (1996) *Cartography: Thematic Map Design* (4th edn). Boston: WCB McGraw-Hill.

Dent, B. D. (1999) *Cartography: Thematic Map Design* (5th edn). Boston: WCB McGraw-Hill.

Harley, J. B. (2001) *The New Nature of Maps: Essays in the History of Cartography*. Laxton, P. ed. Baltimore: The Johns Hopkins University Press.

Harley, J. B., and Woodward, D. eds. (1987) *The History of Cartography, Volume 1: Cartography in Prehistoric, Ancient, and Medieval Europe and the Mediterranean*. Chicago: The University of Chicago Press.

MacEachren, A. M. (1995) *How Maps Work: Representation, Visualization, and Design*. New York: The Guilford Press.

Monmonier, M. (1996) *How to Lie with Maps* (2nd edn). Chicago: The University of Chicago Press.

The International Cartographic Association (2008) ICA Mission Statement. http://cartography. tuwien.ac.at/ica/index.php/TheAssociation/Mission [accessed August 6, 2009].

Wood, D. (1992) *The Power of Maps*. New York: The Guilford Press.

## Additional Resources

Crampton, J. (2010) *Mapping: A Critical Introduction to Cartography and GIS*. New York: Wiley-Blackwell. A much needed introductory text that summarizes post-1990 developments in cartographic theory and technology and presents case studies that help readers understand how to use maps and GIS critically.

Harmon, K. (2004) *You Are Here: Personal Geographies and Other Maps of the Imagination*. New York: Princeton Architectural Press. A book that invites readers to explore how people past and present see and map their bodies, worlds, imaginations, and values. Very useful in broadening anyone's geographic imagination.

Monmonier, M. (2002) *Spying with Maps: Surveillance Technologies and the Future of Privacy*. Chicago: The University of Chicago Press. In a world of handheld GPS devices, high resolution satellite imagery, geo-coded public and private databases, and surveillance cameras, this book introduces the concept of locational privacy and asks if these mapping technologies constitute a significant threat to the right of privacy.

Pickles, J. (2004) *A History of Spaces: Cartographic Reason, Mapping, and the Geo-coded World*. London: Routledge. A valuable history of the development of Western cartography and analysis of how the cartographic gaze shapes our world.

Thongchai, W. (1994) *Siam Mapped: A History of the Geo-body of a Nation*. Honolulu: University of Hawai`i Press. An excellent and often-cited work detailing how a society's understanding of its geography, as mapped, changes through time and with interactions of other societies.

## Exercise 16.1   Analyzing Maps

Find a published map to analyze. It can be historic or contemporary, in print or electronic form, and have any thematic basis (e.g., political, socio-demographic, biogeographic, climatic). Based on the material presented in this chapter, write a short essay on the selected map that answers the following questions:

1   Where did you find the map, and why did you select it for analysis? What is the larger context (e.g., planning/administration, profit, public persuasion) that led to its production?

2  What is the map's stated or presumed purpose?
3  How does the map use projection, scale, generalization, and symbolization to convey geographic information in support of its purpose?
4  Whose interests (political, economic, cultural, administrative, etc.) are served by the way the map represents its geography?
5  What prior knowledge might a reader bring to this map? In other words, how is this map intertextual? To answer this question, think about: the style and content of this map in relation to other maps; the histories of the places shown on the map; important demographic, economic, or political information; other representations of the places in popular culture (film or television, for example); and/or how nature is represented in the map. How might these aspects affect how the map is interpreted?

# Chapter 17

# Descriptive Statistics

## *Sent Visser and John Paul Jones III*

- Introduction
- Frequency Distributions
- Graphs
- The Shape of a Frequency Polygon
- Statistics for Describing Frequency Distributions
- Describing Relationships between Variables
- Conclusion

Keywords

| | |
|---|---|
| Central tendency | Parametric/non-parametric |
| Co-variation | Skewness |
| Frequency distribution/frequency polygon | Variation |
| Kurtosis | Z-scores |

## Introduction

This chapter and the next cannot teach you how to do statistical analyses of data. As in GIS and remote sensing, most geography departments offer semester-long or even year-long courses in statistics. We hope, however, to introduce you to statistical reasoning and to make you a more discriminating consumer of other researchers' statistical results. The chapters presume that you have some understanding of algebra and its symbols (like the summation sign), that you can read Cartesian graphs, that you understand the meaning of variables and relationships, and that you have a good grasp of levels of measurement (see Chapter 4).

A statistic is a number that summarizes a variable or relationship between variables in some way. These numbers describe the distribution of the values of a variable, and the nature and degree of relationships between variables. Because condensing data into a single number or equation loses a lot of information, tables and graphs should be employed so that you can actually visualize the distributions and relationships, especially since statistical software can create these so easily.

The ultimate task of statistics is to unearth and describe relationships between variables. Describing averages, variation, and simple relationships – which are topics of this chapter – are a means to this end. Often a relationship is described as a difference between the averages of a variable for two sub-groups of observations within the sample; e.g., average incomes of men and women in the same occupations. Variation is relevant because we describe relationships as **co-variation** between variables, and because the nature of the variation within each variable affects whether the co-variation between two variables is meaningful.

Statistics can also be categorized as parametric and non-parametric (Norcliffe 1982). **Parametric** statistics refers to those techniques that are only applicable to metric or inter-val/ratio variables. **Non-parametric** techniques apply to ordinal and nominal variables. Parametric statistics are more "powerful" in the sense that they depend on knowledge of the magnitude of the difference between the values of a variable. Non-parametric statistics only consider the ordering of observations on a variable, or which category of a variable that an observation falls in. Sometimes parametric statistics are used on ordinal data, but the results are garbage.

Another categorization of statistics is into simple and multivariate. Multivariate statistics encompass the techniques that try to uncover the independent relationships existent between more than two variables (Johnston 1978). Simple statistics look at the interdependence of only two variables at a time, and ignore the influence of other variables on the distributions of the values of those two variables.

## Frequency Distributions

Tables and graphs of frequency distributions can be utilized to quickly see how the values of a variable are distributed. The values of the variable have to be grouped to do this, although in the case of a nominal variable, the unit of measurement is a group or category. Frequency refers to the number of observations in a group.

In Table 17.1, functional type is a variable generated by analyzing the employment structure of US cities in 1950 using a technique called cluster analysis. This technique groups cities according to similarities in their employment structure. The column labeled "f" (frequency) is the number of cities in each category of the variable. Relative frequency (% f) is the percentage or proportion of the total number of observations that are in each group. (A proportion is a number expressed as a ratio of the number 1; 13.10 percent expressed as a proportion is .1310.)

It is desirable to show both frequencies and relative frequencies in this kind of table to ease comparison with similar types of tables. For example, this table might be compared with tables for other decades in which the number of cities classified is different from that in 1950.

**Table 17.1**   Functional type of US SMSAs, 1950

| Functional type | f | % f |
|---|---|---|
| Diversified | 19 | 13.10 |
| Diversified/manufacturing | 44 | 30.34 |
| Federal government | 1 | 0.69 |
| Manufacturing | 12 | 8.28 |
| Mining | 3 | 2.07 |
| Other government | 7 | 4.83 |
| Public admin./regional centers | 6 | 4.14 |
| Transportation/wholesale/finance | 25 | 17.24 |
| Service centers | 7 | 4.83 |
| Transportation | 1 | 0.69 |
| Professional service | 20 | 13.79 |
| Number of Cases/Sum | 145 | 100.00 |

*Notes*: SMSAs are Standard Metropolitan Statistical Areas. This designation
was later replaced by MSAs, Metropolitan Statistical Areas.
*Source*: data from Subramaniam-Bryson (1991)

Tables that show frequency distributions for ordinal variables are organized the same way as are tables for nominal variables except that the categories of the ordinal variable would be put in order of rank, from lowest to highest or highest to lowest. In the case of nominal variables the categories may be placed in any order (see Chapter 4).

If the ordinal variable is one in which there are not many observations tied at each rank, then the table of the variable must be prepared as for a ratio variable. The ranks will need to be grouped. To make tables of variables that are measured on an interval or ratio scale the variable values must be grouped or categorized such that several observations fall into a group. This is also true of an ordinal variable in which most of the observations have a unique rank.

Table 17.2 shows the frequency distribution for the variable "VTO," which is the number of people who voted in the 1980 presidential election as a percentage of the total population aged 18 and over in all Texas counties in 1980. There are 254 counties, and so the variable has been grouped into 14 mutually exclusive classes (groups). As the number of observations in a study decreases the number of classes should be smaller. The class width chosen for VTO is 5 percent. Generally speaking, the classes should be of the same width; otherwise, the distribution of the variable values will be distorted. There are instances, however, in which you might want to have classes of different width (e.g., grouping a population in this way: children, 0–17 years; adults, 18–64; elderly, 65+).

The column labeled "f" is called the **frequency distribution**. It shows how the values of VTO are distributed. Is there a concentration of values? Around what value are they concentrated? Is the distribution of values symmetrical? That is, are there approximately equal numbers of exceptionally high and exceptionally low values?

The column labeled "% f" is the relative frequency distribution. It is the frequency as a percentage of the total number of observations (n). The label "F" denotes the cumulative frequency distribution. It is the total number of observations that have values smaller than

**Table 17.2**  Voter turnout, Texas counties, 1980 presidential election

| VTO | f | % f | F | % F |
|---|---|---|---|---|
| 22.5–27.5 | 2 | 0.79 | 2 | 0.79 |
| 27.5–32.5 | 3 | 1.18 | 5 | 1.97 |
| 32.5–37.5 | 4 | 1.57 | 9 | 3.54 |
| 37.5–42.5 | 14 | 5.51 | 23 | 9.06 |
| 42.5–47.5 | 55 | 21.65 | 78 | 30.71 |
| 47.5–52.5 | 85 | 33.46 | 163 | 64.17 |
| 52.5–57.5 | 31 | 12.20 | 194 | 76.38 |
| 57.5–62.5 | 28 | 11.02 | 222 | 87.40 |
| 62.5–67.5 | 22 | 8.66 | 244 | 96.06 |
| 67.5–72.5 | 6 | 2.36 | 250 | 98.43 |
| 72.5–77.5 | 0 | 0.0 | 250 | 98.43 |
| 77.5–82.5 | 2 | 0.79 | 252 | 99.21 |
| 82.5–87.5 | 1 | 0.39 | 253 | 99.61 |
| 87.5–92.5 | 1 | 0.39 | 254 | 100.00 |
| Number of cases/sum | 254 | 99.97 | | |

*Source*: Data from Tebben (1990)

the class upper limit. The percentage cumulative frequency provides the percentile of the upper class limit. The *kth* percentile of a variable is that value of the variable such that at least *k* percent of the counties have values that are smaller than or equal to it. 67.5 is the 96.06th percentile of VTO, and thus only 3.94 percent of the cases have values greater than 67.5. Percentiles also apply to ordinal variables, but have no meaning in the case of nominal only variables.

# Graphs

Although you might be able to imagine the distribution of VTO from Table 17.2, graphs are actually better than tables in providing an immediate visual impression of the distribution of a variable. The frequency distribution for a nominal variable may be depicted via a pie diagram (circle graph) or bar chart. In the bar chart frequency or percent frequency is on the Y-axis and the categories of the variable are on the X-axis. Each category has a bar with the length of the bar proportional to the frequency of the category. The bars have the same width and are not contiguous because there is no continuity of values between the categories of the variable.

A histogram is a bar chart used to portray the distribution of a variable measured at the interval or ratio level. The bars are contiguous to show continuity in values of the variable. Each bar portrays the frequency for a class of a grouped variable. The bars are the same width because the classes all have the same width.

Figure 17.1 shows a histogram for the voter turnout data in Table 17.2, except that the class widths have been reduced to only 2.5%. The area of a bar in a histogram is proportional to the class frequency. That is, the area of the histogram between any two values of VTO is calculated as a percentage of the total area of the histogram. This provides the

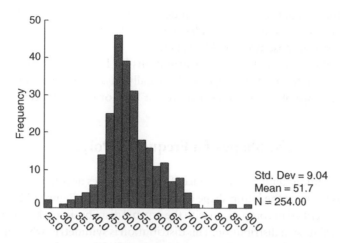

**Figure 17.1**  Histogram for voter turnout (VTO) in Texas counties

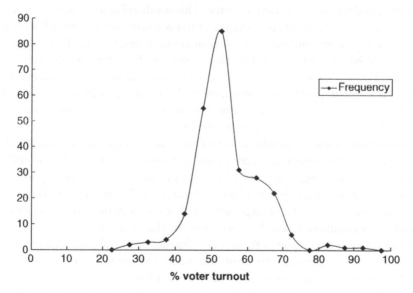

**Figure 17.2**  Frequency polygon for voter turnout in Texas counties

percentage of all observations that have values in that class interval. This is only true if the class widths are all the same.

A frequency distribution can also be portrayed as a line graph called a frequency polygon. Figure 17.2 shows a frequency polygon for the voter turnout data using the same classes as in Table 17.2. The frequency value for a class is plotted against the midpoint of the class, and the plots are then connected with a line.

Drawing a frequency polygon is equivalent to joining the midpoints of the tops of the bars of the histogram, with the proviso that empty classes must be added at both ends of the distribution. The zero frequency in each of these classes is plotted so that the polygon is brought down to the X-axis and is enclosed. This makes the area of the polygon equal to the area of the histogram, and the area under the curve between two values of the variable, as a percentage of the total area under the curve, is equal to the relative frequency of observations between those two variable values.

Line graphs of cumulative frequency distributions (% F in Table 17.2) are called ogives. These slope from bottom left to upper right usually in an S-shape, and show the total number or percentage of observations that are less than or equal to a certain value.

## The Shape of a Frequency Polygon

Frequency distributions tend to have a concentration of values near the average for the variable. This tends to create the hump shape of a distribution as in Figure 17.2. The shape of a distribution is described in comparison with the normal distribution, which is also called the bell curve or the Gaussian distribution. This distribution is symmetric, which means that the left side of the graph is the mirror image of the right side. Those values of the variable that are below the "average" vary in the same way as the values that are above the average.

The uniform distribution is also symmetric. This is a distribution where all values of the variable occur with equal frequency, and the graph is a horizontal straight line. If a single dice is thrown repeatedly, and the values thrown are recorded, then the frequencies of each number should be approximately equal. That is, the numbers thrown will have a uniform distribution. This is what most people think of as random; namely an equal chance of occurrence. In a random sample of observations, every observation has an equal chance of being selected, but the likelihood of getting certain values of a variable measured on those observations depends on the shape of the frequency distribution for that variable.

The term **skew** means the distribution of values is not symmetric. Positive or right skew means that most of the observations are clustered around some value that is smaller than the average of all the values. There are a lot of small values, and a few exceptionally large ones. The frequency polygon has a tail to the right. Negative or left skew is the opposite of right skew. There is a cluster of large values with a few exceptionally small values. Right skew is far more common in social science variables than left skew.

**Kurtosis** refers to the degree of flatness of a frequency distribution. A leptokurtic distribution has a high degree of clustering of observations around a central value with a great degree of variation among a few small values and/or a few large values. A platykurtic distribution indicates a low degree of clustering of values. The uniform distribution is very platykurtic. The normal distribution is mesokurtic. It is the standard by which other distributions are considered leptokurtic or platykurtic. The distribution shown in Figure 17.2 is neither uniform nor highly skewed. It has some tendency toward a normal shape, but it is not symmetric. It is slightly right skewed and leptokurtic.

## Statistics for Describing Frequency Distributions

### Central tendency

Averages summarize the central location of a frequency distribution. They are called measures of **central tendency** because they tend to be near the center of a variable's distribution and tend to be the value around which most of the observations cluster. Thus averages are also referred to as the "norm."

The mode is the most frequently occurring value of a variable. It is only useful for nominal or ordinal variables since many ratio variables either do not have a mode, or have several modes.

If the values of the variable are arranged in order of size (including those observations that have a variable value of zero), the median is the value of the variable of the middle observation. So the median is also the 50th percentile. A median cannot be calculated for nominal data, but it is suitable for ordinal variables. The median of VTO is 49.86. Of the counties, 50 percent have a value of VTO that is less than or equal to 49.86, and 50 percent of the counties are above that value.

Percentiles in general do not measure central tendency, but they do provide a measure of location in a frequency distribution. SAT scores are scaled in such a way that the variable cannot be considered an interval/ratio scale. Your percentile, then, provides what a score of 1,100 out of 1,600 really means. If the 80th percentile is 1,100, 80 percent of those who took the exam had a score lower than or equal to 1,100 and 20 percent scored more.

Other names for percentiles are: the first decile (the 10th percentile), the first quartile (the 25th percentile, or bottom 25 percent), and the third quartile (the 75th percentile). The first quartile for VTO is 46.63 and the third quartile is 56.07. The third quartile is farther from the median than is the first quartile, indicative of the variable's right skew.

The common understanding of average is actually called the arithmetic mean, or just mean. The symbol for the mean of variable X, calculated from sample data, is $\bar{X}$. The symbol for the mean where variable X has been collected for the entire population (called the population parameter) is $\mu_x$ (pronounced "mu").

The mean is calculated by adding all the values of the variable and dividing by the number of observations.

$$\bar{X} = \frac{\sum_{i=1}^{i=n} X_i}{n} \tag{17.1}$$

The mean applies to interval/ratio data, although the calculation of a mean rank (an ordinal variable) is commonly done, for example, grade point average.

Often the median and mean are quite different in value. If the variable is right skewed the mean will be larger than the median. The mean will be smaller than the median if the data is left skewed. This is because the mean is greatly influenced by exceptionally large or exceptionally small values, and the median is not. If the distribution is symmetric, they will be equal. In skewed data, the median may be a preferable measure if the intent of the "average" is to describe a "typical" value, but otherwise the mean is the preferred measure of "average" because it contains more information (every value of a variable contributes to the value of the mean), and because it is mathematically defined, its variation between samples drawn from a single population is known. The mean for VTO is 51.67, which is larger than the median, indicative of the right skew in the variable.

If several different samples of the same size are drawn from a single population, the different sample means will vary in value as will the different sample medians. If the distribution of the population variable is fairly symmetrical the sample means will vary less around their true population mean than will the sample medians around their true population median. Furthermore, the degree of variation of these sample means is known. The sample mean is, therefore, a more reliable estimator of the population mean than is the sample median of the population median.

## Variation, spread or dispersion

The mean offers information about a distribution's central tendency, but it tells us nothing about the amount of variability in a variable. For that we need statistics that describe **variation**. The value of most measures of variation depends on the size of the numbers of the variable, i.e., on the unit of measurement of the variable. Comparing the variation of two variables with very different measurement units is therefore fruitless unless these variation measures are standardized in some way so that the variations are expressed relative to the average size of the numbers.

One important use for measures of variation is in statistical inference. If several samples of size $n$ are drawn from a single population, the means of those samples will vary more as the variation in the variable increases. Therefore the degree of variation in the values of a variable affects how reliable a single sample statistic (like the sample mean, $\bar{X}$) is as an estimator of the population parameter ($\mu_X$).

The range is the simplest measure of variability. It is found by taking the difference between the smallest and largest values of a variable. The smallest value of VTO is 23.92, and the largest is 91.14. The difference of 67.22 is the range. The range depends entirely on the two extreme values of a variable. So while it is simple, it is relatively useless. The inter-quartile range is the range between the 25th and 75th percentiles ($56.07 - 46.63 = 9.44$ for VTO). The decile deviation is the range between the 10th and 90th percentiles. These measures remove the effect of extreme values, but there are more useful measures of variation in a variable that utilize the information contained in all of its values.

A deviation is the distance that a variable value is from a statistic, usually the mean. Algebraically it is written as: $x = X - \bar{X}$ for variable $X$. Values that are below the mean will have negative values. The average distance of each value from its mean on both sides of the mean is an intuitively appealing measure of variation. But the sum of the deviations around the mean is zero. The deviations around the mean can be made positive by squaring them. The average (or mean) squared deviation around the mean is called the variance. The sum of the squared deviations is called the total variation of the variable or the sum of squares. The first formula below is for the sample variance ($s^2$), and the second formula is for the population variance ($\sigma^2$). Division by $n - 1$ to calculate the sample variance is because that provides the best estimate of the population variance. A sample is unlikely to include an unusual extreme value of the variable, which inflates the variance, and so the sample estimate of it is inflated by using $n - 1$ as the divisor.

$$s^2 = \frac{\sum(X - \bar{X})^2}{n-1} \text{ and } \sigma^2 = \frac{\sum(X - \mu)^2}{N} \tag{17.2}$$

The variance for VTO is 81.79. Its metric is squared percent. To convert variance to a metric that is the same as the original variable, we use the standard deviation, which is the

**Box 17.1**   Chebyshev's Inequality

Chebyshev's inequality (also known as Chebyshev's theorem) indicates that, no matter what the shape of the frequency distribution, at least $\left(1-\dfrac{1}{k^2}\right)100$ percent of the values are within $k$ (for $k > 1$) standard deviations either side of the mean. So at least 88.89% of the values of a variable must be within three standard deviations of the mean. Usually almost 100 percent of the values are within 3 standard deviations of the mean.

If a variable has a normal distribution, exactly 68.26 percent of the values fall within one standard deviation from the mean, 95.44% within two standard deviations of the mean, and 99.74% within three standard deviations. So values that are more than 3 standard deviations from the mean are very unusual, and may indicate a data error.

In the data for VTO, three standard deviations below and above the mean lie at 24.54 and 78.80, and two standard deviations of the mean results in values of 33.58 or 69.75. In the actual data, one value of VTO is less than 24.54 and four are greater than 78.80. So 98.03 percent of the values are within three standard deviations of the mean. Five values are smaller than two standard deviations below the mean, and 7 values are more than two standard deviations above the mean. So 95.28 percent are within two standard deviations of the mean. But one value is more than four standard deviations above the mean, and so the distribution cannot be regarded as normal.

square root of the variance. Its symbols are $s$ for the sample statistic, and $\sigma$ (sigma) for the population variance. It is usually much smaller than the variance, unless the variance is less than one. The standard deviation for VTO is 9.04 (percent).

The standard deviation is only applicable to interval/ratio data, and it is the most common measure of variation for such data, even though its intuitive meaning is hard to grasp. Its value is sensitive to extreme values in the data because of the squaring of the deviations. We do, however, have a good idea of how many values of a variable should be within a certain number of standard deviations from the mean (see Box 17.1).

## Coefficient of variation

Often it is desirable to compare the variation in two variables that have the same units of measurement but very different means. For instance, we may want to see if income inequalities, as measured by the standard deviation for per capita county incomes in a state, are decreasing or increasing over time. Over a 30 year time span, however, per capita incomes have increased 3-fold on average, and so the standard deviation should have increased three-fold over time even though relative variation in incomes may not have changed.

The coefficient of variation solves this problem of comparisons. It is calculated as the standard deviation divided by the mean ($s/\bar{X}$). This calculation standardizes the standard deviation for the average size of the variable values, and so enables a comparison of the kind described. If the coefficient of variation increased over time, then regional inequalities of income within a state have increased over time.

## Z scores

Standard scores or **z-scores** express a value of a variable as the number of standard deviation units that the value is from the mean. That is,

$$z = \frac{X - \bar{X}}{s} \text{ or } z = \frac{X - \mu}{\sigma}, \tag{17.3}$$

and so

$$X = \bar{X} + zs \text{ or } X = \mu + z\sigma \tag{17.4}$$

This states that a value equals the mean plus z standard deviations. If the value is below the mean, z will be negative, and the value will equal the mean minus z standard deviations. For example, a value of −1.5 for z means that the X value for which it was calculated is 1.5 standard deviations less than its mean. The transformation of the values of a variable into z-scores allows comparison of variables that not only have different means, but also have different units of measurement. Sometimes z-scores are referred to as normalized scores, but this a misnomer, because if the variable does not have a normal distribution, neither do the z-scores.

The smallest value of VTO is 23.92. $(23.92 - 51.67)/9.044$ equals −3.068, which is the z-score for 23.92. As noted above, three standard deviations below the mean is $51.67 - 3(9.044)$, which equals 24.54. 23.92 is the only value smaller than three standard deviations below the mean. Three standard deviations above the mean is calculated as $51.67 + 3(9.044)$, which equals 78.80. The largest value of 91.14 is $(91.14 - 51.67)/9.044$, which equals a z-score of 4.364. So it is 4.364 standard deviations above the mean, and it is one of four values that are more than three standard deviations above the mean.

The discussion above illustrates that Z-scores can be used to determine whether a variable's frequency distribution approximates that of a normal distribution. Z-scores, calculated for a variable that has a normal distribution, have a mean equal to 0 and a standard deviation equal to 1, and the frequency polygon for these z-curves has a total area equal to 1. The discussion of frequency polygons noted that the area between two values of X expressed as a total proportion of the area under the frequency polygon is equal to the proportion of all values of X that are between those two values. The area between any two z-scores in a normal distribution with a mean of zero and a standard deviation equal to one is equal to the proportion of all z-scores that are between those two values. The areas, and therefore the proportions, that are between any z-score and the mean of 0 are provided in all introductory statistics books. They also tabulate these proportions (areas under the curves) for other important frequency distributions including the t-distribution, the F, the binomial, and the Chi-square ($\chi^2$) distributions.

## Describing Relationships between Variables

Analyzing relationships between variables can be done with tables, graphs, and statistics. Different procedures have been developed for different types of data. We often use contingency tables for nominal and ordinal data, and scatter diagrams and correlation coefficients for ratio and interval data.

### Contingency tables

A bi-variate contingency table shows how the values of a variable are distributed depending on the values of another variable. The number of observations in a category of one variable is "contingent" on which category of the other variable to which each observation belongs. More than two variables can be shown, but generally the limit is three, because a large number of observations are needed to show the distribution of a variable contingent on the values of two other variables.

In Table 17.3, the functional types of Table 17.1 have been regrouped into three city types. The population growth rates of the SMSAs from 1950 to 1980 have also been grouped. The cell frequencies (33, 11 etc.) are the number of observations. Thus, 33 of the 52 cities with growth rates below 40 percent were classified as manufacturing or diversified/manufacturing.

Functional type is a nominal variable. Grouping of the growth rates, a ratio variable, has made it into an ordinal variable. Tables such as these are used to show the relationship between two variables in which a large number of observations fall in each category of the

**Table 17.3**  US urban growth rates, 1950–80, by urban functional type

| Population growth | Functional type, 1950 | | | Total |
|---|---|---|---|---|
| | *Manufacturing* | *Service* | *Government/Prof* | |
| <40% | 33 | 11 | 8 | 52 |
| | (58.93) | (36.67) | (13.56) | (35.86) |
| 40.01%–75.0% | 20 | 11 | 21 | 52 |
| | (35.71) | (36.67) | (35.59) | (35.86) |
| >75.0% | 3 | 8 | 30 | 41 |
| | (5.36) | (26.67) | (50.85) | (28.28) |
| Total | 56 | 30 | 59 | 145 |

*Notes*: Proportions (%) within categories are shown in parentheses under numeric counts.
*Source*: Data from Subramaniam-Bryson (1991)

variables. Therefore, contingency tables are generally used to show relationships among nominal and ordinal variables.

Growth rate is the dependent variable in Table 17.3, and functional type is the independent variable (see Chapter 4). This is because functional type in 1950 could not be "caused" by subsequent growth rates. The categories of the dependent variable should always be down the side of the table with the categories of the independent variable across the top (just like graphs). This makes the relationship shown in the table much more apparent.

The percentages in Table 17.3 are calculated to see if there is a difference between the functional types in terms of how fast the cities grew. Given that the dependent variable is down the side, the cell frequencies are calculated as a percentage of the column totals. So 33 as a percentage of 56 is 58.93. The percentages show that manufacturing cities grew the least as reflected by the 59 percent in the lowest growth category and the 5 percent in the highest growth category. The government and professional cities tended to grow the fastest, with service cities having a more balanced pattern. Functional type in 1950 seems to have had a strong influence on subsequent growth rates (see Box 17.2).

It is possible that the relationship is spurious. Regional location in the US can influence both functional type and urban growth rates. To assess the relative impact of regional location, the country could be divided into three regions (say north, south and west), and region could then be included in a three-way table to see if functional type influences growth rates when region is controlled. As an independent variable, the three categories of region would also be on the top of the table. Each category of region is then divided into the three functional types, resulting in nine categories across the top of the table, and with three growth categories, there would be 27 total cells. At this level of detail, some cells might record no observations. Relative frequencies could still be calculated as a percentage of the column totals. If a high percentage of manufacturing cities have a low growth rate in every region, then functional type is exerting an independent influence on growth rates. If the low growth rate manufacturing cities tend to be in only one region, then it is region that is influencing the growth rates. In fact, both variables have had strong independent effects on urban growth rates.

## Scatter diagrams and correlation coefficients

Tables are valuable when we are examining relationships between ordinal and nominal variables. But when we have ratio or interval data, scatter diagrams (also called scatter plots) and simple correlation coefficients are more appropriate. Scatter diagrams are graphs that take advantage of the fact that interval/ratio variables contain information about the magnitude of differences between all pairs of observations. These graphs plot the variable values of each observation in two dimensions. By convention, in graphing these variables we put the dependent variable, Y, on the vertical axis, and the independent variable, X, on the horizontal axis. Visualization of these plots is a straightforward way to assess whether or not relationships exist. It can also tell us something about the form of the relationship (positive, negative, linear, non-linear). If the points in the scatter diagram appear to follow a straight line, then a line that describes the average relationship can be fitted to the points. We take up that technique in Chapter 18; for purposes of illustration,

## Box 17.2   Cramer's V

Cramer's V is a Chi-square based correlation coefficient, which in turn is based on the difference between the actual cell frequencies and the expected frequencies if the two variables were completely independent of each other. In Table 17.3, 35.86 percent of all the cities had growth rates lower than 40 percent. If the two variables are uncorrelated, then 35.86 percent of each city type should have growth rates that are less than 40 percent. Of the 56 cities in the manufacturing group, 35.86 percent is 20.08, and so 20.08 of those cities should have had growth rates below 40 percent. The service group should have had 10.76 with growth rates below 40 percent, and the expected frequency for the government/prof group is 21.16. A direct formula to calculate expected frequencies is:

$$f_e = ((\text{row total for the cell})(\text{column total for the cell}))/n \qquad (17.5)$$

where n is the total number of observations. For the first cell $f_e = (52)(56)/145 = 20.08$ as before. Actually, 33 of the cities in the manufacturing group have low growth compared to the expected value of 20.08, and only 8 of the government/prof cities have low growth rates where 21.16 was expected if the two variables are independent of each other. Calculating all the expected frequencies reveals that government/prof cities tend to be in the third category (highest growth rate) of the dependent variable, and the manufacturing cities tend to be concentrated in the first category of growth rate. That describes the "nature" of the relationship.

Chi-square ($\chi^2$) is a statistic that derives from the differences between the observed frequencies and expected frequencies in all the cells.

$$\chi^2 = \sum \frac{(f_o - f_e)^2}{f_e} \qquad (17.6)$$

where $f_o$ and $f_e$ are observed and expected frequencies in each cell, and the summation is over all the cells. The equation states that the difference between observed and expected frequencies are squared and then divided by the expected frequency for the cell. These quantities are then added for all the cells. In Table 17.3 they add up to 37.572.

Cramer's V, then, is an index between 0 and 1 (from no relationship to a perfect one), and is calculated as:

$$V = \sqrt{\frac{\chi^2}{(n)\min((r-1),(c-1))}} \qquad (17.7)$$

where r is the number of rows, and c is the number of columns. The denominator under the radical means that the number of observations (n) is multiplied by the smallest of $r - 1$ and $c - 1$. Cramer's V can be calculated for non-symmetric tables, i.e., where r does not equal c. So Cramer's V is the square root of $37.572/((145)(2))$, which equals .36. The relationship is not quite as strong as the table suggests because the expected frequencies and observed frequencies are only markedly different in the four corner cells of the table. Otherwise functional type is not a good predictor of which category of growth an observation falls in.

however, the scatter diagrams in this chapter show the "best fit" line through the pattern of points.

There are a number of different correlation statistics for variables of differing levels of measurement. The most commonly reported for interval and ratio data is Pearson's product moment correlation coefficient. It is a simple, single measure that tells us the degree to which two variables "co-vary," or move about their means, in similar or dissimilar ways (or not at all). The measure, known as $r$, ranges from $-1.0$, through 0, and on to $+1.0$. The negative and positive endpoints tell you that you have a perfect correlation between the two measures; this is when the two variables are actually the same, and if you were to encounter such a result you would want to re-check your data. In between these extremes, however, are degrees of correlation between variables: for example, between snowfall in November to February and river heights in March and April (hypothesis: positive); between poverty rates and levels of education (hypothesis: negative); between elevation and the density of trees (hypothesis: negative); between price per square foot of retail space and distance to the downtown (hypothesis: negative); between hours studying geography and geography GPA (hypothesis: positive).

The formula for the Pearson product moment correlation is:

$$r = \frac{\sum (X - \bar{X})(Y - \bar{Y})/n}{s_x \cdot s_y} \tag{17.8}$$

This is one of the most important measures in all of science, and for that reason it is worthwhile taking a few paragraphs to see how it works. First, recall that X and Y are the independent and dependent variables, respectively, and $s_x$ and $s_y$ are their standard deviations. Recall, second, that the measure varies between $-1.0$ and $+1.0$, with positive values of $r$ indicating a positive association and negative values of $r$ showing an inverse relationship. How does the formula work to create this outcome? For each individual observation's value on X and Y, we take its deviation from the mean of X and Y, and make a product of that value. The product will be zero if either of the values fall on their respective mean; if both values are greater than their means, then the value returned is positive; the same holds true if both values are negative (since two negative products produce a positive value); and if one is higher and the other lower than their means, then the returned value will be negative. In terms of relationships, when positive values are being consistently returned it means that the dataset is characterized by observations in which high values on one variable are found with high values of the other, and that when low values appear, so too do low values in the other variable. This is a case of a positive, or direct, relationship. When negative products dominate, then the situation is reversed: low values on one variable are found alongside high values of the other. Note that "positive" and "negative" refer to the direction of the relationship, and not its strength. A correlation of $+0.6$ is weaker than a correlation of $-0.7$.

We follow this calculation for each observation in the data set, keeping a running total as indicated by the summation sign. The resulting sum tells us the direction of the relationship under study: (a) if positive, then the correlation coefficient is positive; (b) if negative, then the relationship is negative; and (c) if the value is zero, or very low, then throughout the procedure the positive results have been cancelled out by the negative ones, and there is no consistency in the extent to which the two variables co-vary with respect to their individual means.

The remaining operations are essentially aimed at scaling $r$ to a dimensionless number that can be compared across variables and, indeed, data sets. First, the numerator's division by $n$ creates an average of the case-by-case product deviations. This value's metric is in X·Y, however; to convert it to $r$ we simply divide through by $s_x \cdot s_y$, the standard deviations of X and Y which are already in the metric of X·Y. This operation thus divides a product of X and Y by another product of the same, giving us a measure that is independent of the values of X or Y. Re-scaled to remove the variability in both variables, $r$ therefore ranges from $-1.0$ to $+1.0$. Generally speaking, for analyses with $n > 50$, values of $r$ greater than $+0.25$ or less than $-0.25$ tend to be "statistically significant" – that is, large enough for us to assume that a relationship really exists, even though it might be hidden by the operation of other factors working on causing variation in Y (the missing variable problem).

The relationship between scatter diagrams and correlation coefficients is shown in Figure 17.3, below. The best fit lines (described in Chapter 18) are shown for convenience. They summarize the scatter of points, but as you can see they should prove useful should you want to predict values of Y (vertical axis) based on values of X (horizontal axis).

Finally, you should keep in mind that while researchers often key in on linear relationships, in which case observations are oriented around a straight line through them (as in the top two panels of Figure 17.3), in actuality relationships can be much more complex. This is why you need to visualize data, so that you do not rely on a single measure like the correlation coefficient (Clark 1951). Scatter diagrams can help reveal all sorts of other relationships. For example, as the X-values increase the Y-values could tend to decrease, and then increase again as the X-values further increase. Such a U-shaped relationship is called a quadratic. The relationship could also be an inverted-U, a quadratic relationship in which Y-values tend to first increase and then decline. Scatter diagrams are also impor-

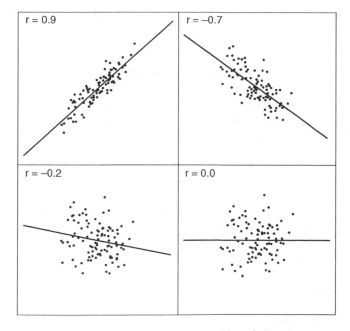

**Figure 17.3**  Scatter diagrams, correlation coefficients, and best fit lines

tant for identifying J-shaped or inverted J-shaped curves, and for detecting the common S-shaped curve, such as we find in studies of diffusion, among others.

## Conclusion

Descriptive accounts of numerical data combine tabular arrays, graphs, and single statistical measures (means, standard deviations, correlation coefficients). Knowing your data descriptively is the first and most important step in data analysis. It enables you to spot problems in data collection (such as extreme values, or outliers); it gives you an intimacy with the observations and their variables (individually and in combination with other variables); and it helps guide you in posing higher order questions about causality or explanation, including those that recognize a world that is largely multivariate – or multiply caused. The first step in the explanation of this complexly interrelated world, however, is to know your variable's tendencies, variability, and simple relations with other variables.

## References

Clark, C. (1951) Urban population densities. *Journal of the Royal Statistical Society, Series A* 114: 490–6.

Johnston, R. J. (1978) *Multivariate Statistical Analysis in Geography*. London: Longman.

Norcliffe, G. B. (1982) *Inferential Statistics for Geographers* (2nd edn). London: Hutchinson.

Subramaniam-Bryson, C. (1991) *The Measurement of Urban Functional Type: An Assessment of Criterion-Related Validity*. Unpublished Masters paper, Southwest Texas State University, San Marcos, TX.

Tebben, E. (1990) *Get the Vote Out*. Unpublished Masters paper, Southwest Texas State University, San Marcos, TX.

## Additional Resources

Anyone familiar with Microsoft's Excel can do any of the operations described in this chapter. Another approach for descriptive statistics is to bookmark one of any number of quick and easy interactive websites for calculating means, variances, and standard deviations. This is a relatively easy and quick one: http://www.easycalculation.com/statistics/standard-deviation.php (last accessed August 25, 2009).

There are a number of websites with interactive scatter diagrams and correlation analyses. Some let you enter data, others allow you to play around with different sample sizes and simulate your own correlation coefficients and scatter plots. This one is a good example: http://argyll.epsb.ca/jreed/math9/strand4/scatterPlot.htm (last accessed August 25, 2009).

Field, A. (2009) *Discovering Statistics with SPSS* (3rd edn). London: Sage. While you can find free software for all sorts of statistical research, there is still nothing easier than SPSS for the personal computer. This book is introductory and takes you step by step through the program's many capabilities. It's nice to have a statistics book and a computer manual in the same text.

Rogerson, P. (2006) *Statistical Methods for Geography: A Student's Guide* (2nd edn). Thousand Oaks, CA: Sage. This introductory text takes students from descriptive to explanatory (e.g., multiple regression) statistics in one semester. Intended for advanced undergraduates and beginning graduate students, this is a relatively concise but nonetheless broad ranging introduction to a wide range of statistical techniques.

Silk, J. (1979) *Statistical Concepts in Geography*. London: George Allen and Unwin. One of a number of older introductory statistical texts for geographers likely to be in your college or university library. Clear exposition and worked examples.

Taylor, P. (1977) *Quantitative Methods in Geography: An Introduction to Spatial Analysis*. Prospect Heights, IL: Waveland Press. This well selling text is decades old now, but it still bears a read for the way that Taylor describes, in logical and deeply geographic ways, some of the intricacies of spatial analysis. The volume covers a host of operations not addressed in this chapter, including the weighted mean center and the standard distance, point pattern analysis, and other operations used by geographers.

Tufte, E. (2002) *The Visual Display of Quantitative Information* (2nd edn). Cheshire, CT: Graphics Press. The classic treatise on the graphical display of quantitative data. It includes the famed graph showing the annihilation of Napoleon's army during the retreat from Moscow. Tufte is the expert on showing how graphs can both inform and mis-inform about what the data is indicating. For example, by appropriate scaling of both axes of a graph that shows the value of something over time, you can make growth look rapid, slow, or non-existent. The book is also a popular coffee table choice.

## Exercise 17.1    Scatter Diagrams and Simple Correlation Coefficients

The data below is for the 15 counties of Arizona (US Bureau of the Census 2000). The first variable is the percent of total persons below the poverty line. The second variable is the

**Table 17.4**    Data for Exercise 17.1

| *County* | *(Y)* | *(X)* | *(Y − Ȳ)* | *(X − X̄)* | *(Y − Ȳ)·(X − X̄)* |
| --- | --- | --- | --- | --- | --- |
| | *Poverty* | *<High school* | | | |
| Apache | 37.8 | 36.4 | | | |
| Cochise | 17.7 | 20.5 | | | |
| Coconino | 18.2 | 16.2 | | | |
| Gila | 17.4 | 21.8 | | | |
| Graham | 23.0 | 24.4 | | | |
| Greenlee | 9.9 | 17.5 | | | |
| La Paz | 19.6 | 30.7 | | | |
| Maricopa | 11.7 | 17.5 | | | |
| Mohave | 13.9 | 22.5 | | | |
| Navajo | 29.5 | 28.8 | | | |
| Pima | 14.7 | 16.6 | | | |
| Pinal | 16.9 | 27.3 | | | |
| Santa Cruz | 24.5 | 39.3 | | | |
| Yavapa | 11.9 | 15.3 | | | |
| Yuma | 19.2 | 34.2 | | | |
| *Totals* | | | | | |

percent of the over 25 population with less than a high school diploma. Assume for this problem that poverty is dependent (Y) and that education is independent (X). We want to visualize the relationship between these two variables and compute a correlation coefficient.

Using a piece of graph paper, plot the 15 counties' values in Y and X space. Does it appear that this data offers a positive, negative, or no relationship? Based on the scatter diagrams in Figure 17.3, what do you predict the correlation coefficient to be?

Now calculate the correlation coefficient for this set of data. You can use the space in the table provided or make a copy of the page and use that as a worksheet. A hand calculator is permitted, but the point of the exercise is to get a feel for the calculations involved. After calculating *r*, does there appear to be a strong or weak relationship in the data? If you were an official of the state of Arizona, what would you conclude on the basis of this short exercise?

# Chapter 18

# Explanatory Statistics

*Sent Visser and John Paul Jones III*

- Introduction
- Thinking Causality Statistically
- Experiments on Humans?
- Simple Linear Regression and Correlation
- Multivariate Linear Models
- Causality Issues in Multiple Regression
- Assumptions
- Conclusion

Keywords

Coefficient of determination
Dependent/independent variables
Dummy variable
Homoscedasticity

Misspecification
Multicollinearity
Spatial autocorrelation

## Introduction

In this chapter we continue our look into statistical reasoning that we began in Chapter 17. Here, however, we take up the topic of explanatory as opposed to descriptive statistics. As in that chapter, we introduce only a modest number of tools. The main focus nonetheless remains: to give you a clearer idea about how researchers think through their problems. In this chapter the problem is causality. In what follows we first give an overview of some of the basics about causality in experimental and non-controlled research settings. We then cover simple and multivariate regression analysis (see Johnston 1978).

Regression is the workhorse of statistical explanation. Learning how it works is a tremendous aid in thinking causally about the world: about how things vary and why, and about how we can be fooled into making incorrect conclusions if we don't pay sufficient attention to our data and to the assumptions we make in our analyses.

## Thinking Causality Statistically

A major purpose of statistics is to describe the strength and nature of the relationship between two variables. This is done with the purpose of providing evidence that one variable "causes" the other. The causal variable is considered the **independent** variable (X) and the response variable is the **dependent** variable (Y). The independent variable is the one whose values are given, or can be controlled in an experimental situation. The dependent variable is the one hypothesized to vary with, or to respond to, the independent variable.

For scientific analysis of this sort to be successful, there is a requirement that X has some independent effect on Y – independent in terms of other potentially causal variables. Often this is not the case. Many times different X variables share similar patterns of variation, and sometimes they are so similar that it is difficult if not impossible to sort out their independent effects on Y. This is an extraordinarily important issue in validating cause. If $X_1$ and $X_2$ share half of their total variation in common, how can we sort out which of the Xs is causing Y? This is known as the **multicollinearity** problem; we return to it below. Equally important, the values of the dependent variable should, theoretically speaking, not influence the values of the independent variable. In practice, especially in human geography, this may not be the case (see Chapter 2). If the relationship between X and Y is recursive (i.e., $X \leftrightarrow Y$), then how much of what we uncover is X causing Y, and how much is Y causing X?

The complications of independence can be obviated in some disciplines through the experimental method. This method pivots on the idea of independent effects, how they are measured, and the hazards of assigning the "cause" label to an X variable even if it is clearly an independent effect. Most importantly, a controlled experiment is the best way to remove the effects of other confounding variables that confuse our assessment of the relationship between X and Y. For example, to determine whether chemical X causes cancer, two groups of rats can be randomly selected from a single pool of genetically identical rats. One group is the control group and the other is the experimental group. The only difference between the two groups is that every rat in the experimental group will get X on a regular basis, and none of the rats in the control group get it. If after one year, 20 out of 100 rats in the experimental group have contracted cancer while only 4 out of 100 rats in the control group contracted it, then the X in the diet must have caused it because that was the only difference that existed between the two groups. If we make X vary from small to medium to high doses, and the progress of cancer responds to these different amounts, then we have even more evidence for the relationship between X and Y. Varying the dosage rate might also suggest to us different threshold amounts of the chemical, beyond which cancer begins to emerge.

Another possibility is that X is carcinogenic only when it interacts with another substance, which may have been present in the laboratory situation but is not normally present in the real world. When two substances interact to generate an effect, which neither can produce by themselves, this is called synergy. It is more usual that X does not cause cancer in the experimental situation, but in the real world it does, because of synergy with some unknown substance that it is more likely to meet. Positive synergy is when the combined effect is multiplicative, rather than the effect being that of adding the two independent effects. Negative synergy is when the presence of two factors negates the effect of one or both, or the combined effect of the two factors is less than the sum of their independent effects. Positive and negative synergies are concepts that apply to effects measured as ratio variables, and each variable is having an effect on the response variable. But, science teaches us that there can also be a catalyst variable. Something may be present in the experimental situation that does not induce an effect on Y by itself, but it does induce an interaction between X and Y when present, and there is no interaction or relationship between X and Y when it absent. Synergy and catalyst variables are a huge problem in establishing cause and effect, especially in medical research, because all combinations of causative factors must be tried to reveal it, and this is monetarily not practical (see Table 18.1).

Further complicating the issue is the fact that if X does cause cancer in rats, this does not mean it will cause cancer in humans. It probably will because rats are not that genetically different from humans, but while most factors that cause cancer in rats probably do so in humans, not all will, because of our genetic difference with rats, and because the environments we live in (and the synergy and catalyst factors we experience) are also somewhat different. Note that this problem is not one of false analogy. False analogies are logically flawed, and are a separate problem in measurement and interpretation.

Finally, the relationship between X and cancer may simply be coincidence. A major task for statistics in the above study is to decide if the difference between 4 out of 100 (4%) and 20 out of 100 (20%) is statistically significant. If neither group of rats had received X,

**Table 18.1** Types of causal relations

| Type of relation | Graphic representation |
|---|---|
| Simple causal relation with independence | $X \rightarrow Y$ |
| Multivariate causal relation with independence | $X_2$ <br> $\searrow$ <br> $X_1 \rightarrow Y$ |
| Recursive relations | $X \leftrightarrow Y$ |
| Multicollinearity | $X_1 \leftrightarrow X_2$ <br> $\searrow \swarrow$ <br> $Y$ |
| Multiplicative synergy | $X_2 \cdot X_1 \rightarrow Y$ |
| Catalyst effect | $X_2$ <br> $\downarrow$ <br> $X_1 \rightarrow Y$ |

what is the chance that there were 4 incidences of cancer in one group and 20 in the other; that is, just by luck? This chance or probability can be calculated, and in fact its value is 0.0004 or 0.04%. Thus the difference could have happened by chance, but it is extremely unlikely. The researchers would therefore conclude that the differences in the occurrence of cancer were because of X, and that if the experiment were run on one million rats, there would be a higher rate of cancer in those that received X than in those that did not. This is statistical inference – the work of inferring from a sample of observations to the population of observations.

A population of observations is all possible observations relevant to a study (Chapter 6). In the above instance it is all rats at all times. One requirement of a random sample is that the chosen observations must be independent of one another (a third issue with respect to the word independence). If that is not true, the independent effect of X on Y cannot be identified. In the described experiment this requirement has actually been flouted, because the rats are genetically identical, and therefore not representative of all rats. So these rats may lack a gene that is needed to acquire cancer as a result of exposure to X, or they may have a gene, which most lack, that causes cancer as a result of exposure to X. So our ability to infer to the population is partly the result of the independence of the observations.

These issues are more easily controlled for in experimental situations, but are less so in social contexts. When households in a neighborhood are sampled for a study, the people interviewed are not independent; they share a "neighborhood effect" (e.g., similar class, ethnic, and national origin backgrounds). When surveys are administered over land lines, you miss those people who only carry cell phones. When counties in a state are the units of observation, the ones closer to one another will tend to be more alike, and therefore not independent. The same can be true of weather stations, river gauges, and the pixels that record vegetative cover through a remote sensing device. Statistics was invented for a world of independence; but outside of experimental situations, a great number of our observations are interdependent upon one another. This problem is called autocorrelation; if the units involved are dependent because they share a common geography (e.g., common boundary, propinquity, spatial connection), then the problem is called **spatial autocorrelation**.

## Experiments on Humans?

What if we turn to the laboratory to design an experiment for humans that both identifies the independent effect of a variable and removes confounding variables? One problem we might have is ethical: there are some experiments scientists will conduct on animals that they won't sanction for human populations (note that many animal rights activists and ethicists are against animal experimentation as well). When researchers do conduct experiments on humans, say to see what substances or treatments may cure disease, one group usually receives a treatment (stimulus) and another group doesn't. If the treatment is effective, members of the group that didn't receive it may die or die sooner than the treatment group. This sounds terrible, but it is the only way to really determine if a treatment is effective. Many people have died by not getting the right treatment, and many have received damaging treatments that were of no benefit.

Another problem in human experiments is that there are no identical individuals to place into groups. So people are randomly selected into two groups. The individuals in each group are not identical to each other, but hopefully the variation in attributes of individuals within the groups is the same. In this case independence among observations is controlled by random sampling.

A remaining issue, however, may be the contamination of experiments because the groups think they know they are, or are not, receiving the treatment. This is the placebo effect. In a double-blind experiment, the staff that administers a drug to patients does not know whether an individual is actually getting the drug or the placebo. Otherwise they may affect the outcome. They may feel sorry for those who are getting the placebo and somehow communicate a sense of "no hope," which depresses the patient and lowers their chance of improvement. In a triple-blind experiment the researchers who measure whether the patients have "improved" or are "cured" also do not know who got the drug and who got the placebo. Otherwise, their interpretation of improvement may be biased.

As you can see, even in the ideal research setting of experiments, the measurement of an independent relationship between two variables is actually fraught with error, and open to misinterpretation. And although scientists are well aware of the influence of confounding variables on the independent relationship between two variables, they sometimes fail to question whether the relationship depends entirely on context (i.e., the presence of a catalyst).

For most geographic and social science research problems the experimental method cannot be used. One group of counties cannot be asked to try things one way while another group of similar counties tries things some other way. To identify an independent relationship between two variables, therefore, requires that we attempt to control the variation in a large number of variables so that we can focus in on the variable whose causal efficacy is of primary interest. Analytically, this can present great difficulties; there are simply very few natural experiments out there in the "real world."

## Simple Linear Regression and Correlation

Regression describes the co-variation of two variables measured on an interval/ratio scale. A simple regression means there is only one independent variable. Correlation describes the closeness or strength of that co-variation (Chapter 17). If the points in the scatter diagram appear to follow a straight line, then such a line that describes the average relationship can be fitted to the points. Linear regression is finding the equation of the line that best fits the scatter.

Figure 18.1 shows a weak positive relationship between the Democratic party vote in 1980 (Y, or DEMVOTE) among selected Texas counties and the percentage of the county population below the poverty line (X, or PERPOV), with the regression line plotted on the scatter (Chapter 17 and Tebben 1990). The equation for a regression line has the form:

$$\hat{Y}_i = a + bX_i \tag{18.1}$$

$\hat{Y}$ is the predicted value given the value of X. The subscript "i" refers to the $i^{th}$ observation of X and Y. The parameters "a" and "b" are called the constant and the regression

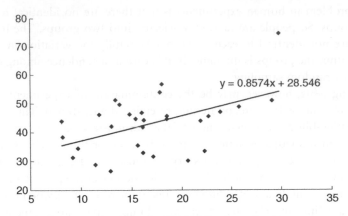

**Figure 18.1**   DEMVOTE (vertical axis) and PERPOV (horizontal axis)

coefficient, respectively. The constant is the predicted value of Y where X is zero; it is also called the intercept and can be read off the Y axis at that point. The regression coefficient is also called the slope, because it is the slope of the line. It also is a measure of the predicted change in Y if the value of X is increased by one unit. The regression coefficient would be negative if Y decreases as X increases, describing an inverse relationship.

The regression line for the 30 county set of observations is:

$$DEMVOTE = 28.55 + 0.86 \cdot PERPOV \qquad (18.2)$$

For all of the state's 254 counties, the equation is:

$$DEMVOTE = 29.71 + 0.81 \cdot PERPOV \qquad (18.3)$$

The latter equation indicates that, on average, a county that has a PERPOV value one unit higher than another county should have a vote for the Democrat that is 0.81 units (percent) higher. The inference is that counties with more poor people are more likely to vote Democrat. PERPOV is measured in percent, as is DEMVOTE, so in this case an increase of one percent in PERPOV increases DEMVOTE by 0.81 percentage points.

To estimate the Y value for a particular observation given its X value, the X value is simply plugged into the regression equation. The vertical or Y-axis difference between an observation's actual value on Y and the value predicted from the regression equation is the prediction error for that observation. For example, one county has a value for PERPOV of 20.6 percent. Putting this value into the regression equation for the 30 counties results in a DEMVOTE prediction of 46.27 (28.55 + 0.86(20.6)). The actual value of DEMVOTE for that county was 30.1 percent. The prediction error is written as $e_i = Y_i - \hat{Y}_i$, and was $30.1 - 46.27 = -16.17$ for the example observation. Observations below the regression line will have negative errors (actual less than predicted), and those above the regression line will have positive errors (actual more than predicted). The prediction error is usually called a residual.

If the prediction errors of all the observations are squared, and then the squared deviations are summed, the total is called the error sum of squares or residual sum of squares (RSS).

$$RSS = \sum e_i^2 = \sum \left(Y_i - \hat{Y}_i\right)^2 \tag{18.4}$$

The computer calculates values for the regression coefficient (b) and constant (a) that minimizes the error or residual sum of squares. The regression equation is therefore called the least-squares linear regression equation, and this equation is also considered the best-fit line. The regression line that results is unique. There is only one possible value for "a" and only one possible value for "b" for which the sum is minimized. Visually, if you were to imagine increasing or decreasing the slope of the line in Figure 18.1, you should be able to see that it would increase overall squared error; likewise if you were to raise or lower the line (increasing or decreasing the intercept), the error will also appear to increase. Thus, the sum of the squared errors around the line is less than the sum of the squared errors around any other line.

Notice how the linear regression line tracks Y for all observations that have a given value of X. So the regression line is really a moving average of Y that is a linear function of the X variable.

The regression errors are equivalent to deviations around this moving mean. Computationally, the errors around the regression line are handled in the same manner as variation in a variable around a mean (Chapter 17). The errors are squared, summed across all observations, and averaged, just as in the calculation of variance. Division of the residual sum of squares by the number of observations provides the regression variance. The best estimate of the population regression variance, however, requires that the residual sum of squares be divided by $n - 2$. The standard deviation of the residuals is therefore calculated by taking the square root of the regression variance. But this term is not called a standard deviation. It is called the standard error of the estimate or just the standard error. Its symbols are $s_e$ for sample data, and $\sigma_e$ for the population.

$$s_e = \sqrt{\frac{\sum \left(Y_i - \hat{Y}_i\right)^2}{n-2}} \text{ and } \sigma_e = \sqrt{\frac{\sum \left(Y_i - \hat{Y}_i\right)^2}{N}} \tag{18.5}$$

The standard error of the estimate for the regression of DEMVOTE against PERPOV is 10.14. The standard deviation of DEMVOTE is 11.72. PERPOV has therefore "explained" some of the variation in the DEMVOTE variable. To find out how much statistical explanation has occurred, we go back to the total variance in the variable. The difference between a Y value and its mean $(Y_i - \bar{Y})$ can be viewed as being composed of two parts: the variation that can be explained (or is "caused") by the X value $\hat{Y}_i - \bar{Y}$, and the error $Y_i - \hat{Y}_i$, which is that portion that X cannot account for. The residual or error sum of squares (RSS) was explained above. The total variation or total sum of squares was introduced in Chapter 17. Here we write it as:

$$TSS = \sum (Y_i - \bar{Y})^2 \tag{18.6}$$

The explained sum of squares is ESS $= \Sigma\,(\hat{Y}_i - \overline{Y})^2$. It has been proven that the explained sum of squares plus the residual (or unexplained) sum of squares equals the total sum of squares. If the explained sum of squares is then divided by the total sum of squares, the result is the proportion of the total variation in the Y values that is "explained" by the X variable. This result is the **coefficient of determination**, or $r^2$: the proportion of the total variation of Y explained by the linear relationship with the independent variable.

Since $r^2$ is a proportion, its values vary between 0 and 1. A value of 1 means the explained sum of squares equals the total sum of squares; so all the scatter diagram points would be on the straight line, and there would be no errors. An $r^2$ of zero means that the unexplained sum of squares equals the total sum of squares, and the regression line will be horizontal and equal to the mean of Y, or $\overline{Y}$. Also, the regression coefficient will be zero, and the constant equal to the mean of Y.

The square root of $r^2$ is $\pm r$, which in the simple regression case is the same as Pearson's correlation coefficient, which was described in Chapter 17. The $r^2$ for DEMVOTE regressed against PERPOV is .255 for all 254 counties. The Pearson's is $+.505$. Pearson's always has a larger absolute value than $r^2$, and so reporting it alone makes the relationship look stronger than it really is, in addition to the fact that it describes the strength and direction of a linear relationship with a single number. If r is .20 the $r^2$ is .04, which indicates virtually no linear relationship, but the r doesn't look that bad. Research that only reports the correlation coefficient must be viewed with skepticism. Is a linear relationship the most appropriate? Does the relationship exist because of one or two observation values? Do the errors have a normal distribution?

There is meaning to $r^2$. An $r^2$ of .8 is twice as good as a .4. But this is not true of r. R is only an index of the closeness of a linear relationship and .5 is better than .25 but not twice as good. If there is no relationship between X and Y then $r = r^2 = b = 0$. But, if $r = r^2 = b = 0$, this does not mean that there is no relationship between X and Y; it only means there is no linear relationship between them.

In the case of bivariate (simple) regressions, the correlation coefficient is also the standardized regression coefficient (beta). If the X value is increased by 1 standard deviation then the Y value on average increases by .505 standard deviations of Y (using r = .505). The regression coefficient tells us how much DEMVOTE (in its unit of measurement) changes if PERPOV is increased by one of its units of measurement (both percentages in this case).

## Multivariate Linear Models

An interval/ratio scale variable can be related to many independent variables simultaneously by expressing it as a linear function of several variables. The form of the equation is:

$$\hat{Y} = b_0 + b_1 X_1 + b_2 X_2 + --- + b_n X_n \tag{18.7}$$

The Xs are the different independent variables measured on an interval/ratio scale, although they may be dichotomous nominal variables (**dummy variables**). These are variables that have only zero and one as possible values. A value of one means that a characteristic is present, and a zero means that it is not. The constant is $b_0$, which is the predicted value of

Y when all the X variables equal zero, an improbable event. The b values are the partial regression coefficients describing the effect of the respective X variable on Y. The coefficients for the dummy variables indicate how different the means of Y are for the two groups that make up the dummy variables. For example, if yield in bushels is a dependent variable and the dummy variable is Entisols = 1, and not Entisols = 0, and the regression coefficient is 10, then that indicates that average yields are 10 bushels higher on Entisol soils versus other soils, all other things being equal.

The word "partial" in partial regression coefficient probably derived from "partial derivative" in calculus, because it describes the independent effect of any given X on Y; that is, it is the effect of $X_1$ on Y when $X_2$ and all the other Xs in the equation have been controlled for. In turn, that means the b values associated with the Xs provide the influence of an X variable on Y when none of the other X variables vary. For example, assume an equation with two independent variables:

$$\hat{Y} = b_0 + b_1 X_1 + b_2 X_2 \tag{18.8}$$

In principle, the computer regresses Y and $X_1$ against $X_2$, and calculates the errors or residuals of each, and then regresses these residuals against each other. So the effect of the variation of $X_2$ on Y and on $X_1$ is removed, and therefore, the relationship between Y and $X_1$ is independent of the variation of $X_2$. Similarly the regression coefficient for $X_2$ describes the effect of $X_2$ on Y without any variation in $X_1$. The mathematics for calculating the regression coefficients are not actually sequential as described, but they are equivalent. The technique is called ordinary least-squares (OLS) multiple regression.

Just as in simple regression, the sum of the squared errors in multiple regression is minimized. If a simple regression with one independent variable is said to minimize the sum of squares of all the points around a line that goes through them, then a multiple regression with two independent variables minimizes the sum of the squared deviations of all the points around a plane (visualize the plane as floating in three dimensions with Y on the vertical axis, $X_1$ and $X_2$ on the horizontal axes, and the data points situated above and below). The plane has a height measured by the multiple regression equation intercept, and partial correlation coefficients can be calculated for each independent variable; these are also the slope of the plane in the dimension of the variable in question. Finally, the results of an OLS multiple regression include a multiple coefficient of determination, $R^2$, which is the proportion of the total sum of squares explained by all of the independent variables together.

## Causality Issues in Multiple Regression

Multiple regression is a powerful technique. But it cannot solve problems inherent in either data or careful thinking about causality. To illustrate, let's return to the problem of multicollinearity, briefly raised at the beginning of this chapter.

Recall the data reproduced in Table 4.2 (Chapter 4), taken from Visser (1979; 1980). The problem was one of determining what causes variation in agricultural intensity. Intensity is the capital and labor expenditures per acre of farmland. According to traditional agricultural theory, counties that are more distant from the market for agricultural produce should have lower intensities because they must spend more on transportation costs and less on growing costs. A more significant influence on intensity, however, is fertility, which may be measured as average annual rainfall, length of growing season, and type of soil. As intensity is increased the increases in yield are greater on more fertile land. The effect of fertility on intensity must therefore be controlled to see if there is an independent effect of distance to market on intensity.

But complicating matters is the fact that the counties that are farthest from the market also have the lowest rainfall on average. There is a close inverse relationship between the two variables. Both lower rainfall and higher transport costs are likely to reduce intensity, but what are the relative independent impacts of the two variables? Distance and rainfall are very closely correlated, such that there is little variation in distance and intensity that is independent of rainfall. Because of the high degree of multicollinearity, multiple regression cannot tell us if it is the rainfall or the distance that is influencing intensity. The variables co-vary so closely that their independent effects simply could not be separated, especially over the small geographic scale of the study. For all intents and purposes, they are one and the same variable.

Another problem can arise if relevant independent variables are left out of the analysis. In this case, the partial regression coefficients may not be true, in the sense that the independent effect of an X variable on the Y variable has not been discovered. This is the specification (or **misspecification**) problem, of which missing variables are a part. Missing variables are a very important and overlooked problem in trying to establish causality between X and Y, for the introduction of a variable previously left out of the analysis may cause the relationship between a given X and Y to shift dramatically. For example, in a spurious correlation an omitted variable may have a common correlation with both X and Y, and more importantly may be the cause for the common variation in both variables. The addition of the new, previously missing independent variable may cause a relationship between Y and one of the X variables to disappear as it now accounts for the variation in both. In other cases where there is no correlation between two variables, a correlation may appear if a new variable is added.

## Assumptions

Given a particular value of a population parameter (either known or hypothesized), inferential statistics provide the probability that a sample statistic, calculated for a random sample drawn from this population, will be greater than or less than some value. These statistics don't prove or disprove or explain anything: they only indicate how likely you are to get a range of sample statistic values of that magnitude if the population parameter has a certain value. But at the same time, these tests of the probability that an estimated statistic, such as a regression coefficient, is of a certain value (say $\neq 0$) are extremely important in assessing the strength of relationships, and that's why we are interested in them. If the probability is extremely low that a calculated statistic of a given size could have occurred just by the chance sampling from a population without a relationship, then we are fairly confident in saying that a relationship exists.

To make statistical inferences in regression, however, requires that certain assumptions be addressed. But, frequently researchers do not check to see whether the assumptions about the nature of the data are in fact met. Certainly, they rarely report whether the assumptions have been satisfied, and it is wrong to assume that they are, because the social world usually does not provide variables with symmetric distributions. As we will see, there should be no exceptionally large or exceptionally small values of a variable, and usually there are. Many of the relationships reported in the social science literature, therefore, may be viewed with skepticism. Specifically, statistical significance tests for linear regressions have several assumptions. If these assumptions are violated, the statistical tests associated with them are invalid. These assumptions include:

1   the regression residuals ($e_i$ or $Y_i - \hat{Y}_i$) must have an approximately normal distribution;
2   there must be no outliers;
3   the variance of the residuals should be relatively constant for all values of the independent variable;
4   the model must be properly specified as linear;
5   the residuals must be independent of one another (no autocorrelation).

If both or one of the variables has a skewed distribution these assumptions will tend to be violated. **Homoscedasticity** is the condition where the variance of the Y-values tends to be the same for all values of X. Often, in a positive relationship between two variables, the higher Y-values tend to vary more at higher X-values. This is called heteroscedasticity. The scatter diagram of the residuals against the X values will tend to have a megaphone or funnel shape. Since linear regression minimizes the sum of the squared regression errors, the few higher values of Y and X will have a greater influence than the smaller variable values on where the regression line lies. That is, the location of the regression line becomes more dependent on just a few of the observations, and so less reliable, invalidating the significance test.

An outlier is when the residual for an observation is exceptionally large. It may be more than three standard errors from the regression line, and because regression minimizes the sum of the squared errors around the line, just one or two observations may have a major influence on where the line is located. Figures 18.2 and 18.3 illustrate how one observation may negate a strong linear relationship between two variables. Figure 18.2 shows a close relationship between X and Y. The correlation coefficient is very high at .905 and the

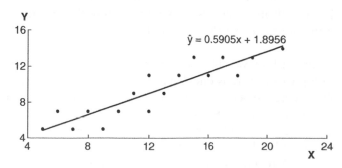

**Figure 18.2**   A good relationship

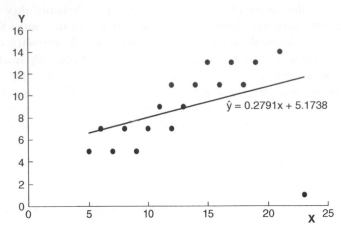

**Figure 18.3**    The effect of an outlier

regression analysis shows statistical significance (the details of these tests are not discussed here; see Rogerson 2006). In Figure 18.3, the same data have had an outlying observation (X = 23, Y = 1) added to it. This single observation has changed the location of the line dramatically, and the correlation coefficient has been reduced to .407. Based on the statistical tests performed (not reported here), we would not be able to conclude that the sample was drawn from a population in which there was a linear relationship between the variables. But, clearly there is, except for the outlier. One observation has completely negated a very nice relationship.

When X is 23, the equation $\hat{Y}_i = 5.174 + .279X$ predicts that the Y value should be 11.591. The actual value of Y is 1; so it is 3.15 standard errors from the regression line ($s_e$ was 3.36 in the second regression analysis). A value that large is by definition an outlier. A plot of the errors – standard fare on most statistical packages – would reveal that they do not have a normal distribution, and so the statistical significance test is not valid. Of interest is the fact that this outlier is not the result of a severe skew in either the X or Y variable; it results from their particular combination (large X, very small Y). This result points to the importance of looking at a frequency distribution of residuals to determine whether the regression assumption is violated.

Outliers must not be discarded simply because they are outliers. That is falsifying the data. What to do about them, though, is beyond the aegis of this chapter. The example is presented to simply emphasize that checking on the assumptions of a statistical test really matters.

Severe skew in one or both variables can create relationships where there are none, or negate a relationship where there is one. Figure 18.4 illustrates the former case. In it a strong relationship is detected because of one observation, where there should be no relationship. Both variables in Figure 18.4 have a severe right skew, and it happens that the same observation (X = 23, Y = 17) has an exceptionally large value for both variables. This observation is not an outlier, because it is quite close to the regression line (.783 standard errors below the line). A frequency plot of the residuals would not reveal a non-normal distribution. But a frequency distribution of each variable (as discussed in Chapter 17)

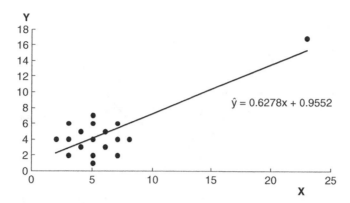

**Figure 18.4** The problem of skew

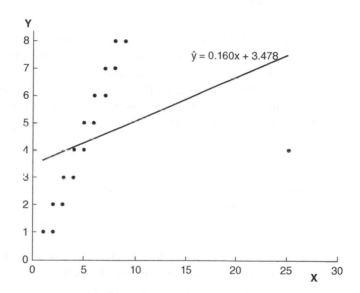

**Figure 18.5** Another case of skewness

would show the skew. When the single observation is excluded, the $r^2$ is 0. When it is included the value is .675. If the regression assumptions are not checked you would conclude that you've discovered a strong relationship.

The next example (Figure 18.5) illustrates how severe skew in the X variable may negate an otherwise strong linear relationship between two variables. Again, the example created a problem with just one observation, and again, the problem observation might not be revealed as an outlier. Although the illustration shows a right skew in the X variable, it can be imagined how a severe left skew in X could also negate a strong linear relationship between two variables. The relationship in Figure 18.5 has an $r^2$ of .142, and the regression analysis is not significant. If the observation with X = 25 were absent, the $r^2$ climbs to .955, which is of course highly significant. A frequency plot of the residuals is needed to detect this sort of effect, although here it is clearly evident on the original scatter diagram.

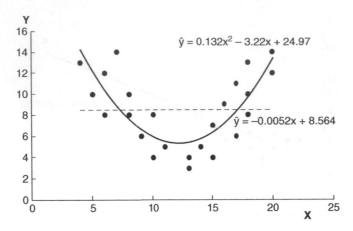

**Figure 18.6**    Specifying the relationship

Scatter diagrams are also helpful in detecting non-linear relationships; one can plot Y and X individually or each residual can be plotted against each X in a multiple regression. Non-linear relationships are common in the natural and social sciences, so these are important exercises. The effect is shown in Figure 18.6. Here the linear regression (shown as the dashed line) shows no relationship. There is, however, an excellent curvilinear relationship that is described by a quadratic equation.

$$\hat{Y} = b_0 + b_1 X_1 + b_2 X_1^2 \tag{18.9}$$

The quadratic relationship has an $r^2$ of .627. The regression coefficients for both X and $X^2$ are significant. Failure to plot the data would result in the conclusion that there was no relationship between the two variables, when in fact there was just no linear relationship.

There are various remedies for massaging the data when it does not satisfy the assumptions needed for testing the significance of linear relationships. Many variables in the social sciences have right skew. Often taking the logarithm of the variable, and then regressing the logarithmic values against another variable can remove the right skew. This is called transforming the variable, and is quite acceptable, although it will change the nature of the relationship. For example, regressing the natural logarithm of Y against the natural logarithm of X generates the linear equation:

$$\ln(Y) = \ln(a) + b(\ln(X)) \tag{18.10}$$

This actually describes a curvilinear relationship between the two original variables of the form

$$Y = aX^b \tag{18.11}$$

Finally, residuals are supposed to be independent of one another (not autocorrelated). Given the variability (i.e., the information content) that these measures contain as $e_i$ values (i.e., $Y_i - \hat{Y}_i$), it becomes clear how this assumption can easily be violated with spatial data. Specifically, $e_i$ values represent the original data ($Y_i$) "purged" from (or accounting for) the explanatory effects of all the Xs in the regression model ($\hat{Y}_i$). Hence, they can only contain four kinds of information: (a) unaccounted-for random variability that existed in the original Y variable; (b) information from misspecifications present in the model (e.g., non-linear effects such as those shown in Figure 18.6); (c) missing variable information in $Y_i$ not picked up by the model; or (d) some sort of spatial interdependence in the observations $Y_i$ that found their way onto $e_i$ because they were not removed from the model. This latter dependence, introduced earlier as spatial autocorrelation, is a violation of the regression assumptions. The easiest way to detect it is to map the residuals from a regression analysis. If you find that positive (above the line) residuals cluster geographically, while negative ones are also spatially associated, then there is evidence of spatial autocorrelation. Autocorrelated residuals can result from spatial patterns of missing variables or from genuine spillover effects caused by the fact that spatial processes do not obey the boundaries of observational units. In either case the mapped residuals contain interesting informa-

**Box 18.1   Spatially Varying Parameters**

Another issue to contend with in regression analyses is the stability of relationships across different spatial units. Imagine that you run a regression for the over 3,000 counties in the United States. Then you divide those counties into three regions (e.g., North, South, and West) and re-run the regression. Would you anticipate getting the same result? The first analysis offers a national portrait of the relationships among the variables. The partial regression coefficients describe what's going on, for that X, for the entire country. But what if that portrait is simply a generalized view of parameters that in fact vary regionally? Wouldn't you want to know if the relationships among your variables differed across the spatial units in your data set? That would certainly be in keeping with the idea that geography matters in the explanation of processes.

There are techniques developed to assess what has been called "spatial parameter variation" (Jones 1984). One of them builds on the "expansion method," an approach invented by the geographer Emilio Casetti (1972) that incorporates sub-models for regression parameters within regression models. In the expansion approach developed by Jones (1984), the parameters are redefined as a function of trend surface coordinates (Cartesian north-south soundings in $p$, $q$, $p^2$, $q^2$, etc.). These provide regional scale estimates of the variability in parameters. Another approach to parameter variation developed by Casetti is termed "Drift Analysis of Regression Parameters" (DARP) (Casetti and Jones 1983; Casetti and Can 1999). In it, a central point is selected and the spatial units in the regression are weighted in accordance to their distance from the point. Observations that are farther away from the pivot point are discounted, and the result is a local regression. Fotheringham et al. (2000) have extended and popularized DARP under the name of "geographically weighted regression."

tion that can bear on causality. Geographers have been at the forefront in addressing spatial autocorrelation, typically by trying to identify and remove the effect through advanced statistical procedures (Anselin et al. 2004).

## Conclusion

This chapter has endeavored to provide you with an understanding of the problems of measuring independent effects or causes of one variable on another. Much has been left out, particularly the details of hypothesis testing, probability, and significance. In these topics, the point is to determine whether you can infer that a relationship between two variables found in a random sample of data is reflective of a relationship in the population of observations from which that sample was drawn. The powerful parametric tests, which enable you to make these inferences, have several assumptions that must be checked. Non-parametric tests have no assumptions about the distributions of the variables, but they are not as powerful, and they do not incorporate all of the information that may be in the data.

The natural sciences unabashedly think of the social sciences as the soft sciences. The causes of individual or group behaviors are extraordinarily complex, and the relationships that provide evidence of cause are often not directly measurable, or other influences on a dependent variable cannot be controlled via the experimental method. Even in the experimental method it is easy to contaminate the results, and we cannot know whether a relationship found in a laboratory experiment will be stable outside the laboratory. In non-experimental research correlations may be spurious or coincidental. A hypothesized relationship may not be found because the researcher did not control for a variable that suppresses the independent relationship between the two variables of interest.

And even when the majority is satisfied that a relationship does exist, that it is causative, and that it is enduring, we find that it isn't. Most social relationships are not stable with respect to time and other contexts, such as space or culture (Box 18.1). Sometimes these contexts can be measured as variables, but often they cannot. Economists use the term *ceteris paribus*, meaning all other things being equal, to denote that they are talking about the independent effect of one variable on another. Unfortunately it is not a *ceteris paribus* world. Their relationships have not allowed for all other influences, they are not deterministic, and they are not true of all times (they are not stable), cultures, and other contexts. Thus, reliance on these relationships leads to predictions of future economic events, which are actually less accurate than random guesses. Statistical inference makes no statements about causality. It just indicates that a relationship probably exists. Whether your independent variable actually causes the dependent variable to vary is another issue.

## Acknowledgment

This chapter is dedicated to Adriana, 1976–88.

# References

Anselin, L., Florax, R., and Rey, S. (2004) *Advances in Spatial Econometrics*. Berlin: Springer.

Casetti, E. (1972) Generating models by the expansion method: applications to geographical research. *Geographical Analysis* 4: 81–91.

Casetti, E., and Can, A. (1999) The economic estimation and testing of DARP models. *Geographical Systems* 1: 91–106.

Casetti, E., and Jones III, J. P. (1983) Regional shifts in the manufacturing productivity response to output growth: sunbelt vs. snowbelt. *Urban Geography* 4: 285–301.

Fotheringham, S., Brunsdon, C., and Charlton, M. (2000) *Quantitative Geography: Perspectives on Spatial Analysis*. Thousand Oaks, CA: Sage.

Johnston, R. J. (1978) *Multivariate Statistical Analysis in Geography*. London: Longman.

Jones III, J. P. (1984) A spatially-varying parameter model of AFDC participation: empirical analysis using the expansion method. *The Professional Geographer* 36: 455–61.

Rogerson, P. (2006) *Statistical Methods for Geography: A Student's Guide* (2nd edn). Thousand Oaks, CA: Sage.

Tebben, E. (1990) *Get the Vote Out*. Unpublished Masters paper, Southwest Texas State University, San Marcos.

Visser, S. (1979) *The Spatial Dynamics of Agricultural Intensity*. Unpublished PhD dissertation, The Ohio State University, Columbus.

Visser, S. (1980) Technological change and the spatial structure of agriculture. *Economic Geography* 56: 311–19.

# Additional Resources

Field, A. (2009) *Discovering Statistics with SPSS* (3rd edn). London: Sage. While you can find free software for all sorts of statistical research, there is still nothing easier than SPSS for the personal computer. This book is introductory and takes you step by step through the program's many capabilities. It's nice to have a statistics book and a computer manual in the same text.

Griffith, D. A., and Amrhein, C. G. (1997) *Multivariate Statistical Analysis for Geographers*. Upper Saddle River, NJ: Prentice Hall. Excellent resource for some of the higher order methods, including cluster analysis, canonical analysis, principal components and factor analysis.

Hoskings, P. L., and Clark, W. A. V. (1986) *Statistical Methods for Geographers*. New York: Wiley. For many years the standard text in upper division undergraduate and beginning graduate courses in geography and statistics. Covers everything from basics in research design to multivariate regression and extensions.

## Exercise 18.1 Assessing Multiple Regression Analysis

Take an hour or so to peruse the back pages of some of the more empirically oriented journals in your favorite area of geography. Look for examples of multiple regression in practice. Some of the journals you might look through include: *Applied Geography, Area, Economic Geography, Geografiska Annaler (Series A Physical Geography), Geomorphology, International Journal of Population Geography, Journal of Applied Meteorology and Climatology, Journal of Transport Geography, Progress in Physical Geography, Social Science and Medicine, Urban Geography,* and *The Professional Geographer*. Not every issue will have an example of a paper using multiple regression, but it is a common enough technique that you eventually find an article that uses the technique. Once you do, read it and answer the following questions:

1 What are the dependent and independent variables used in the analysis?
2 Did the selection of variables tend to follow a strongly grounded theoretical rationale, or was the objective more exploratory?
3 What specific causal effects was the author(s) searching for?
4 Is there evidence that the author(s) gave careful attention to the quality of the data?
5 How many models appear to have been estimated in the process of the research?
6 What assumptions about regression were discussed? Is it clear that tests were performed to ensure that the assumptions were not violated?
7 What were the major findings? Were there any surprises emerging from the analyses?

# Chapter 19
# Mathematical Analysis

*Sandra Lach Arlinghaus*

- Introduction
- Euclidean Geometry and Visualization
- Arithmetic, Algebra, and Data
- A View of the Future
- Conclusion

## Keywords

Antipodal points
Assignment algorithm for simple closed
   curves
Composite number
Distributive Law
Equator
Eratosthenes's measurement of the Earth
Euclid's Parallel Postulate
Factorization
Four color theorem
Fundamental Theorem of Arithmetic
Geodesic
Great circle
Jordan Curve Theorem

Large scale map
Latitude
Longitude
Meridian
Non-Euclidean geometry
Parallel
Prime Meridian
Prime number
Representative fraction
Small circle
Stereographic projection
Transformation

# Introduction

Geographers often view mathematical analysis as a collection of tools that can be used to solve a problem or generate some result. While it is true that mathematics, when used carefully, does have that capability, it is also far more than a mere toolkit. Mathematics is a rich theoretical subject based on the power of logic; it is an art as well as a science. To make good choices about how to use mathematics it is important to have some understanding of its broad conceptual structure. Such understanding, however, can be quite difficult and time-consuming to gain because, unlike geography, mathematics curricula are quite linear in character and prerequisites have to be fulfilled before it is possible to move to higher levels, making it difficult to sample the subject matter. For this reason, the mathematical ideas introduced in this chapter are either drawn from those areas of mathematics in which a linear buildup of prerequisites is not an absolute necessity, or are drawn from those areas that the reader should already be familiar with (the material is not difficult but it may require careful reading to digest).

Both mathematics and geography can be visual in nature. Geometry is a branch of mathematics that is visual, and geometric analysis can be used to examine some visual components of geography. Euclidean geometry, for example, is employed to coordinatize Earth using **latitude** and **longitude** (Coxeter 1961; Loeb 1976). Eratosthenes (276–194 BC), the chief librarian of the great library of Alexandria in ancient Egypt, employed the mathematics of Euclid to solve an important geographic problem: how big is Earth? Starting with the work of Eratosthenes and trigonometry, this chapter moves from a consideration of analysis using geometrical figures to analysis using numerical figures. We shall also examine Eratosthenes' prime number sieve (an algorithm for finding prime numbers up to a specified integer) and its implications for doing calculations of various sorts. In so doing the importance of both mental agility and accuracy in performing calculations in geographic research is underscored. The topics we shall cover include: mathematical **transformations** and the creation of maps; the **Four Color Theorem** and the coloring of maps; and the **Jordan Curve Theorem** and the problem of assignment.

## Euclidean Geometry and Visualization

Consider Earth as a sphere. While Earth is not actually a sphere, a sphere is a good approximation to its shape and is amenable to the classical mathematics of Euclid.

### Latitude and longitude

Given a sphere and a plane, there are only a few logical possibilities about their relationship to each other:

- The sphere and the plane do not intersect.
- The plane is tangent to the sphere.
- The plane intersects the sphere. If this, then:

   – The plane does not pass through the center of the sphere (the circle of intersection is called a **small circle**).

   – The plane does pass through the center of the sphere (the circle of intersection is as large as possible and is called a **great circle**).

Great circles are the lines along which distance is measured on a sphere:

- In the plane, the shortest distance between two points is measured along a line segment and is unique.
- On the sphere, the shortest distance between two points is measured along an arc of a great circle.
  - If the two points are not at opposite ends of a diameter of the sphere, then the shortest distance is unique.
  - If the two points are at opposite ends of a diameter of the sphere, then the shortest distance is not unique (one may traverse either half of a great circle).

Points at opposite ends of a diameter of a sphere are called **antipodal points**. More generally, to reference measurements on the Earth sphere in a systematic manner we can introduce a coordinate system.

- One set of reference lines is produced using a great circle in a unique position (bisecting the distance between the poles), the **Equator**. A set of evenly spaced planes, **parallel** to the equatorial plane, produces a set of evenly spaced small circles, commonly called parallels.
- Another set of reference lines is produced using a half of a great circle, joining one pole to another, that has a unique position: the half of a great circle that passes through the Royal Observatory in Greenwich, England (three points determine a circle). Here it is historical consideration that produces the uniqueness in selection. Choose a set of evenly spaced halves of great circles obtained by rotating the diametral plane along the polar axis of the Earth. These lines are called **meridians**. The unique line is called the **Prime Meridian**.

This particular reference system for the Earth is not unique; an infinite number is possible. There is abstract similarity between this particular geometric arrangement and the geometric pattern of Cartesian coordinates in the plane (in the Cartesian coordinate system a point in the plane is represented by a pair of numbers).

    To use this arrangement, one might describe the location of a point, *P*, on the Earth sphere as being at the 8th parallel north of the Equator and at the 14th meridian to the west of the Prime Meridian. While this might serve to locate *P* according to one reference system, someone else might employ a reference system with a finer mesh (halving the distances between successive lines) and for that person, a correct description of the location of *P* would be at the 16th parallel north of the Equator and at the 28th meridian to the west of the Prime Meridian. Indeed, an infinite number of locally correct designations might be given for a single point: an unsatisfactory situation in terms of being able to replicate results. The problem lies in the use of a relative, rather than an absolute, location system.

To convert this system to an absolute system, which is replicable, we employ some commonly agreed upon measurement strategy to standardize measurement. One is the assumption that there are 360 degrees of angular measure in a circle. Suppose $O$ is the center of the sphere. Thus, $P$ might be described as lying 40 degrees north of the equator (angle AOP in Figure 19.1 with angular measure about $O$ in the clockwise direction away from the Equator), and 70 degrees west of the Prime Meridian (angle BOA in Figure 19.1 with angular measure about $O$ in the counterclockwise direction away from the Prime Meridian). The point $P$ is located at 40 degrees north latitude (north meaning north of the equator) and 70 degrees west longitude (west meaning west of the prime meridian). Often, instead of using "north" or "west", a "plus" or a "minus" symbol is used. When the rotational direction is clockwise the value of the associated numeral is positive; when the rotational direction is counterclockwise the value of the associated numeral is negative.

Most points are not conveniently located at multiples of 10 degrees. Parts of degrees may be noted as minutes and seconds, or as decimal degrees. Conversion from one form

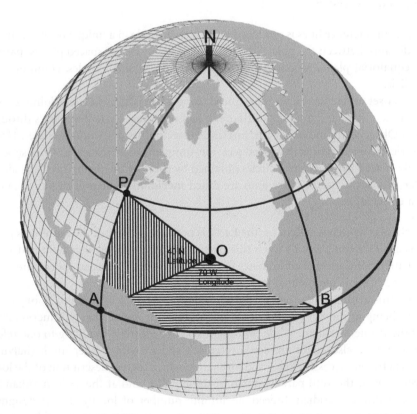

**Figure 19.1**  Determining location on a sphere. The point $P$ lies at 40 degrees north latitude, north of the equator, and at 70 degrees west longitude, west of the prime meridian. Often, north and south are replaced by + and − as are east and west. In that case, $P$ would have coordinates 40 degrees latitude and −70 degrees longitude

to another is simple to execute using a calculator. For example, 42 degrees 21 minutes 30 seconds converts to 42 + 21/60 + 30/3600 degrees = 42.358333 degrees; notice the powers of 60.

The geometric arrangement described above permits the creation of a systematic methodology for locating points on the Earth in a unique fashion. Analysis of these points often employs trigonometry. Figure 19.2 offers a visual review of trigonometric functions shown in the unit circle with axes labeled "axis" and "co-axis" (for complementary axis). This geometric analysis is based on Euclidean geometry, assuming **Euclid's Parallel Postulate** (given a line and a point not on the line, through that point there passes exactly one line

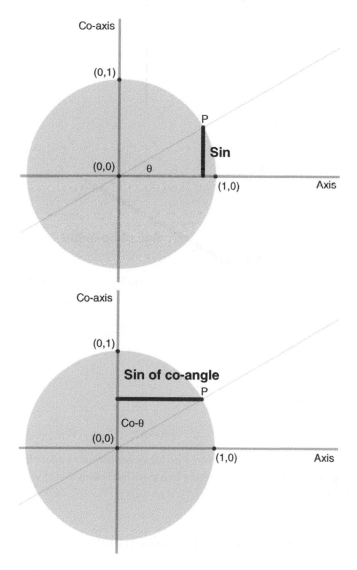

**Figure 19.2**   Visualizing trigonometry. Analyzing the angle theta and its complementary angle, co-theta (after *Spatial Synthesis*, Arlinghaus and Arlinghaus (2005) and used here with permission)

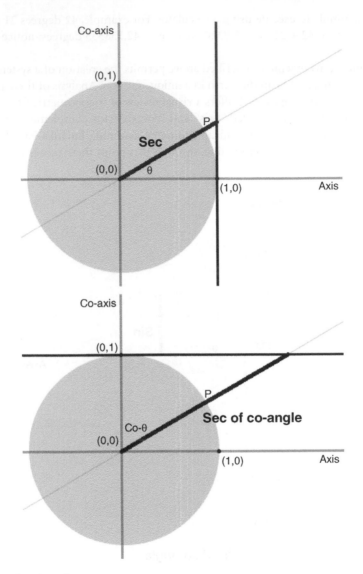

**Figure 19.2** *Continued*

that does not intersect the given line). Non-Euclidean geometries violate this Postulate (try to imagine the geometry of the Earth sphere in a non-Euclidean world where parallel lines intersect at infinity!).

## Eratosthenes and Earth's circumference

**Eratosthenes** worked out how to **measure the circumference of the Earth** without ever having traversed it. To do so, he used Euclidean geometry and simple measuring tools. His

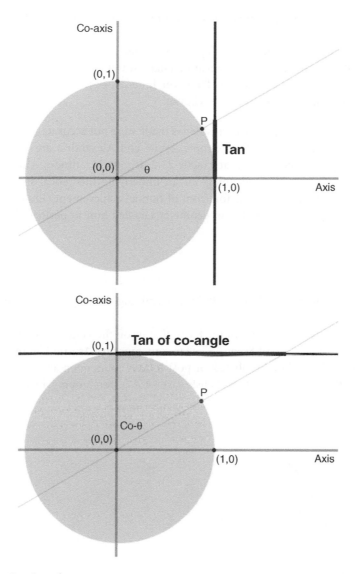

**Figure 19.2** *Continued*

research methodology rested on a clear understanding of the mathematics of his times. That understanding enabled him to tackle a problem of gigantic proportions.

Eratosthenes Assumptions

- Earth is a sphere.
- The circumference of the sphere is measured along a great circle on the sphere.
- Rays of the Sun are parallel to each other.

- The Sun's rays are directly overhead, on the Summer Solstice (*c.* June 21), at 23.5 degrees N latitude.
- Syene is located at about 23.5 degrees N latitude. Hence, on the Summer Solstice, the reflection of the sun will appear in a narrow well (and it will not on other days).
- Alexandria is north of Syene. Thus, on June 21, objects at Alexandria will cast shadows whereas those at Syene will not.

In fact, many of the assumptions Eratosthenes made were not accurate (the errors balance out to produce a good result; for example, Syene and Alexandria are not on the same meridian and Syene is not located at exactly 23.5 degrees N latitude). However, research methodology that takes clear advantage of an underlying theoretical structure, such as mathematics, has often been at the forefront of human achievement (Eratosthenes understanding of the concepts of Euclidean geometry enabled him to perform his remarkable calculation).

Eratosthenes' Method

- Find the circumference of the Earth by determining the length of the intercepted arc of a small central angle.
- Find two places on the surface of the Earth that lie on the same meridian (Eratosthenes chose Alexandria "*A'*" and Syene "*S*," near contemporary Aswan: Figure 19.3a).
- Eratosthenes focused on an obelisk or post located in an open area in Alexandria. He measured the shadow that the obelisk cast (*A'A''*), functioning in the manner of a

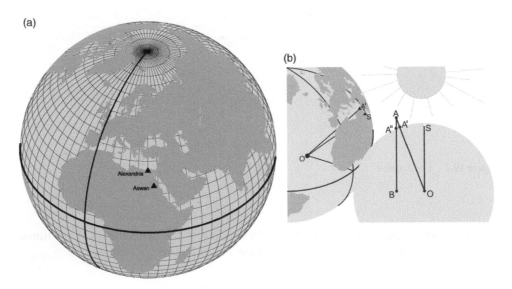

**Figure 19.3**   a. Relative location of Alexandria and Aswan. They are close to lying on the same meridian (half of a great circle). b. The obelisk (post) at Alexandria is *AA'* and the well at Syene is at a depth *S* below the surface of the Earth. This figure shows a detailed, geometric view of the relations among Alexandria *A'*, the obelisk *AA'*, the well at Syene *S*, the center of the Earth *O*, the shadow of the obelisk *AA''*, and the extension *AB* of that shadow into the Earth

gnomon on a sundial, and then measured the height of the obelisk (*AA* ʹ) (perhaps using a string anchored to the tip of the obelisk, Figure 19.3b).

* According to Euclid, two parallel lines cut by a transversal have alternate interior angles that are equal. The Sun's rays are the parallel lines. One ray, at Alexandria, touches the tip of the obelisk and extends earthward toward the tip of the shadow of the obelisk, *AA* ʺ. It is extended to *AB* in Figure 19.2b. The other ray, *SO*, at Syene, goes into the well and extends abstractly to the center of the Earth, *O*. The obelisk, *AA* ʹ, also extends abstractly to the center of the Earth, *O*; thus, the line, *AO*, determined by the tip of the obelisk and the center of the Earth is a transversal cutting the two parallel rays, *SO* and *AB*, of the sun.

* Angles (*BAO*) and (*SOA*) are thus alternate interior angles in geometric configuration described above; therefore, they are equal.

* Use the length of the obelisk shadow and the height of the obelisk to determine angle *BAO*; triangle *ΛΛ ʹΛ* ʺ is a right triangle with the right angle at *A* ʹ. Thus, we would note, tan (*A* ʹ*AA* ʺ) = (length of shadow)/(height of obelisk). Eratosthenes' measurements of these values led him to conclude that the measure of angle (*A* ʹ*AA* ʺ) was 7 degrees and 12 minutes.

* The value of 7 degrees and 12 minutes is 1/50th of the degree measure of a circle. Since he assumed that Alexandria and Syene both lay on a meridian (half a great circle), it followed that the distance between these two locations was 1/50th of the circumference of the Earth.

* Eratosthenes calculated the distance between Alexandria and Syene using records involving camel caravans. The distance he used was 5,000 stadia. Thus, the circumference of the Earth is 250,000 stadia, which translates to somewhat less than 25,000 miles (depending on how ancient units convert to modern units). This value is remarkably close to current values.

## Arithmetic, Algebra, and Data

Today, thanks to space exploration, miniaturization, and computers, we are not only able to envision the entire Earth but also to see it in satellite and other images. These methods of envisioning Earth are based, fundamentally, in mathematics. Geometry lies behind images; arithmetic and algebra lie behind data analysis. Often it is easy simply to let a spreadsheet do all of our calculations or to use a calculator to handle summations or multiplications. We should not, however, lose sight of the need to use our brains to work alongside the computer.

### Eratosthenes' prime number sieve and unique factorization

Eratosthenes demonstrated not only a firm grasp of geometric principles as a basic research methodology but also of arithmetic principles critical in successful data management. Integers are the numbers we use much of the time in looking at measurements of the Earth. Therefore, it is important to understand some of the conceptual structure of the integers.

Consider only the set of positive integers. One can abstractly partition this set into two kinds of numbers: those whose only integer factors are themselves and 1, and those whose integer factors include at least one number other than themselves and 1. The former is called a **prime number** and the latter is called a **composite number**. The number 7 is a prime number: its only integral factors are 1 and 7. The number 8 is a composite number: in addition to 1 and 8, 2 and 4 are also factors. The set of positive integers is infinite; the set of primes, contained within the positive integers is also infinite.

How do we know if a given positive integer is a prime or a composite number? We test it to see if it has factors other than itself and 1. How many integers need to be tested? The largest factor a number can have is its square root. Thus, it is necessary only to test numbers, as candidate factors, up to the greatest integer contained in the square root of the number under consideration. Eratosthenes demonstrated a systematic methodology for casting out candidate factors, commonly referred to as his 'prime number sieve'. In it, successive multiples of primes drop through the sieve so that only primes remain in the sieve. In Table 19.1, primes that remain in the sieve are less than 169 (13 squared). The boldest numbers are primes less than 13; the faintest are composite, and the medium bold numbers are primes greater than or equal to 13 and less than 169. The boldest ones were used to eliminate all the faintest numbers. That is, go through and remove all multiples of 2; then go through and remove all multiples of 3; and so on, up to all multiples of 11. What remains will be precisely the set of all primes less than 169.

The sifting of data is an important strategy that computers often use; you may, for example, be familiar with the use of filters in a spreadsheet to hide/reveal important data

**Table 19.1** Sieve of Eratosthenes. Primes are sifted out that are less than $169 = 13^2$. Thus, only multiples of 2, 3, 5, 7, and 11 need be eliminated. Base materials from *Spatial Synthesis*, Chapter 3 with selected associated textual material used here with permission (Arlinghaus and Arlinghaus 2005).

| 1 | **2** | **3** | 4 | **5** | 6 | **7** | 8 | 9 | 10 |
|---|---|---|---|---|---|---|---|---|---|
| **11** | 12 | **13** | 14 | 15 | 16 | **17** | 18 | **19** | 20 |
| 21 | 22 | **23** | 24 | 25 | 26 | 27 | 28 | **29** | 30 |
| **31** | 32 | 33 | 34 | 35 | 36 | **37** | 38 | 39 | 40 |
| 41 | 42 | **43** | 44 | 45 | 46 | **47** | 48 | 49 | 50 |
| 51 | 52 | **53** | 54 | 55 | 56 | 57 | 58 | **59** | 60 |
| **61** | 62 | 63 | 64 | 65 | 66 | **67** | 68 | 69 | 70 |
| **71** | 72 | **73** | 74 | 75 | 76 | 77 | 78 | **79** | 80 |
| 81 | 82 | **83** | 84 | 85 | 86 | 87 | 88 | **89** | 90 |
| 91 | 92 | 93 | 94 | 95 | 96 | **97** | 98 | 99 | 100 |
| **101** | 102 | **103** | 104 | 105 | 106 | **107** | 108 | **109** | 110 |
| 111 | 112 | **113** | 114 | 115 | 116 | 117 | 118 | 119 | 120 |
| 121 | 122 | 123 | 124 | 125 | 126 | **127** | 128 | 129 | 130 |
| **131** | 132 | 133 | 134 | 135 | 136 | **137** | 138 | **139** | 140 |
| 141 | 142 | 143 | 144 | 145 | 146 | 147 | 148 | **149** | 150 |
| **151** | 152 | 153 | 154 | 155 | 156 | **157** | 158 | 159 | 160 |
| 161 | 162 | **163** | 164 | 165 | 166 | **167** | 168 | | |

or with sorting methods that can handle data sets efficiently. All are variants, some more complex than others, of this type of methodology.

One enduring data-sifting problem in geography involves partitioning data to portray on a map. Geographic Information Systems (GIS) software often partitions data in default mode by "natural breaks." When the data are sifted in some other fashion, however (such as by using "equal intervals"), an entirely different pattern often emerges (Monmonier (1996) provides a more detailed look at issues of this sort).

An important research method in mathematical analysis is the creation of "theorems" (statements made about a broad class of mathematical, or other, objects that can be proved based on an underlying logic system). One of the most important theorems concerning integers is the **Fundamental Theorem of Arithmetic**. It states that every positive integer can be written (factored) uniquely as a product of primes. This uniqueness extends only to the primes involved in the **factorization**. For example, the fact that $12 = 2 \cdot 2 \cdot 3 = 2 \cdot 3 \cdot 2 = 3 \cdot 2 \cdot 2$ does not violate uniqueness. $12 = 2^2 \cdot 3$. By convention, the number 1 is a product of no primes. It is the desire for unique factorization that led to the avoidance of 1 as a prime; obviously, 1 can be included in a factorization of a number as often as desired. The desire for unique characterization of integers in terms of primes is similar abstractly to the desire for unique characterization of places on the Earth in terms of coordinates: clarity and care in system development are critical.

## Factorization and the distributive law

Mathematical laws and theorems can be easy to understand. It would be a mistake, however, to assume that because something is easy to understand that therefore it is not worth much as a research methodology or as an important tool in support of a research methodology. Consider the **Distributive Law** that links the operations of addition and multiplication. This law tells us that $2 \cdot (5 + 7) = (2 \cdot 5) + (2 \cdot 7)$. Stated in general algebraic terms, viewing arithmetic and algebra along a continuous spectrum of mathematical evolution, the law holds that: $a \cdot (b + c) = (a \cdot b) + (a \cdot c)$ (the parentheses in the right hand side of the equation are inserted for emphasis ... they are not necessary using the conventions involving order of operations). Consider some of the implications of this law: what is $(a + b)(a - b)$? Using the Distributive Law, it is simply $a \cdot (a - b) + b \cdot (a - b)$ and that is $a \cdot (a + (-1)b) + b \cdot (a + (-1)b)$. Using the Distributive Law again, on each of the terms in the previous expression, we now have $a^2 + (-1)ab + ba + (-1)b^2$ which is simply $a^2 - b^2$. Using the simple application demonstrated here it is possible to see how the Distributive Law might be used to check on selected elements of complex data management problems. Box 19.1 provides some examples (try to find others based on algebra you have already learned, then use them when handling data sets to see that the results of analyses you perform do make sense).

Mathematical analysis in geography often makes use of large numbers: thousands or tens of thousands of miles (kilometers) may separate places under consideration in a single study. Skill in the mental manipulation of numbers, and in assessing (independent of calculators or computers) whether the answer makes sense is critical. When one is lost in the numbers, one can become lost in the corresponding terrestrial space. The art and science of location theory is an important and complex source of research methods in

**Box 19.1　Does the Answer Make Sense?**

#### What is 24 · 26?

If one recognizes that it is simply the product of $(25 - 1)$ and $(25 + 1)$ then it is clear that that product is the same as $25^2 - 1$.

#### What is $25^2$?

The Distributive Law again provides the answer. State the last problem a bit more generally so that the method that is produced applies more universally – that is, to more than to just 25. Let's multiply $(10x + 5)$ by $(10x + 5)$. The Distributive Law shows that it is $100x^2 + 100x + 25$ which is also, again using the Distributive Law from right to left, $100x(x + 1) + 25$. When $x = 2$, as in the case of 25, then clearly $25^2$ is simply 2 times 3 times 100, or 600, plus 25. That is, $25^2$ is 625; thus 24 times 26 is simply $625 - 1$ or 624.

#### What is 45 · 45?

It is 4 times 5 times 100 or 2000 plus 25 or 2025.

#### What is 98 · 92?

It is $(95 + 3)$ times $(95 - 3)$ or 9025 – or 9016.

Try some for yourself – amaze your friends with newfound number-crunching prowess! Use your brain to check your computer (you are smarter than it is!).

geography (Puu 2003). When numerical and geometrical complexity are linked in a geographic setting, still more questions can be asked and problems solved. Consider, for example, some of the fractal forms in geography (Arlinghaus 1985; Goodchild and Mark 1987; Batty and Longley 1996).

## A View of the Future

At the time when Eratosthenes did his work it was based on high-powered mathematical analysis. Today, the required mathematics is taught in secondary schools. Thus, a modern-day Eratosthenes would need to be abreast of current developments in mathematics.

### Mathematical transformations: the case of stereographic projection

Loosely stated, a mathematical transformation sends elements of one set to elements of another set. The set of positive integers might be transformed to the set of positive even integers by sending each integer to twice its value. The concept of transformation has dominated research-level mathematics since World War II. One can choose to study the objects being transformed, or the transformation itself. To do the latter is universal.

The mathematical transformation of **stereographic projection** is viewed briefly here as it relates to maps. Maps are flat; globes are spheres. Flat maps are convenient and useful for many purposes. How can a globe be flattened to the plane? If uniqueness is required, then each point on the globe must correspond to exactly one point in the plane: the mathematical transformation sending the globe to the plane must be one-to-one. (Think of an orange; can you flatten the skin smoothly? No, it cannot be done.) One method of flattening the globe involves a transformation known as a stereographic projection. In Figure 19.4, the globe has been placed on a plane. The South Pole, *S*, is the point of tangency of the plane to the sphere. Use the North Pole, *N*, as the center of projection. Choose a point *P* on the globe and join it, inside the sphere (dashed line) to *N*. Then, extend the line segment *NP* to pierce the plane and label that as point *P'*. Continue this process for all points on the globe. Each point on the globe goes to a single point in the plane and each point on the plane comes from a single point on the globe. The transformation is a one-to-one transformation – or is it? If it were, then stereographic projection would offer a perfect map of the globe in the plane. Take a closer look: where does the point *N* project to in the plane? Join it to itself and extend the line segment: the resulting line is tangent to the globe and does not intersect the plane. Thus, this mapping does not send all the points from the sphere to the plane; it fails at one point. Conversely, to think about rolling the plane up into a sphere, one needs only to add one point to make the nice compact surface of a sphere. Stereographic projection failed at *N* because the geometry in which it is constructed is Euclidean; once again, Euclid's Parallel Postulate is controlling the analysis from behind the scene (Coxeter 1961). A theorem in more advanced mathematics tells us that in fact, this is the best we can do (there is no transformation that sends all points on the sphere to points in the plane in a one-to-one fashion).

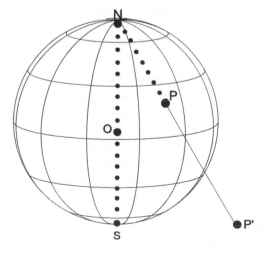

**Figure 19.4**  Stereographic projection of the sphere to the plane

One implication of this fact, for mapmakers, is that there is no perfect map. There is an infinite number of ways of transforming the sphere to the plane. At first glance, this appears to be a disastrous situation. In fact, however, what it does mean is that one needs to exercise care in selecting map projections and choose one that retains desired elements of the globe-view on the paper map (see Chapter 16). The art and science of cartography is devoted to such considerations and many mathematical and geographical treatises have been written on this topic. What non-Euclidean cartography might look like (where parallel lines do meet at infinity) is an interesting topic to speculate on, as are other related issues in cartography (Snyder 1993; Yang et al. 2000).

A traditional use of **non-Euclidean geometry** in geography involves the use of the Manhattan metric in urban geography (Krause 1975). In Euclidean geometry there is a single shortest path between any two points, called the **geodesic**. In Euclidean geometry, a geodesic is unique. Consider, however, the map of an urban downtown arranged on a grid street pattern. Blocks are sets of rectangles arranged, perhaps, along the axes of the cardinal points of a compass. Consider the path from the northwest corner of a block to the southeast corner of a block along the street grid. There are two distinct geodesics: one around the northeast corner of the block and another around the southwest corner of the block. There are two shortest paths; the geometry is not Euclidean. The presence of multiple geodesics has implications in planning and in urban geography: how might one route traffic through the downtown to allow swiftness of passage through the area yet retain flexibility for motorists to gain access to the variety of merchants present there? One-way streets offer a solution and they do so only because of the presence of multiple geodesics; otherwise, the motorist would lose access to the variety of shops present as he/she is routed efficiently through the downtown. Multiplicity of geodesics solves the problem: knowing the mathematics behind the geography is critical.

## Four color theorem

Once we have a map in the plane, in an appropriate projection, showing various regions of interest on the Earth, how many colors does it take to color the map in such a way that adjacent polygons can be distinguished from each other? If one were coloring a map of countries of the world and had as many different colors as there were countries, there would be no problem. That sort of approach, however, does little to reveal pattern and is not an effective use of color. First, consider what is meant by adjacency of regions. For our purposes two polygons will be said to be adjacent if they share a common line segment (touching only at a point does not constitute adjacency). Consider: a map with two adjacent regions. Clearly two colors suffice to color this map, say red and green. Consider a map with three regions. If the three regions are three parallel stripes, like the flag of France, then two colors also suffice: say red on the two end stripes separated by a green stripe in the middle. If, however, the three regions are arranged in an overlapping pattern, as in laying bricks, then three different colors are required, say red, green, and yellow. If one introduces an island in this last map, as in Figure 19.5, then four colors are needed to make the appropriate distinctions between adjacent regions.

If there were five regions in the map, what would be the maximum number of colors one might be forced to use in order to make the required distinctions? One might be

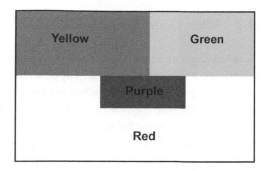

**Figure 19.5**   The four color theorem. The island in the middle requires the introduction of a fourth color; purple, to distinguish it from three adjacent regions

tempted to say five. However, for all maps with five regions, four colors are enough. In fact, recent research has shown that four colors are always enough to color any map in the plane, independent of its size or number of regions! Try coloring the map of the contiguous United States, 48 states, using four colors. While poor choices in color assignment may force the use or more than four colors, the four color theorem says that there exists at least one coloring scheme using no more than four colors for any map in the plane. But it does not tell us how to find that coloring; human thought is required for that.

If one never needs more than four colors to color a map in the plane, then why does cartographic (or GIS) software offer so many color choices? There are a number of answers to this question. One is that often the data being mapped are split into intervals of increasing intensity. If the color scheme is a graduated one of a single color, with intensity increasing with data magnitudes, then the color pattern imparts meaning about the data being mapped. For example, the higher the concentration of pollutants, the deeper the color of red; thus, one can pick out hot spots of pollution. Had the coloring scheme consisted of independent colors such as red, green, yellow, and purple, then one could not locate hot spots from the color because the meaning of the color would not be tied to the meaning of the data being mapped; instead, the meaning of the color would be tied to adjacency patterns of polygonal regions. If there are more than four data gradations then more than four distinct color gradations will also be required to have the desired display in which color intensity corresponds to data density. Thus there are good reasons for using more than four colors on a map.

Another question is: how many colors will always assure the desired coloring pattern on the sphere? It may surprise you to learn that it is the same number as on the plane. The proof, determined long ago, lay in the use once again of stereographic projection. Consider any map on a sphere. Poke a hole (remove a point) in the interior of any region of the map. Now use that hole as the center of a stereographic projection to map all the other points to the plane. The map in the plane can be colored (using four colors) and, because the hole was made in the interior of a region, the projected map in the plane has a sea around its rim. Reproject the four color map in the plane back to the sphere and color the added point the same color as the "sea." Thus, the map on the sphere also has four colors.

**Box 19.2**   What Does "**Large Scale Map**" Mean?

Maps are sometimes called large or small scale. These terms can be confusing but are not when the underlying mathematics is clear. Suppose, that on one map, the scale is 1 inch on the map represents 50,000 inches on the surface of the Earth. Another way to capture that idea is to express it as a ratio, or fraction, as 1/50,000, independent of units of measure (they are the same in numerator and denominator). This form of expression for scale is often called the "**representative fraction.**" Suppose a different map has a scale of 1/100,000. Which map is said to have "larger" scale? Which fraction is larger? The fraction 1/50,000 is larger than 1/100,000 (would you rather have a piece of pie from a large pie cut into 50,000 pieces or from a large pie of the same size cut into 100,000 pieces). Thus, the map with the scale 1/50,000 is larger scale than the one with scale of 1/100,000. That is the simple explanation in terms of the mathematics. The implications in terms of the geography is that larger scale maps give a more local view and may show more detail than smaller scale maps showing a more global view with less detail. The best solution may be simply not to use the terms "large" and "small," opting instead for clearer terms like "local" and "global"; nonetheless, others do use them so it is important to understand both what is meant and that the meaning derives directly from relative sizes of representative fractions. Therefore, to understand what "large" means, one needs to know how the scale is measured and have at least two measures of that scale in order to make a comparison.

## Jordan Curve Theorem

A simple closed curve is one that is topologically equivalent to a circle – that is, any curve that can be snapped back to a circle … like a rubber band. This theorem is due to Camille Jordan (1838–1922), who first presented it in his *Cours d'Analyse*. It is often convenient to think of the different domains as the "inside" of *J* and the "outside" of *J*. A common curve that is not a simple closed curve (or Jordan curve) is the numeral 8. To label "inside" and "outside," walk around the curve in a counterclockwise direction and label the domain to which your left arm points as the "inside" and the domain to which your right arm points as the "outside." In the case of the numeral 8, once the cross-over is passed, the inside becomes the outside and vice-versa! Curves that cross themselves are not simple closed curves.

It is important for mapmakers to know what kinds of curves they are working with. If they do not, they might assign an address to the wrong side of the street. If, for example, all even numbered addresses are assigned to the "inside" and all odd ones to the "outside" of a street, and they are dealing with a street curve that is equivalent to a figure 8 (as are many routes or parts of routes in gridded urban street patterns), then the even numbers on the map might appear "inside" the 8 on the top and "outside" on the bottom, contrary to the corresponding real-world situation. The solution is simple: split the 8 apart at crossovers into two distinct simple closed curves. The solution is simple once one understands the concepts involved!

Fill packages in modern software will fill, when using a paint bucket, only one half of the numeral 8. (The software developers know of the situation with the Jordan Curve Theorem but the casual user might not know why the bucket only fills half the figure 8.)

In the case of many simple closed curves it is an easy matter to tell if two arbitrary points lie on the same or on opposite sides of the curve. In complex cases, it may not be easy at all. Using Figure 19.6 we can demonstrate how to obtain a solution in a complex case. It can be done for a few positions for *a* and *b* (the *Fleur-de-Lis* is a simple closed curve; it is somewhat complex, although nowhere near as complex as urban street networks, but it has no crossovers).

Given any two points *a* and *b* in the plane, not on a simple closed curve *J*. How can you tell if *a* and *b* lie on the same side of *J* or on opposite sides of *J*?

- Pick a fixed direction.
- Join *a* and *b* with a line segment (this may be achieved uniquely because we are in the plane rather than, for example, on the sphere).
- Count the number of intersections the segment *ab* makes with *J*.
  - If the number of intersections is even (0 is even), then *a* and *b* lie on the same side of *J*.
  - If the number of intersections is odd, then *a* and *b* lie on opposite sides of *J*.

The Jordan Curve Theorem thus:

- Permits correct assignment of addresses on either side of streets. One must have the Jordan Curve Theorem built into the software if geocoding is to work on matched addresses.
- Permits visually appropriate coloring of polygons.
- Illustrates the need to split complex curves apart at nodes where the curve crosses itself in order to ensure that the properties above will hold on maps. This fact is important in digitizing maps (and elsewhere).
- Permits the development of a topological coding system for maps that maintains the characteristics of direction and connectivity between objects, without regard for distance. The Topologically Integrated Geographic Encoding and Referencing (TIGER) system of the US Census Bureau and US Geological Survey was developed in preparation for the 1990 decennial census of the population. Topologically improved, the system enabled one to distinguish inside/outside and street address positions. This had not been the case universally in the Geographic Base File/Dual Independent Map Encoding (GBF/DIME) system previously in use.

In this way, the Jordan Curve Theorem reveals some contemporary uses of mathematics as a research methodology in geography.

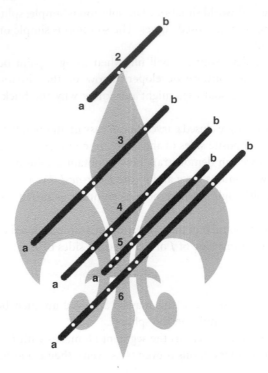

**Figure 19.6  Assignment algorithm** related to the Jordan Curve Theorem determines when *a* and *b* lie on the same or on opposite sides of a simple closed curve (the *Fleur-de-Lis* in this case)

# Conclusion

Applications of mathematics have a rich history, and an exciting future, in geography and related earth sciences. In this chapter we have reviewed Eratosthenes' innovative use of the contemporary mathematics of his time (to measure the circumference of Earth), considered how geometry helps drive mapping forward, and emphasized the importance of systematic and thoughtful approaches to numbers and the handling of data (by again drawing on the work of Eratosthenes). We have also examined the Jordan Curve Theorem and the four color theorem. Eratosthenes, Jordan, and others created great conceptual theorems and algorithms that led to advances in applications. Above all, these approaches demonstrate the need to understand the mathematics and logic behind maps and the decisions we make. Maps influence our decisions; decisions influence our maps. Clear understanding of how maps and mathematics are related is therefore a critical part of the geographic enterprise. The volumes edited by James R. Newman (2003) provide a perspective on the work of other mathematicians, and offer valuable guides to primary and secondary sources in the mathematical literature. These may help to expand your level of knowledge as you seek to apply mathematics to real world problems.

# Acknowledgment

The author wishes to thank Professor Jonathan D. Phillips, Department of Geography, University of Kentucky, for careful reading of a preliminary draft and for his constructive comments that helped to improve the chapter.

# References

Arlinghaus, S. (1985) Fractals take a central place. *Geografiska Annaler* B 67: 83–8.

Arlinghaus, S., and Arlinghaus, W. (2005) *Spatial Synthesis*. Ann Arbor: Institute of Mathematical Geography, http://www.imagenet.org [accessed August 8, 2009].

Batty, M., and Longley, P. (1996) *Fractal Cities*. London: Academic Press.

Coxeter, H. S. M. (1961) *Introduction to Geometry*. New York: John Wiley & Sons.

Goodchild, M. F., and Mark, D. M. (1987) The fractal nature of geographic phenomena. *Annals of the American Association of Geographers* 77: 265–78.

Krause, E. F. (1975) *Taxicab Geometry*. Menlo Park CA: Addison-Wesley.

Loeb, A. L. (1976) *Space Structures: Their Harmony and Counterpoint*. Boston: Addison-Wesley.

Monmonier, M. (1996) *How to Lie with Maps*. Chicago: University of Chicago Press.

Newman, J. R., ed. (2003) *The World of Mathematics*. New York: Dover Publications.

Puu, T. (2003) *Mathematical Location and Land Use Theory*. Berlin: Springer-Verlag.

Snyder, J. P. (1993) *Flattening the Earth: Two Thousand Years of Map Projections*. Chicago: University of Chicago Press.

Yang, Q., Snyder, J. P., and Tobler, W. (2000) *Map Projection Transformation: Principles and Applications*. Oxford: Taylor and Francis.

# Additional Resources

Arlinghaus, S., and Arlinghaus, W. (1989) The fractal theory of central place hierarchies: a Diophantine analysis of fractal generators for arbitrary Löschian numbers. *Geographical Analysis* 21: 103-21. An article that uses fractal geometry and number theory to prove a set of theorems about classical central place theory in geography. Requires a substantial background in mathematics.

Arlinghaus, S., Arlinghaus, W., and Harary, F. (2002) *Graph Theory and Geography: An Interactive View (eBook)*. New York: John Wiley and Sons. Wiley's first eBook offers readers an opportunity to see mathematical theorems come to life as animations on maps and in a variety of other contexts. The level of mathematical difficulty ranges from beginning to advanced.

Mandelbrot, B. F. (1983) *The Fractal Geometry of Nature*. San Francisco: W. H. Freeman. An important contribution showing the power of the computer to visualize complex mathematics and spatial associations that had previously been relegated only to the imagination. Take a look at the images and read the associated text to learn how the images were created. This book can be read here and there; it is not necessary to begin at the beginning and read straight through to the end.

Nystuen, J. D. (1966) Effects of boundary shape and the concept of local convexity. *Papers, Michigan Interuniversity Community of Mathematical Geographers* (unpublished). Reprinted, Ann Arbor: Institute of Mathematical Geography, http://www.imagenet.org [accessed August 8, 2009]. One of a number of classic papers from the MICMG group. All are archived on the url above. The complete set has been scanned and is maintained on the website. Browse through the foundations of mathematical geography in the twentieth century! Difficulty level varies.

*Solstice: An Electronic Journal of Geography and Mathematics*. Ann Arbor: Institute of Mathematical
    Geography, http://www.imagenet.org [accessed August 8, 2009]. An online journal since 1990
    devoted to the interaction between mathematics and geography. Newer issues contain a wide
    variety of animations and interactive applications such as Java™ Applets and Virtual Reality
    models.
Thompson, D. W. (1961) *On Growth and Form*, abridged and edited by Bonner, J. T., Cambridge:
    Cambridge University Press. An important contribution showing how transformations carry
    spatial form, interpreted here as biological objects. Watch as a slender fish transforms into a
    puffy fish. See plants transform from one shape to another. In all cases, keep your eye on the
    underlying graph paper: the grid tells the story here as it does with maps.

## Exercise 19.1   Mathematical Geography in Action

To test your knowledge of both the geographical and mathematical aspects of coordinatiz-
ing the Earth, construct an increasingly difficult sequence of problems involving latitude
and longitude, that include questions such as:

- How long (in miles) is one degree of latitude?
- How long (in miles) is one degree of longitude at 42 degrees of latitude?
- At what latitude is the length (in miles) of one degree of longitude exactly half the value
  of one degree of longitude at the equator?

Color a real-world map (try altering some of the fundamental assumptions, such as those
about adjacency or about the surface being mapped, to create a modified coloring scheme).

Describe some real-world situations in which paths that are not simple closed curves could
lead to erroneous mapped results (show how to correct the pattern so that correct results
will be achieved).

# Chapter 20

# Regional Analysis

## *Gordon F. Mulligan*

- Introduction
- Economic Base Analysis: The Fundamentals
- The Marginal Multiplier
- Conclusion

## Keywords

## Introduction

Social scientists design and update models that often assist decision-makers in the private and public sectors. These models must however satisfy two conditions before practitioners will adopt them. First, they must be firmly grounded in theory. Theory, which is continually contested, allows practitioners to understand the causal mechanisms that underlay visible events. Second, these models must resemble reality to a satisfactory degree. Repeated empirical testing allows practitioners to accept or discard models based on their accuracy or usefulness. This is all part of the enterprise of verifying or falsifying hypotheses.

This larger theme is explored in this chapter on regional analysis. Rather than examine many different areas of research and practice, I focus on economic (or export) base analysis. The concepts and techniques of **economic base** analysis have a long tradition, dating from the middle of the twentieth century (Blumenfeld 1955; North 1955; Tiebout 1955, 1962), and they continue to be widely used by geographers, planners, and economists in the United States and other countries to study the performances of regional and local economies. The "small areas" examined in economic base analysis include towns, sub-metropolitan cities, and sparsely settled counties. The basic proposition underlying the theory is that the health of a small-area economy largely depends upon the performance of its export industries. Economic base analysis is closely related to several other topics of interest to students of regional development, including staples theory, input-output analysis, and export-led industrialization.

The main purpose of this chapter is to show how traditional economic base analysis was modified as theoretical concerns changed, as data sources improved, and as society as a whole was transformed between the mid-1960s and the mid-1990s. Making use of a unique data set on southwestern US towns, I focus on economic base **multipliers** in telling this story. These multipliers, which are frequently mentioned in the media, simply measure the propensity of export-oriented jobs to induce locally-oriented jobs in regional econo-mies. Later on we will see how the estimates of multipliers always depend upon the various assumptions and subjective choices that practitioners make in their studies.

Day-to-day research and scholarship in the social sciences is largely guided by theory that already has been accepted in the scientific community. This activity leads to *positive* social science, which is largely involved with the formulating and testing of hypotheses. Moreover, inquiry largely proceeds according to certain "rules of the game" that all schol-ars more-or-less accept, and most inquiry is therefore called *normal* social science. But existing theory sometimes proves to be so inadequate that it must be radically modified or even completely abandoned. During those periods of time when scientists are trying to fashion a new paradigm to direct their inquiry, the community of scholars can be divided along the lines of methodology. Eventually, though, some new body of theory becomes acceptable to a younger cohort of scholars and yet a new round of normal activity ensues (Kuhn 1970). However, in the social sciences our research and scholarship is also guided by other important concerns that are directly related to human welfare. This concern leads to *normative* social science. Often in the social sciences we must confront fundamental moral, ethical, or ideological issues that can also divide the community of scholars. We live in a world that is experiencing rapidly changing economic and social circumstances and the social sciences have a special responsibility not only to make sense of all these changes, but also to assist public policy-makers in making informed and responsible deci-sions in the face of these changes.

To help us in understanding research and scholarship in the social sciences we should take another moment to think about the production and uses of knowledge. I find it helpful to visualize a *meta-model* to characterize many of the issues and problems that I study in the field of regional development (Figure 20.1). In fact this meta-model can be visualized as a triangle whose three vertices are *theory* (abstract concepts, mathematical statements, etc.), *evidence* (different types of primary and secondary data, etc.) and *policy* (the appropriate scale for study, concerns of social justice, etc.). As in any triangle, a direct linkage exists between any one vertex and the other two vertices. But in our case the link-ages can be very complex because they ultimately involve people and their diverse beliefs and interests. Nevertheless, when a change occurs at any one vertex of this triangle there

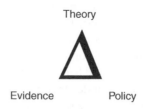

**Figure 20.1**  A meta-model for research in regional development

are direct implications – and, later, indirect implications – for change at the other two vertices.

As we will see, external circumstances during the 1960s and 1970s forced analysts in the field of regional development to reformulate their theories. This led to a series of remarkable improvements in economic base analysis – first during the 1980s and later during the 1990s. Among other things, entirely new types of data were needed to test or confirm a whole series of improved models. But along the way practitioners found they were also able to address a number of public-policy concerns in a much more satisfactory way. For instance, they could now assess not only the expected size but also the expected distribution of new jobs in small-area impact studies. This meant that public officials could now gauge the degree to which disadvantaged social groups might or might not participate in expanding labor markets. In short, "new" moral and ethical concerns merged with "old" theoretical concerns to redirect inquiry regarding the nature of regional economies. The triangular meta-model of theory, evidence, and policy captures these shifts, and I use it to guide and inform the discussion that follows.

## Economic Base Analysis: The Fundamentals

This chapter makes extensive use of the Arizona Community Data Set (ACDS; Vias 1996), which shows the *basic*, *nonbasic*, and *total* employment of nearly 50 towns. These data were derived from numerous field surveys undertaken between the mid-1970s and mid-1990s. All data were collected at the level of the establishment (i.e., factory, office, store) and virtually all establishments were surveyed when a town was targeted. Given the present purposes, we will examine a sample of only 15 of the Arizona communities.

**Basic employment** refers to export-oriented jobs and **nonbasic employment** refers to locally-oriented jobs. This bifurcation can be achieved in a variety of ways (see below), but survey results clearly are the most accurate. The employment of each establishment is divided into these two components based on its annual revenues. Basic employment reflects those jobs that are generated by sales to agents (firms and households) found outside the town's designated base area. By contrast, nonbasic employment serves the needs of local firms and households.

To clarify this distinction, consider a small plant that produces a widely distributed roof-coating material. Management knows both the plant's annual revenues and employment figures, and it also has a good estimate of the proportion of annual revenues coming from beyond the plant's sales region (i.e., from shipments beyond the town or county).

With that information, regional analysts can determine the amount of the plant's total employment that is export oriented. Given the nature of the product, we would expect that proportion to be quite high. Now consider the corner grocery store across the street from the plant. With the exception of some sales to tourists or others just "passing through," almost all of the store's sales (and hence its employee mix) will be local in nature, or nonbasic. This distinction is key to understanding economic sustainability in a small-area economy: that economy cannot survive solely on nonbasic activities; it must bring in money from the outside in order to prosper, that is, in order to support local employment in stores that sell groceries. Moreover, in the real world not all **basic activities** are equal in their ability to support a healthy range of **nonbasic activities**. For these reasons, analysts want to know both the structure of employment in a small area and the economic leverage that the basic employment has in producing local prosperity.

## The data set

The ACDS basic–nonbasic distinction is consistent over three scales: the establishment itself (e.g., a store or office), the major industry group (e.g., manufacturing or retailing) identified by the SIC code, and the entire town economy. For our purposes, jobs are consolidated into four general sectors: $i = 1$, primary (agriculture; mining), $i = 2$, secondary (construction; manufacturing), $i = 3$, trade (transportation, communications, and public utilities; wholesale trade; retail trade), and $i = 4$, services (finance, insurance, and real estate; other private services; public administration).

Table 20.1 shows basic (BE) and nonbasic (NE) employment figures for each of the 15 towns, which are listed in alphabetical order. The reader is encouraged to consult an atlas to find their locations. All basic employment, BE, represents the so-called *economic base* or *export base* of the economy. Sector-by-sector averages for both basic and nonbasic employment are shown at the bottom of the table.

The first thing to notice about the various towns is that they have different *types* of economic bases. In other words regional economies typically *specialize*. Some places, like Bisbee and Globe, have most of their basic employment in primary activities; other places, like Sedona and Wickenburg, have most of their basic employment in trade and services; and other places, like Douglas and Snowflake, have most of their basic employment in secondary activities. Further scrutiny also reveals that these towns have somewhat different *degrees* of economic specialization. These categories of **functional specialization** are important in explaining why we find different responses to economic stimuli from one place to the next.

## The traditional multiplier

In economic base analysis we make use of two important equations. The first equation is an accounting identity that simply recognizes that total employment always equals basic employment plus nonbasic employment. For the entire economy, note that:

$$TE = BE + NE \qquad (20.1)$$

**Table 20.1**  Basic and nonbasic employment for 15 Arizona towns

| Town | B1 | B2 | B3 | B4 | BE | N1 | N2 | N3 | N4 | NE |
|---|---|---|---|---|---|---|---|---|---|---|
| Bisbee | 1,266 | 5 | 155 | 252 | 1,678 | 4 | 46 | 367 | 507 | 924 |
| Casa Grande | 749 | 962 | 363 | 888 | 2,962 | 111 | 264 | 846 | 1,061 | 2,282 |
| Clifton | 2,754 | 6 | 52 | 108 | 2,920 | 0 | 11 | 440 | 358 | 809 |
| Douglas | 46 | 1,194 | 333 | 488 | 2,061 | 3 | 98 | 561 | 1,026 | 1,688 |
| Globe | 2,388 | 37 | 313 | 578 | 3,316 | 6 | 109 | 1,030 | 1,367 | 2,512 |
| L Hav City | 2 | 1,356 | 322 | 665 | 2,345 | 28 | 515 | 774 | 1,039 | 2,356 |
| Holbrook | 3 | 30 | 580 | 622 | 1,235 | 5 | 18 | 401 | 530 | 954 |
| Parker | 49 | 121 | 283 | 823 | 1,276 | 32 | 54 | 561 | 624 | 1,271 |
| Payson | 1 | 53 | 167 | 169 | 390 | 2 | 53 | 278 | 253 | 586 |
| Safford | 227 | 131 | 356 | 889 | 1,603 | 13 | 176 | 878 | 868 | 1,935 |
| Sedona | 15 | 13 | 217 | 234 | 479 | 5 | 53 | 275 | 240 | 573 |
| Show Low | 2 | 87 | 237 | 210 | 536 | 0 | 19 | 263 | 178 | 460 |
| Snowflake | 65 | 550 | 96 | 85 | 796 | 0 | 22 | 224 | 373 | 619 |
| Superior | 685 | 4 | 58 | 73 | 820 | 0 | 7 | 145 | 140 | 292 |
| Wickenburg | 4 | 17 | 200 | 229 | 450 | 13 | 39 | 225 | 290 | 567 |
| Average | 550.4 | 304.4 | 248.8 | 420.9 | 1,524.5 | 14.8 | 98.9 | 484.5 | 590.3 | 1,188.5 |

*Notes:* The average refers to the statistically representative town in the data set. B is for basic employment (total: BE); N is for nonbasic employment (total: NE). Sector 1 is primary employment, 2 is secondary, 3 is trade, and 4 is services. For further details, consult text

*Gordon F. Mulligan*

The second equation introduces the *traditional* economic (export) base multiplier, which was in dominant use until the 1980s. Here the so-called *multiplier effect* is calculated by simply taking the ratio of total jobs to basic jobs across the entire regional economy:

$$m = TE/BE = 1 + NE/BE \qquad (20.2)$$

Throughout the chapter we use the lower case m to designate any multiplier *estimate* that is computed by the employment ratio approach.

Using equation (20.2), the multiplier concept is simple to explain. Each basic job brings money into the community from outside and some of this money is spent locally, thereby creating nonbasic jobs. The first expression on the right-hand side of equation (20.2) indicates how *on average* each export job is directly transformed into each total job. But the second expression in equation (20.2) is more informative. Here we see that each basic job actually induces or supports other jobs that exist in the nonbasic portion of the economy. In fact, it is precisely this ratio of locally-oriented employment to export-oriented employment that determines the size of the multiplier effect. If the ratio NE/BE is large, then the multiplier is large; but if the ratio NE/BE is small, then the multiplier is small. The second expression also indicates that as long as just one nonbasic job exists in the economy, its economic base multiplier must be greater than unity (i.e., m > 1).

Now examine Table 20.2, which shows a number of other attributes of the 15 towns, including total employment TE and the estimate m of the traditional multiplier. Note that

**Table 20.2**  Additional attributes and estimates for 15 Arizona towns

| Town | TE | m | TR | m* | ER% | $m_a^*$ | $m_q^*$ | Type |
|---|---|---|---|---|---|---|---|---|
| Bisbee | 2,602 | 1.55 | 414 | 1.44 | 7.6 | 1.74 | 2.46 | A |
| Casa Grande | 5,244 | 1.77 | 895 | 1.59 | 11.3 | 2.06 | 4.27 | B |
| Clifton | 3,729 | 1.28 | 184 | 1.26 | 1.6 | 1.32 | 1.71 | A |
| Douglas | 3,749 | 1.82 | 1,106 | 1.53 | 18.9 | 1.98 | 2.45 | B |
| Globe | 5,828 | 1.76 | 874 | 1.60 | 10.0 | 1.96 | 2.98 | A |
| L Hav City | 4,701 | 2.00 | 770 | 1.76 | 13.6 | 2.05 | 2.70 | B |
| Holbrook | 2,189 | 1.77 | 211 | 1.66 | 6.6 | 8.99 | 2.92 | C |
| Parker | 2,547 | 2.00 | 333 | 1.79 | 11.7 | 4.89 | 3.37 | C |
| Payson | 976 | 2.50 | 491 | 1.67 | 49.7 | 2.44 | 2.16 | C |
| Safford | 3,538 | 2.21 | 928 | 1.76 | 25.6 | 3.03 | 3.06 | C |
| Sedona | 1,052 | 2.20 | 660 | 1.50 | 46.7 | 2.29 | 1.91 | C |
| Show Low | 996 | 1.86 | 285 | 1.56 | 19.2 | 3.26 | 2.66 | C |
| Snowflake | 1,415 | 1.78 | 126 | 1.67 | 6.6 | 2.02 | 3.00 | B |
| Superior | 1,112 | 1.36 | 160 | 1.30 | 4.6 | 1.49 | 1.96 | A |
| Wickenburg | 1,017 | 2.26 | 432 | 1.64 | 37.8 | 2.87 | 2.16 | C |
| *Average* | *2,713.0* | *1.87* | *524.6* | *1.58* | *18.0* | *2.83* | *2.65* | |

*Notes*: TE is total employment; m is the traditional multiplier; TR is basic employment generated from transfer payments; m* is the transfer payment adjusted multiplier; ER% is an estimate of the inflation of m relative to m*; $m_a^*$ is the transfer payment adjusted, assignment-based multiplier; $m_q^*$ is the transfer payment adjusted, location-quotient derived multiplier; Type refers to the specialization of the economy (A: mining; B: manufacturing; C: services) the estimate of the traditional multiplier ranges between a low of 1.28 (Clifton) and a high of 2.50 (Payson), where the average value of m is 1.87. Therefore, in the *statistically average* or *representative* town of the ACDS – not that this town really exists – each basic job is responsible for inducing another 0.87 nonbasic jobs, thereby generating 1.87 total jobs (including itself).

## Two changes in economic base analysis

Since the 1960s two significant changes have forced analysts to change their thinking regarding the operation of local and regional labor markets. First, more economically advanced nations all have assumed **postindustrial** attributes (Bell 1973). Innovation, production mode, and access to both risk capital and human capital have become very important concerns of firms and access to natural and human-made **amenities** has become an especially important concern of households (Diamond and Tolley 1982; Glaeser et al. 2001). Primary and secondary activities have become relatively less significant and service and (some) trade activities have become relatively more significant, at least in employment terms. Moreover, people are living longer, they are more mobile than ever, and they increasingly form household types that are very different from those that were dominant just a few decades ago.

Second, government tax agents in these advanced nations collect and then redistribute funds in support of a variety of public programs. Some of these monies flow directly back to households in the form of public transfer payments. In the US, such monies include social security, unemployment insurance, payments for the disabled, and the like. These payments are distinguished from private monies like rents, interest payments, and dividends that many households also receive. Together, analysts often talk about households receiving **non-earnings income**, which is comprised of payments that are received for their investments, for their past services in the workplace, and for the entitlements they presently enjoy from government (Gibson and Worden 1981; Manson and Groop 1990).

In light of these remarkable transformations, both academics and practitioners in regional development began to ask important questions like the following: What are the implications of increasing female participation in the labor force? How does human capital (reflected in the health, education, and skills of workers) affect regional job growth? What will happen in rural areas as large numbers of retirees (who substitute high wages/low amenities for high amenities/low wages) move there? How will these in-migrants alter the demography and affect the delivery of public goods in these rural areas?

By the late 1970s the economic base model was itself being updated in light of these and other important questions. Two very general changes in the model are worthy of some mention at this time. First, it was realized that the economic base of many postindustrial regional economies was often much different than what was being captured by traditional thinking. Many small-area economies thrived not because they exported primary or manufactured commodities but because they specialized in trade or in business or personal services. The economic base logic of the 1960s and 1970s was now seen as dated because it no longer correctly identified all the sources of employment-related income flowing into the economy. Second, it was recognized that many small-area economies were now dependent upon changes in non-earnings income for significant amounts of local job creation. In fact, the ACDS gives special attention to the issue of *public* transfers and data covering these payments were carefully collected at a very intimate geographic scale. If these transfer payments were disregarded, an important part of each town's economic base would be omitted and any study of the regional labor market would be flawed.

Returning to my meta-model introduced earlier, changing external circumstances (involving new empirical evidence but also new policy issues) forced general inquiry (economic theory) to shift. This in turn led to the updating of a specific model (economic base) that we use in regional development. Later we will see how this model assumed even more complex features after the initial updating of the early to mid-1980s. And we will also see how these newer versions of the economic base model have allowed those responsible for formulating public policy to make more accurate and informed decisions.

## Public transfer payments

But before moving on to these other topics we must say a few more words about public transfers. For the sake of illustration, suppose a town receives $10 million a year in transfer payments from various levels of government and the average wage or salary in that town is $25,000 per year. Clearly these transfer payments represent the *equivalent* of adding an extra 400 (i.e., $10 million/$25,000) export-oriented or basic jobs to the regional economy. Following this logic, employee-equivalent transfer payments (TR) can be computed for regional economies that have been surveyed in different years. In fact, these values have been computed for all towns in the ACDS and Table 20.2 shows the levels of TR for the 15 towns that are of interest to us. Astonishingly, these transfer payments represent the equivalent of nearly 525 basic jobs in the statistically representative town of our data set, ranging from a low of 126 equivalent basic jobs in Snowflake to a high of 1,106 equivalent basic jobs in Douglas.

Now let's return to equations (20.1) and (20.2) above and *adjust* the various terms in the traditional model accordingly. Note first that the adjusted employment figures are $BE^* = BE + TR$ and $TE^* = TE + TR$, where nonbasic employment (NE) remains the same. The adjusted economic base multiplier is

$$m^* = TE^*/BE^* = 1 + NE/BE^* \tag{20.3}$$

Throughout the rest of the chapter an asterisk is used to represent any multiplier estimate that is adjusted for transfer payments.

From the last expression in equation (20.3) we see that the adjusted multiplier $m^*$ is always less than its unadjusted counterpart m. In other words, whenever practitioners fail to adjust for the presence of transfer payments, their multiplier estimates will always be too high. The importance of this discrepancy for public policy will be examined in more detail below. Suffice to say here that since most economic base studies undertaken before 1980 failed to make this adjustment, they were characterized by a degree of multiplier inflation.

Table 20.2 provides some evidence about the multiplier discrepancies that probably arose. For each of the 15 towns this discrepancy can be computed as a percentage error ER% where $m^*$ is assumed to be the "correct" estimate: $ER\% = 100 \times (m-m^*)/m^*$. We see that the unadjusted multiplier is on average 18 percent greater than the adjusted multiplier, where these errors range widely, from a low of approximately 2 percent in Clifton to a high of nearly 50 percent in Payson. In other words, most studies of regional labor markets undertaken during the 1960s and 1970s probably inflated their multiplier estimates by one-fourth or even one-third simply because the role of public transfer payments in local job creation was not sufficiently appreciated.

## Indirect or shortcut methods

It is expensive and time-consuming to survey individual businesses, even in small towns like those of the ACDS having at most 5,000–6,000 persons and 200–250 establishments (i.e., factories, offices, and stores). A reasonable guess is that it would cost at least $5,000 to update any one of the nearly 50 surveys that were undertaken in creating the ACDS. Furthermore, this cost figure would rise very quickly for surveys undertaken in somewhat larger places, such as small cities having 25,000–50,000 persons. Consequently, both prac-titioners and public officials have been long interested in devising indirect or shortcut methods of estimating economic base multipliers (Mulligan 2008). Typically these methods manipulate sector-specific total employment data, which are made available by various state and federal sources. In fact most of the economic base studies that were undertaken during the 1960s and 1970s used indirect methods and we now can assess how accurate those studies were.

The most common approach is to assign entire sectors of total employment either to the basic or to the nonbasic category. Some difference of opinion exists about exactly how to do this, but most analysts would assign the total employment in sectors 1 and 2 to basic and the total employment in sectors 3 and 4 to nonbasic. For an example, let's examine Lake Havasu City. Adding the two components (B1+N1, B2+N2, etc.) in the survey data (Table 20.1), we see that T1 = 30, T2 = 1,871, T3 = 1,096, and T4 = 1,704. This means that 1,901 jobs are assigned to basic and 2,800 jobs are assigned to nonbasic by the assignment method. From equation (20.2) the new multiplier estimate is $m_a$ = 2.47. The assignment-based multiplier also can be adjusted for transfer payments, an operation that reduces the estimate downward to $m_a^*$ = 2.05. The adjusted assignment multipliers $m_a^*$ are shown for all 15 towns in Table 20.2.

A second popular method involves designating the "excess" total employment in each sector as basic, and the remaining total employment in that sector as nonbasic. This method enjoys some popularity in the postindustrial era because it at least recognizes that service and trade activities can perform an important role in a regional economy's export base. A percentage allocation Ti%, where Ti% = 100 × Ti/TE, is first calculated for each of the four sectors of each town, and then the average allocation Ti% is computed across the entire sample of 15 towns. These sector-specific averages are then used as references or benchmarks in applying location quotients (LQi) to the actual employment in each town. In the average town of the ACDS we see that T1% = 17.51, T2% = 13.00, T3% = 30.97, and T4% = 38.52. The town-specific percentages for Lake Havasu City are T1% = 0.64, T2% = 39.80, T3% = 23.32, and T4% = 36.25. Taken together, these two sets of percentages indicate that Lake Havasu City has excess employment in sector 2, where LQ2 = 39.80/13.00 = 3.06, which is greater than unity. But in each of the other three sectors the value for LQi is less than unity, so no excess employment exists. The excess employment in sector 2 is T2 × (LQ2-1)/LQ2 = 1,871 × (39.80-13.00)/39.80, which means that 1,260 jobs are assigned as basic and the remaining 611 jobs as nonbasic in sector 2. All of the remaining 2,830 jobs in sectors 1, 3, and 4 are assigned as nonbasic, making 3,441 nonbasic jobs in all. This means that the multiplier estimate for the town is $m_q = 3.73$, which is reduced downward to $m_q^* = 2.69$ with the inclusion of transfer payments. Adjusted location quotient multipliers are shown for all 15 towns in Table 20.2.

It is a sobering exercise to examine how the estimates based on these two indirect methods compare to those based on the more expensive survey approach. Upon examining the descriptive statistics we see that the *average* multiplier estimates for the assignment method are $m_a = 7.80$ and $m_a^* = 2.83$. These are somewhat higher than the *average* multiplier estimates for the location quotient method, which are $m_q = 4.46$ and $m_q^* = 2.65$. But, upon recalling that the *average* survey-based multipliers are $m = 1.87$ and $m^* = 1.58$, it is clear that both indirect methods severely inflate either the unadjusted or adjusted multiplier estimate. In fact, this problematic inflation of the multiplier holds for all towns in the sample. The average error (computed as before) for the adjusted multiplier using the assignment method is 75.2 percent and the average error using the location quotient method is 66.7 percent. Furthermore, the *correlation* between the survey and location quotient multipliers (r = 0.57) is somewhat stronger than the correlation between the survey and assignment multipliers (r = 0.42). Evidently, the location quotient method provides multiplier estimates that are superior to those of the assignment method; however, *neither* shortcut approach generates multiplier estimates that are very reliable. Consequently, we should be very skeptical of any approach that does not use some sort of survey data when estimating the economic (export) base multiplier of a regional labor market.

## An application to public policy

Multiplier models are very helpful to public officials who wish to evaluate the impacts of new industries on small-area economies. They can provide guidance with respect to needed new housing and infrastructure, schools and libraries, and other public needs. These same models can also be used to project the economic contractions caused by departing or closing industries. For purposes of illustration, suppose that a new export-oriented factory is proposed for the representative town in the ACDS. From a case study elsewhere, suppose that public officials can gauge that 80 new workers will be added to the economy. These officials might then adopt the average value of m* from Table 2, and estimate the impact-related change in nonbasic employment to be 0.58 × 80 = 46 new jobs and the impact-related change in total employment to be 1.58 × 80 = 126 new jobs. These officials could also use the ACDS to specify the most likely allocation of the 46 new nonbasic jobs across the four employment sectors. Appropriate allocation percentages can be calculated from the averages for N1, N2, N3, and N4 (in relation to NE) shown at the bottom of Table 20.1. Those officials would then arrive at the following distribution for the 46 new local jobs: sector 1, 0.6 jobs; sector 2, 3.8 jobs; sector 3, 18.8 jobs; and sector 4, 22.8 jobs.

If these officials were comparing several proposed new industries for an available site, they could estimate the job-creation impact of each proposal and rank the proposals from best to worst on this criterion. Naturally, though, these officials would want to consider a lot of other pertinent information – including project-, site-, and town-specific data – before they reached a final decision about the future of the site.

This simple example sheds light on the relative usefulness of the numerous economic base impact studies that were undertaken in the US prior to the 1980s. We can now recognize that many of these studies were doubtless flawed, sometimes severely, because either local job creation was overestimated by indirect methods or transfer payments were entirely disregarded. In many cases the overestimation of multiplier effects was doubtless very high because errors arising from these two deficiencies actually compounded one another.

## Some analytical shortcomings

But, even as this one simple application reveals, a number of important issues are not fully addressed by the traditional economic base approach. First, we just *assumed* that the multiplier effect is satisfactorily estimated as the ratio between total employment and basic employment. Second, we just *assumed* that the size of the impact always would be independent of the sector where the impact originates. So we simply supposed that 80 new export jobs in manufacturing would have exactly the same impact as 80 new export jobs in government. And third, we just *assumed* that the impact itself would not have a substantial effect on the very nature of the small-area economy. The first of these three issues is fairly straightforward to address but even here we must eventually modify the logic of traditional economic base analysis. The second and third issues are more advanced and cannot be addressed in this chapter (see Mulligan and Vias 1996).

An additional issue to consider in assessing the quality of economic base studies is that of the leakages to the economy caused by commuting. Some economic base studies recognize the role of household commuting behavior in affecting the size of the multiplier effect. Taking in-commuting into account, for example, recognizes that some people work in the target region but spend most of their earnings elsewhere (i.e., where they live), whereas out-commuting recognizes that some people work elsewhere but spend most of their earnings inside the target region. The neglect or exclusion of households in-commuting (out-commuting) in the study leads the analyst to deflate (inflate) the regional multiplier estimate. This behavior is especially important to take note of when the target region is a rural town or bedroom community located within the commuting shed of a large, metropolitan economy. Related to this issue is the fact that a small area's relative distance (measured in physical, time, or money units) to other regions has an effect on the nature and size of the regional economy, especially in small places like mining towns or resort communities. When these places are relatively isolated and protected by distance, they tend to have more nonbasic activities than is the norm (based on their population size alone), making the economic base multiplier somewhat larger in size than would be expected. On the other hand, when these places are found very close to much larger economies they tend to suffer many leakages (e.g., from high rates of consumer out-shopping) and consequently have fewer nonbasic activities than is the norm, making the multiplier effect smaller than would be expected.

# The Marginal Multiplier

We now must become acquainted with the idea of a *marginal* multiplier, a concept that has been more prevalent in economics than in geography or planning (Armstrong and Taylor 2000). Marginal multipliers can be estimated by using longitudinal, cross-sectional, or pooled data sets. Now we no longer compute 15 town-specific multipliers and then take their mean value to estimate a multiplier for the representative town. Instead we directly calculate a multiplier for this average town that minimizes the sum of the (squared) errors involved in performing the estimation across those 15 towns. In order to accomplish this we use ordinary least-squares (OLS) regression (see Chapter 18; Griffith and Amrhein 1997).

The first model, which we call Model 1, actually retains the underlying economic logic of the traditional analysis. Practitioners used this version of the model during much of the 1980s. We begin by assuming that local jobs and export jobs always exhibit a *linear* or straight-line relationship, which means that we can estimate the four sector-specific relationships as:

$$Ni = a_i + b_i BE \qquad (20.4)$$

and that we can estimate the overall relationship as:

$$NE = a + bBE \qquad (20.5)$$

where $a = \Sigma\, a_i$, $b = \Sigma\, b_i$ and $i = 1,2,3,4$. Next, using calculus, we differentiate the dependent variable by the independent variable in each equation. This informs us how a 1-unit change in BE induces a marginal change in Ni in equation (20.4) and a marginal change in NE in equation (20.5). Using the symbol $\Delta$ to represent a small change, we see that:

$$\Delta Ni/\Delta BE = b_i \qquad (20.6)$$

and

$$\Delta NE/\Delta BE = b \qquad (20.7)$$

The slope coefficient $b_i$ shows how many extra (fractional) local jobs are created in sector i by a 1-unit shift in a town's export jobs. In like fashion, the slope coefficient b shows how many local jobs in all four sectors are created by this same 1-unit shift.

Equation (20.7) suggests a multiplier measure that is somewhat different than the one we saw earlier. Now consider how a *change* in basic employment induces a *change* in total employment. Adding one extra job to both sides of this equation we see that:

$$M_1 = (\Delta NE + \Delta BE)/\Delta BE = \Delta NE/\Delta BE = 1 + b \qquad (20.8)$$

where 1+b is the estimate of the marginal multiplier. In this chapter the upper case is used to indicate all estimates of marginal multipliers. We say that $M_1$ is a marginal multiplier because it relates the final, or resulting, *change* in total employment to the initial, or exogenous, *change* in export employment. In making impact assessments, regional development analysts believe the marginal multiplier is superior to the ratio multiplier because it more accurately represents how the impact process actually unfolds.

The first numerical column of Table 20.3 shows the relevant estimates for Model 1. Note, for instance, that a 1-job shift in BE induces a shift of 0.311 local jobs in N4, 0.609 overall local jobs in NE, and 1.609 total jobs in TE. The slope estimate b has a high t-score (t = 4.71), thereby indicating that in all likelihood the strong, positive relationship existing between NE and TE is not based on chance. And, in terms of its overall statistics, we see that the model performs fairly well – the adjusted R-squared is 0.60 and the standard error of the estimate (SEE) is 482.2. Finally, we see that the estimate of the marginal multiplier for Model 1 is $M_1 = 1.61$, which is somewhat lower than the average value of 1.87 for the 15 ratio multipliers shown in Table 20.2.

**Table 20.3** OLS regression estimates of nonbasic employment

| Sector | Model 1 (BE) | Model 1* (BE,TR) | Model 2 (TE) | Model 2* (TE,TR) |
|---|---|---|---|---|
| NE1 | −9.26 | −9.26 | −7.94 | −9.69 |
|  | 0.012 | 0.009 | 008* | 0.007 |
|  |  | 0.020 |  | 0.009 |
|  | 0.11 | 0.08 | 0.18 | 0.12 |
|  | 26.8 | 27.2 | 25.7 | 26.6 |
| NE2 | 3.47 | −52.44 | −36.49 | 60.80 |
|  | 0.063* | 0.033 | 0.050* | 0.034 |
|  |  | 0.192* |  | 0.128 |
|  | 0.15 | 0.28 | 0.34 | 0.35 |
|  | 123.9 | 114.6 | 109.8 | 109.0 |
| NE3 | 143.25 | 29.11 | 67.17 | 27.51 |
|  | 0.224* | 0.164* | 0.154* | 0.128* |
|  |  | 0.392* |  | 0.209* |
|  | 0.61 | 0.77 | 0.84 | 0.86 |
|  | 173.4 | 134.3 | 112.4 | 102.6 |
| NE4 | 116.58 | −63.37 | 8.14 | −64.18 |
|  | 0.311* | 0.216* | 0.215* | 0.168* |
|  |  | 0.618* |  | 0.381* |
|  | 0.61 | 0.82 | 0.84 | 0.90 |
|  | 242.3 | 165.8 | 154.3 | 123.7 |
| NE | 259.92 | −95.96 | 30.89 | −107.16 |
|  | 0.609* | 0.422* | 0.427* | 0.337* |
|  |  | 1.223* |  | 0.727* |
|  | 0.60 | 0.81 | 0.86 | 0.91 |
|  | 482.2 | 332.4 | 287.8 | 227.0 |

*Notes:* The estimates shown in each of the five rows are as follows: intercept, employment (using BE or TE), transfer payments (if appropriate), the adjusted R-squared, and the standard error of the estimate. In rows 2 and 3, the symbol * indicates a slope estimate for employment and transfer payments that is significant at the 0.10 level

As before, we can re-estimate the marginal multiplier by introducing (employee-equivalent) public transfer payments. Call this Model 1*. But now we have the attractive feature of being able to introduce TR as an entirely separate independent variable where:

$$Ni = a_i + b_i BE + c_i TR \qquad (20.9)$$

and:

$$NE = a + bBE + cTR \qquad (20.10)$$

In other words, we can now disentangle how actual export jobs and equivalent export jobs affect local job creation. Equation (20.10) indicates that a 1-job shift in BE (real export jobs) creates an overall shift of b nonbasic jobs and that a 1-job shift in TR (equivalent export jobs) creates an overall shift of c nonbasic jobs. The various regression estimates for Model 1* are shown in the second column of Table 20.3. Note that the slope estimates $b_i$ and b are pushed downward with the inclusion of transfer payments, indicating (as we would have anticipated) that the results for Model 1 overestimated the importance of basic (real) jobs in local job creation. These results echo our earlier findings, using the ratio multiplier, regarding the relationship between multiplier estimates that have and have not been adjusted for non-earnings income. The regression estimates indicate that the adjusted marginal multiplier estimate is $M_1^* = 1.42$, which is 13% lower than the unadjusted marginal estimate of $M_1 = 1.61$.

## Another improvement

Later in the 1980s a *theoretical* deficiency was noted in the novel regression-based analysis. So a new version of the economic base model was designed that had a more satisfactory underlying logic, one in which nonbasic jobs could actually induce other nonbasic jobs. To accommodate this, we can simply substitute total jobs for basic jobs on the right-hand side of equations (20.4) and (20.5) and call the new specification Model 2. Thus, for each of the four employment sectors consider:

$$Ni = a_i + b_i TE \qquad (20.11)$$

and for aggregate employment consider:

$$NE = a + bTE \qquad (20.12)$$

Do the same for equations (20.9) and (20.10) and then consider:

$$Ni = a_i + b_i TE + c_i TR \qquad (20.13)$$

and:

$$NE = a + bTE + cTR \qquad (20.14)$$

as the appropriate equations for Model 2*. Finally, after making manipulations much like those performed for Models 1 and 1* earlier, we arrive at the following equation for the estimate of the marginal economic base multiplier using either Model 2 or 2*:

$$M_2 = \Delta TE/\Delta BE = 1/(1-b) \tag{20.15}$$

Our change in logic, although subtle, is very important and is worthy of some discussion. First, when using BE (instead of TE) we were completely *missing* some jobs that would be created by the local spending of workers engaged in the town's various nonbasic activities. To take one example, we failed to recognize that at least some of the demand for goods sold in a town's *trade* sector would arise from the needs of other locally-oriented workers in that same town's *service* sector. In omitting NE from the right-hand side of the regression equations, as we did earlier, the importance of BE in local job creation was inflated. In turn this meant the estimate of the multiplier effect was biased upward. Second, we now have a model that has more appealing *behavioral* properties. Much like a Keynesian model, where consumption is viewed as a linear function of overall or total income, we now have a model where local employment has a *structural* relationship with total employment. In fact, $b_i$ and b are said to represent *propensities* for creating nonbasic employment, and each is standardized by the total employment in the economy.

Before we can conclude that this "new" marginal multiplier model is actually better than the "old" one, yet another round of empirical testing must be undertaken. The new estimates are shown in the last two columns of Table 20.3. Note that the overall statistical fits fully endorse the change we made in substituting TE for BE as being the "correct" employment variable driving local employment NE. As indicated by both the adjusted R-squared statistic (which is higher) and the SEE statistic (which is lower), Model 2 outperforms Model 1, and Model 2* outperforms Model 1*. Furthermore, Model 2* is clearly the best of all four possibilities. For the aggregate relationship between NE and TE, Model 2* explains 91 percent of the variation in NE and has by far the lowest standard error (SEE = 227.0) of any of the four cases. The estimates for Model 2* indicate that a 1-job change in TE is (structurally) accompanied by a 0.337-job change in NE and a 1-equivalent-job change in TR induces a shift in NE of 0.727 jobs. The introduction of transfer payments in Model 2* deflates the slope estimate for NE down to 0.337 from 0.427 in Model 2, indicating that (as before) any model excluding transfer payments tends to overstate the ability of export jobs to create local jobs. So we should recognize $M_2^* = 1.51$ as being a better estimate of the marginal multiplier than is $M_2 = 1.73$.

Again we see how theory and evidence are closely intertwined in the meta-model that I use to understand the evolution of inquiry in regional development. The accepted economic base model was improved upon, yet again, by first adapting the logic of the model to better economic theory, and then empirical testing clearly endorsed the change that had been made. Moreover, as the reader should now have anticipated, this all has implications for the accuracy and usefulness of public-policy assessments.

Returning to the impact study outlined earlier, we can now re-estimate both the *number* and the *distribution* of the new locally-oriented jobs that will be added to the impacted regional labor market. Recall that 80 new basic jobs are expected. From equation (20.15) we see that an extra $0.508 \times 80 = 40.6$ nonbasic jobs can be expected from the impact, leading to the creation of 120.6 new jobs in all. The allocations of these nonbasic jobs to

**Box 20.1** Some Other Popular Methods in
Regional Analysis

### Adjustment models

The term refers to a multi-region model that allows employment and population levels (or densities) to adapt or adjust to one another by introducing a time lag. More complex models even incorporate per capita income or land-use absorption as a third interacting variable. Estimation of the two-variable model is done with two-stage least squares regression. In the best-known study, in 1987, Gerald Carlino and Edwin Mills examined the growth of all US counties during the 1970s, while controlling for a variety of amenities and public-policy factors. Some studies have substituted each region's basic employment for its total employment in order to improve specification of the bi-directional model.

### Economic-demographic models

Sometimes called demoeconomic, a model that links together employment projections (often based on the economic base logic) with population projections to simulate the longitudinal attributes of a regional labor market. An employment module represents regional labor demand and a population module represents regional labor supply, and the outputs of the two modules are compared each year through age- and gender-specific labor force participation rates. When demand exceeds supply in-migration of new workers is induced and when supply exceeds demand regional out-migration occurs. These simulation models are used by people like Andrei Rogers to forecast the most likely futures of regional labor markets and serve to inform public-policy decisions regarding local taxes and infrastructure expenses.

### Input-output models

This is a very popular model in regional science that was developed by Wassily Leontief to examine the transactions (purchases or sales) between all industries in a regional or national economy. There is an important distinction made between final demand (due to households, governments, and exports) and intermediate demand (due to industries trading with one another), and the key idea is that each industry has its own multiplier effect. Economic base and input-output models belong to a wider family, called social accounting models (SAMs), which were advocated by Richard Stone. Analysts like Geoffrey Hewings, Walter Isard, Ronald Miller, and Karen Polenske often distinguish between *intra*regional and *inter*regional models. In the latter, export changes in the target region not only impact the industries of that same region but also create spillover effects in the industries of other regions, and then these effects in turn ripple back through the industries of the target region. Linear algebra is used to trace out the complex patterns of direct, indirect, induced, and feedback effects that result.

the various sectors are computed by forming the ratio between the estimate $b_i$ and the overall estimate b, using equations (20.13) and (20.14). So the expected allocations are as follows: sector 1, $(0.007/0.337) \times 40.6 = 0.8$ jobs; sector 2, 4.1 jobs; sector 3, 15.4 jobs; and sector 4, 20.3 jobs. Evidently the ratio multiplier model discussed earlier in the chapter over-predicted the impact by 5.4 nonbasic jobs, or 13.3 percent, and the greatest errors were incurred in sectors 3 and 4. In summary, a better economic base model – one that stands on sounder economic principles – allows public officials to make more accurate predictions about expected employment outcomes in impacted regional economies.

# Conclusion

Rather than review and simply "skim the surface" of the many research perspectives now current on regional economies, this chapter has examined in some detail one prominent line of inquiry. There are many others, a few of which are briefly outlined in Box 20.1. In closing, I want to situate this larger body of work, including the material in this chapter, within a broader cross-disciplinary framework.

The study of regional economies is influenced by various disciplines, but especially economics and geography. In the decades following the 1950s much research has been practiced under the banner of "regional science" (Boyce 2003), a field of study partially founded and popularized by the economist Walter Isard (e.g., see Isard 1960). In using a Venn diagram to support his view of this new interdisciplinary field, Isard was responding to the often aspatial approaches of economists (some of whom have been said to analyze the economy as if it functioned "on the head of a pin") *and* to the often non-analytic nature of much of human geography (which at the time was largely confined to descriptive case studies). Over the past half-century, regional science has had a transformative impact on *both* economics and geography. Moreover, it continues to attract researchers interested in allied areas such as urban and regional planning, international and regional development, transportation and land use, demography and migration, resource management and ecological analysis, GIS and spatial statistics, and others. Though diverse, all of the work done under the heading of regional science recognizes two fundamental truths: (a) we cannot understand economic growth or decline without taking into account the fact that national economies are territorially partitioned, first into regional and then into local economies; and (b) this requires not only that we pay attention to the differential spatial impacts of economic change, but that our models and policies take spatial difference fully into account. It is with these points of reference that both researchers and practitioners have forged ever new ways of integrating theory, evidence, and policy in regional analysis.

# References

Armstrong, H., and Taylor, J. (2000) *Regional Economics and Policy* (2nd edn). Oxford: Blackwell.

Bell, D. (1973) *The Coming of Post-Industrial Society*. New York: Basic Books.

Blumenfeld, H. (1955) The economic base of the metropolis. *Journal of the American Institute of Planners* 21: 114–32.

Boyce, D. (2003) A short history of the field of regional science. *Papers in Regional Science* 83: 31–57.

Carlino, G., and Mills, E. (1987) The determinants of county growth. *Journal of Regional Science* 27: 39–54.

Diamond, D., and Tolley, G., eds. (1982) *The Economics of Urban Amenities*. New York: Academic Press.

Gibson, L., and Worden, M. (1981) Estimating the economic base multiplier: a test of alternative procedures. *Economic Geography* 57: 146–59.

Glaeser, E., Kolko, J., and Saiz, A. (2001) Consumer city. *Journal of Economic Geography* 1: 27–50.

Griffith, D., and Amrhein, C. (1997) *Multivariate Statistical Analysis for Geographers*. Upper Saddle River, NJ: Prentice Hall.

Isard, W. (1960) *Methods of Regional Analysis*. Cambridge, MA: MIT Press.

Kuhn, T. (1970) *The Structure of Scientific Revolutions* (2nd edn). Chicago: University of Chicago Press.

Manson, G., and Groop, R. (1990) The geography of nonemployment income. *Social Science Journal* 27: 317–25.

Mulligan, G. (2008) A new shortcut method for estimating economic base multipliers. *Regional Science Policy and Practice* 1: 67–84.

Mulligan, G., and Vias, A. (1996) Interindustry employment requirements in nonmetropolitan communities. *Growth and Change* 27: 460–78.

North, D. (1955) Location theory and regional economic growth. *Journal of Political Economy* 63: 243–58.

Tiebout, C. (1955) Exports and regional economic growth. *Journal of Political Economy* 64: 160–9.

Tiebout, C. (1962) *The Community Economic Base Study*. New York: Committee for Economic Development.

Vias, A. (1996) The Arizona community data set: a long-term project for education and research in economic geography. *Journal of Geography in Higher Education* 20: 243–58.

## Additional Resources

Abler, R., Adams, J., and Gould, P. (1971) *Spatial Organization: The Geographer's View of the World*. Englewood Cliffs, NJ: Prentice-Hall. Though now somewhat dated, a perusal of this introductory textbook will offer the inquisitive student one of the most comprehensive surveys of how geographers approached space, methodologically and theoretically, in the late 1960s and 1970s. A very short section on economic base studies points out that regions (especially cities), like individuals, tend to specialize in their productive activity. Then, in trading with one another, they generate flows of people and goods that be classified in two different ways: those flowing between economic sectors and those flowing between places.

Berry, B., and Parr, J. (1988) *Market Centers and Retail Location*. Englewood Cliffs, NJ: Prentice-Hall. This is an excellent introductory account of one of the most important early areas of regional analysis: central place theory. Replete with maps, data, and diagrams this is still a very good text for getting students to think about the operation of both inter-urban and intra-urban markets. Even some cross-cultural examples, including that of periodic markets, are provided. It is often forgotten that the economic bases of many communities are comprised, at least in part, of a wide array of ever-changing service and trade (both wholesale and retail) activities.

Florida, R. (2002) *The Rise of the Creative Class*. New York: Basic Books. The author outlines a highly popular (and controversial) theory that focuses on the emergence of a new social class in the workforce. This "creative class," comprised of various knowledge-based workers – including entrepreneurs, scientists, artists, and media and design professionals – is responsible for most of the new ideas and innovations in our city-based economies. Their presence is believed to add a spark to both the cultural vitality and economic prosperity of urban areas. Echoing ideas of Blumenfeld (1955), the long-run sustainability of large cities is thought to depend on the continuous nurturing of this feature of the urban economic base.

Fujita, M., Krugman, P., and Venables, A. (1999) *The Spatial Economy*. Cambridge MA: MIT Press. This is a very demanding analytical book that examines the forces leading to economic concentration and uneven regional development throughout the world. A strong background in mathematics is needed to appreciate much of the material, which largely focuses on the relations existing between increasing returns, transportation costs, and the movement of labor and capital. One interesting argument concerns "circular and cumulative growth," where the authors argue that a qualitative discontinuity can take place in the evolution of a region's economic base.

Hewings, G. (1985) *Regional Input-Output Analysis*. Beverly Hills, CA: Sage. This is Volume 6 in the popular Scientific Geography Series, edited by Grant Thrall. The first part of the book develops conceptual and analytical links between economic base, Keynesian, and input-output models. In many ways, input-output analysis is seen as a disaggregated (i.e., multi-sector) version of economic base analysis. The second part of the book provides various versions, extensions, and applications of the standard linear model, while addressing the issue of general social accounting. Students are actually shown how to construct a regional input-output model.

Power, T. (1996) *Lost Landscapes and Failed Economies*. Washington, DC: Island Press. This book, somewhat controversial when it first appeared, argues that the natural landscape is an essential part of any region's economic base and that it should never be sacrificed in order to create short-run jobs that are not sustainable over the long run. The author makes a persuasive argument for including quality of the natural environment in the economic base, which has obvious implications for public policy. Local shifts in non-earnings income often reflect the enhanced value being placed on the local natural landscape, particularly by retirement-aged households.

## Exercise 20.1   Economic Base and City Specialization

As noted earlier, an important attribute of regional economies is their specialization. In this exercise you are asked to examine specialization in relation to economic base multiplier effects. Towns in the ACDS can be grouped together to form relatively homogeneous classes, and their varying degrees of specialization can be measured. Such a classification was developed for the 15 Arizona towns, and this is shown in the last column of Table 20.2. Cities with a designation A are largely mining oriented; type B refers to manufacturing; and cities marked with a C specialize in retail or service activities. Given this classification, answer the following questions:

1   Select one of the multipliers discussed in this chapter, and calculate its *average* value separately for the cities designated by the different specializations A, B, and C. What are the mean values for these different city types, and what do you think might account for the differences?

2   Select three cities for further study: one with a large multiplier, one with a small multiplier, and one with a multiplier in the middle range. Based on web research for each of the three cities (e.g., www.az.gov), can you come up with some hypotheses about what might lie behind the sizes of the estimated multipliers? Which, if any, of these factors is purely locational, or social, etc.?

# Chapter 21

# Modeling

*Yvonne Martin and Stefania Bertazzon*

- Introduction
- Modeling in Physical Geography
- Modeling in Human Geography
- Examples of Models in Human Geography
- Model Evaluation
- Conclusion

**Keywords**

Conceptual model
Determinism
Empirical model
Mathematical model
Physical model

Process-based model
Spatial diffusion
Spatial science
Stochasticism

## Introduction

Geographers recognize models as powerful tools that can contribute significantly to understanding system behavior, and in aiding policy and decision making that affect our planet and society at scales ranging from the local to global. In this chapter we consider the use of models in both physical and human geography, the contributions modeling can provide, and the associated assumptions and limitations.

A model can be defined as a simplified representation of a natural and/or human phenomenon or system. Models are idealizations of reality, designed to generate outputs from inputs in an attempt to understand system behavior. Various assumptions and simplifications must be made during model development to allow the problem under investigation to remain tractable. However, if fundamental properties of the system can be retained during this process, a model may provide meaningful results and contribute important knowledge to our understanding of system operation, and aid in predictions of the future. Evaluation of models is a critical stage in the modeling process, as it allows for assessment of both the model itself and its predictions, and yet despite its importance there remain many unresolved questions about what constitutes a critical test of a model's performance. Researchers studying the various branches of physical geography generally accept the use of models, while recognizing both their strengths and their limitations. In contrast, researchers for most sub-disciplines in human geography question and/or reject models, or simply do not employ them as a methodology. Therefore, the role of modeling in physical and human geography is very different, and for this reason each is explored in turn within this chapter following a general introduction.

Before considering in detail the role of modeling in geography, let us first examine two fundamental questions, to set the context for further discussion: why are models a valuable tool when applied to certain geographical problems; and what is the role of models in geography? A primary role of models is to provide a speculative formulation of a problem to guide in its investigation. Models can also be used to test how a system responds to different input parameters, to examine system sensitivity to different scenarios, and to test hypotheses or theories that may or may not have been corroborated in other ways. A hypothesis represents an explanation of a phenomenon based on relevant observations, but that has not yet been generally accepted, whereas a theory represents an explanation for a phenomenon that has been more widely corroborated by the research community. Models can provide evidence to support, reject, challenge, or modify our ideas (hypotheses or theories) of how the world works. Model evaluation then takes on the important role of testing the strength of the explanations that have been put forward. However, models alone cannot provide unequivocal proof of a hypothesis or theory. One way of viewing models is as a means of communication of ideas about the functioning of a particular system. In the real world, it is often difficult to control the natural or human environment. Model variables, parameters, and initial/boundary conditions can all be modified readily to ask a series of critical *"what-if"* questions, which constitutes one of the most important aspects of a modeling exercise. Models are often used in policy and decision making. For example, models have been invoked to make predictions of future system states for issues such as groundwater contamination or optimal transportation corridors in a city. Caution is required, however, when applying models for decision making due to limitations associated with model evaluation.

## Modeling in Physical Geography

Physical geography shares many methodologies, including modeling, with other scientific disciplines. Environmental systems, such as those falling under the realm of physical geography, generally involve a large number of variables and processes. The interactions and feedbacks within the system can be very complex, and the underlying processes are often nonlinear in nature. For this reason, scientific investigations of such phenomena are seldom straightforward and provide exciting challenges for the researcher. The process of attempting to meet this challenge makes the study of physical geography very rewarding, and requires careful consideration of appropriate methodologies to tackle a research question (see Chapters 2 and 5).

Experimentation is the gold standard of methodologies in science, and has been given much consideration by historians, philosophers, and practitioners of science (see Chapter 3). During experimentation, certain variables are controlled while others vary, which allows for identification of critical relationships and a deeper understanding of system behavior. Experimentation becomes especially challenging when applied to complex, environmental systems. For example, difficulties may be encountered in controlling key variables in the field, such as climate or geology. Moreover, even if some degree of control is possible in certain situations, field experimentation is limited in its ability to provide insights into system operation over the medium to large spatial and temporal scales so often considered in physical geography. For this reason, physical modeling and mathematical modeling are valuable methodologies for research in physical geography. Within these frameworks, not only can manipulations be undertaken to control certain variables while allowing others to vary, system operation over medium to large scales can also be explored.

Despite the important role of modeling in scientific investigation, modeling studies alone should not be relied on to provide explanations about scientific phenomena. The greatest progress most often occurs when a critical combination of various scientific methodologies is adopted to tackle a research problem. For example, modeling can be used to test and refine our understanding of knowledge gained using other methodological approaches (such as field-based studies). And these other types of studies can, in turn, contribute to the development and refinement of a model. We now turn to a consideration of the three main types of models commonly employed in physical geography: **conceptual models**; **physical models**; and **mathematical models**. Note that some of the information contained in the following sections is relevant to the later discussion of modeling in human geography. For this reason, fundamental descriptions of primary types of models will not be repeated in that section, but rather the discussion will be extended to consider their application to human geography.

## Conceptual models

Conceptual models represent a widely used and valuable tool in scientific enquiry. They can be constructed at any stage in the study of a phenomenon, to explore and summarize the state of understanding at the time. These models take the form of narrative or visual summaries in which system components (processes and forms), and interactions among them, are described. Conceptual models are often visual, taking the form of "boxes and arrows" diagrams, in which components are shown in boxes and interactions are indicated with arrows (see Figure 21.1 for an example). Such diagrams are particularly useful in describing feedbacks in systems, and are often used in undergraduate textbooks to highlight the principal features of system operation.

Conceptual modeling allows a researcher to think through and analyze a problem, to describe essential features and processes, and to synthesize knowledge. These models may

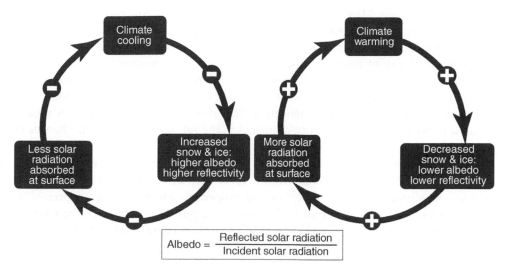

**Figure 21.1** An example of a "boxes-and-arrows" conceptual model for the sea ice-albedo feedback. (a) positive feedback; (b) negative feedback

be developed on the basis of theoretical reasoning/intuition and/or existing theories/data/ knowledge derived by other means (e.g., field measurements). Some conceptual models are quite sophisticated and involve detailed descriptions of process operation, while others, particularly for new areas of research, may be more rudimentary. In either case, new insights about system functioning should be revealed.

Conceptual modeling is often a preliminary step in a scientific investigation, frequently preceding measurement programs, experimentation, physical and/or mathematical (empirical or process-based) modeling. The conceptual model may be used to generate working hypotheses to be tested in various ways. That being said, the relation between conceptual modeling and other methodologies (including other forms of modeling) is not always strictly linear; knowledge gained in other ways can be used to develop new conceptual models or to revise existing ones.

Three basics steps are involved in conceptual modeling. First, the research question, the system to be addressed, and the spatial and temporal scales involved must be defined clearly. Second, the goals of the conceptual modeling exercise should be stated (for example, will the model be used as a basis for further investigation?). Third, the model components, including processes, forms, and interactions/relationships must be identified and any assumptions clearly stated.

## Physical models

Physical modeling (sometimes referred to as hardware or scale modeling) involves construction of an analog for a real-world environmental system, and allows for precise control of variables in a laboratory setting. The latter feature is critical to understanding the particular significance of this type of model, given that our inability to adequately control variables for environmental systems in the field presents a formidable obstacle in studies of physical geography. Many research laboratories, such as the St Anthony Falls Laboratory at the University of Minnesota, have developed sophisticated facilities for the undertaking of physical modeling experiments (http://www.safl.umn.edu/facilities/facilities.html).

Model development involves construction of a physical representation of the target system, which maintains essential attributes and relationships as determined by theoretical reasoning or by understanding of the phenomenon obtained using other methodologies (e.g., conceptual or mathematical modeling, field investigations). The model is usually reduced in both size and complexity compared to the full-scale counterpart. Changes in physical scale may modify the system behavior quite significantly, and formal scaling criteria for the system under examination must be met in order for the system to operate in a meaningful way and to contribute to the interpretability of results. This requires that the relative magnitudes of critical processes and properties, which are often expressed as dimensionless ratios, are preserved in the model (see Box 21.1). Care must be taken as simplification and the change in physical scale affect direct comparison with the full-scale system.

---

**Box 21.1    Physical Models of River Channels**

For over a century river engineers have used small-scale physical models to investigate river processes, such as flow dynamics, sediment transport, and the development of river channel morphology, and to test designs for structures in rivers under controlled experimental conditions. Geomorphologists have adopted these methods and used them to investigate fluvial processes and river dynamics.

Physical models require some validation against real world data, but physical modeling of rivers has a number of advantages compared to making observations in the field, including: a reduced time scale (long-term processes occur faster); control over the river discharge and other conditions, such as the sediment delivery rate and the bed (valley) gradient; and easy observation and measurement of morphology and processes. Physical modeling requires specialized laboratory facilities, such as the flume at the University of Western Ontario shown in Figure 21.2, which has been used to study different aspects of river morphology and processes by P. Ashmore and his collaborators (e.g., Pyrce and Ashmore 2003). This 20 m long, 3 m wide flume has a sand bed with a gradation scaled down to represent medium-sized fluvial gravel. It sits on jacks and can be tilted to different slopes. Sediment that leaves the end of the flume is captured and returned to the upstream end of the flume.

Model rivers may be strictly scaled according to hydraulic scaling laws, so that the morphology, physical processes, relative magnitude of forces, and features in the full-scale "real" river (the prototype) are closely modeled, or they may be analog models for which exact scaling is relaxed, in which case the model provides only an idea of geomorphological processes and responses, without allowing for the same degree of comparison with the full-scale river. For hydraulic scaling, the length scale of the model is initially chosen, often by scaling down the particle sizes of the prototype bed material (for gravel-bed rivers, model scales of 1:30 are typical). The

**Figure 21.2** The flume in the Department of Geography, University of Western Ontario, inset shows a braided river pattern formed in the flume (photo courtesy of P. Ashmore)

channel gradient in the model is the same as the prototype. The discharge scale is calculated by taking the discharge of the prototype and reducing this value by the 2.5 power of the length scale. For example, a prototype discharge of 20 cubic meters per second gives a scaled model discharge of 4 liters per second. The time scalc for the model is the square root of the length scale; a length scale of 30 gives a time scale of between 5 and 6, meaning that flow velocity in the model is about 1/5 to 1/6 of that of the real river. Consequently, equivalent geomorphic processes operate 5 to 6 times faster in the model than in the prototype.

Once the physical model is constructed, it is possible to undertake experiments using the model, in which key variables are controlled in an attempt to understand system behavior. An advantage of physical modeling is the ease with which controlling factors can be manipulated during experiments, relative to the difficulties that are encountered making controlled observations in the field. Since these models are designed to be simpler than real-world systems, data measurement and analysis are relatively straightforward in comparison. Moreover, such modeling allows for intense data measurement procedures. Although construction of these models may involve high initial costs, this type of modeling work, once in operation, may be cost- and time-effective relative to field studies. Physical model experiments allow for accurate and precise control of variables, provide valuable insights to refine and test scientific theories, and can contribute to further investigations using other methodologies. The critical insights that such models can provide to classic and new research topics in physical geography is increasingly being realized, with application of such models to topics covering a full range of temporal and spatial scales.

The following steps are involved in physical modeling. As with conceptual modeling, the system to be studied and the spatial and temporal scales under investigation must be clearly stated. The critical processes and properties of the system must be identified, and those which will be manipulated and those which will remain constant in the experimental phase should be determined. Formal scaling criteria for the system under examination must also be met. During the experimentation phase, variables may be modified as necessary and measurements made. The data are then analyzed and interpreted to identify key processes, properties and relationships. Finally, model results are compared to relevant data for the full-scale system to evaluate model performance, and on the basis of the model results, changes can be made to the physical model itself or to the experimental design.

## Mathematical models

Mathematical modeling is widely used in most branches of physical geography, and represents a powerful tool that has provided many remarkable new insights into the functioning of environmental systems. Simply stated, mathematical models consist of a series of quantitative relations among variables that represent the underlying theory of the system being studied. Mathematical models can vary significantly in their underlying consideration of a phenomenon, with some models attempting to correlate system properties while others focus on mathematical expressions that embody system and/or process functioning and are more explanatory in nature. These two distinct approaches to mathematical modeling will be discussed under two broad headings: **empirical models**; and **process-based models** (which may include physical, chemical and/or biological components). In addition, the manner in which forms and processes are treated may differ in another critical way. Some mathematical models are based on the premise that cause-and-effect is direct, while others incorporate a degree of randomness in their construction. We shall consider this issue under the heading of "determinism and stochasticism."

### Empirical Models

These models can be employed in cases where it may be difficult, or not the goal of a modeling exercise, to create a model that explains and accounts for the underlying system operation. This may arise from a lack of knowledge of processes, or because a modeler may decide that an understanding of system processes is not required for a particular study. In such cases, data may exist for what are thought to be key system variables that can be used in the development of an empirical model (which is sometimes referred to as a black box model). An empirical model does not account for underlying processes, but rather data for two or more variables are used to define functional relationships that capture the main system trends. This may involve, for example, simple linear curve fitting, multivariate regression analysis, or more complex spatial analysis within a GIS framework (see Chapters 18 and 22). Models involving empiricism often have the requirement that some external variables (such as climate) do not change between model generation and application. The relative simplicity of empirical models often makes them desirable in comparison to process-based models, with the latter generally requiring a greater understanding of the system. If a simple empirical model successfully simulates data observations, then the model may provide a powerful summary statement, especially if the processes assumed to underlie the functional relation are real (which it may not be possible to determine). Empirical models are often used to predict behavior when detailed process data and/or understanding do not exist, or when data are available for only certain restricted variables. Empirical models are often the precursors to later, more explanatory, process-based models.

## Process-based Models

These models explain and provide understanding of system operation by defining mathematical expressions for the governing principles and processes. Such models have high explanatory power in comparison to empirical models, and are often preferred for this reason. When a process-based model is derived through careful analysis and understanding of system processes, the model may be accepted even if its predictive capability is not as strong as might be desirable. In this case, the model may be preferred because it attempts to *explain* the inner workings of the system, which may be considered to be as important as empirical confirmation. Nonetheless, many models that are largely process-based, or strive to be so, still involve some degree of empiricism in certain model parameters, as complete explanation is not always possible. For this reason, hybrid models (involving both process-based and empirical components) are common. In such models, certain aspects of the system may be explained by mathematical, physically meaningful statements, with the inclusion of empirically-determined parameter(s). Prior to the computing era, process-based mathematical models were designed so that equations could be solved analytically, often limiting model complexity and making their solution time-consuming. Advances in computing power and numerical techniques now allow complex sets of equations to be solved with ease within a computer modeling framework, and have broadened their impact on the discipline. For example, the development of numerical approximation techniques and corresponding increases in computing power in recent decades now allow for the solution of very sophisticated, process-based models, such as Global Climate Models (GCMs).

## Determinism and Stochasticism

These are concepts that provide a basis for describing two ways in which processes can be treated in a mathematical model. A **deterministic** worldview involves the belief that there is a link between initial conditions and final conditions (that is, if all conditions are known, then the system output can be predicted), while a **stochastic** worldview considers randomness to be inherent to the system (Smart 1979). Models may be constructed that are deterministic, stochastic or some combination thereof. Even if it is believed that a phenomenon is inherently deterministic, it is often the case that stochastic elements must be included to account for the part of the problem that cannot be defined deterministically. Stochasticism may be treated using specialized methods, or may appear as empirical parameters that encapsulate some suitable averaging of the random behavior (in such cases, there is expected to be random error associated with model results). For example, complete deterministic understanding of turbulence in rivers is at present impossible, and impractical. Because of this, a statistical mechanics approach to this problem is often adopted, which relates the behavior of individual atoms/molecules to bulk system properties. While some models may be fully deterministic or fully stochastic, hybrid models, which include both elements, may prove valuable (Vogel 1999); stochastic elements may allow deterministic models to more faithfully preserve relationships between the model and observed empirical data, while deterministic components allow stochastic models to incorporate internal system functioning.

Steps Involved in Mathematical Modeling

Mathematical modeling involves the following steps. In addition to identifying clearly the system to be studied and the spatial and temporal scales under investigation, the type of mathematical model and its elements (empirical or process-based, stochastic or deterministic), as well as relevant processes and properties, must be identified. The parameters to be manipulated and those to remain constant should be defined, and the assumptions and limitations of the model clearly stated. Appropriate data for model input, development, calibration and/or evaluation (see Box 21.2) must be identified and

## Box 21.2   How Do We Evaluate Mathematical Models?

Let us consider a mathematical modeling approach to determining how ice cap morphology and dynamics may respond to climate change. An important question to ask is: how reliable are such predictions? We address this question by using the example of a model that incorporates both ice sheet dynamics and subglacial hydrology, and which has been applied to the Vatnajökull Ice Cap, Iceland (Marshall et al. 2005).

Model evaluation, which often consists of model confirmation by comparing results with real-world measurements, is one way of assessing model performance (and is perhaps the most well recognized). However, there are a series of steps prior to this that may improve the reliability of predictions. First, we need to evaluate the construction of a model itself and recognize its strengths and weaknesses. The strength of a model depends on how well a host of physical processes are understood and simulated; in the case of ice cap modeling, the processes of internal deformation of the ice and basal flow (sliding over the bed and subglacial sediment deformation) must be well constrained and adequately represented in the model, and the subglacial topography must be resolvable at the grid cell dimension used to numerically solve the governing equations. Field studies and remote sensing observations often provide data for constraining model inputs and may be used in the calibration of physical process equations in the model.

Once the model is running, simulations conducted at different resolutions (i.e., changing the size of grid cells making up the glacier and underlying topography) can be used to assess the model's usefulness at different scales and to test consistency of

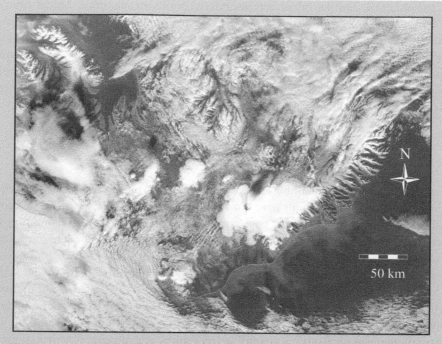

**Figure 21.3**   The Vatnajökull Ice Cap, southeast Iceland 11/07/2004 (Image courtesy of MODIS Rapid Response Project at NASA/GSFC)

results (do the predictions at different resolutions converge?). In addition, modeling experiments with different process parameterizations can illuminate the sensitivity of model forecasts to specific sources of uncertainty. For example, Marshall and his collaborators undertook model runs with and without certain processes included in the model, such as longitudinal stress coupling in the ice dynamics solution, and analyzed the ability of different model constructions to simulate the main features of the Vatnajökull Ice Cap (Figure 21.3). Initial model runs revealed that the predicted average ice thicknesses were too great relative to field observations for Vatnajökull Ice Cap. But further models runs, which included hydrological regulation of basal flow, longitudinal stress coupling, and the subglacial heat flux from geothermal cauldrons, reduced average ice thickness of the ice cap reconstructions to more reasonable values. The process that contributes the greatest amount of error or uncertainty in the prediction may be the best candidate for targeted field studies. In the case of Vatnajökull Ice Cap dynamics, a better understanding of the controls and spatial patterns of basal flow clearly require more constraint.

The ability of model simulations to reproduce control cases, such as a reconstruction of the present-day situation, must be assessed before forecasts (i.e., future predictions) can be seen as being meaningful or useful. If models prove skillful at replicating control cases, they have the potential to become a valuable tool for assessing the future states of complex systems.

collected/obtained, and the initial and boundary conditions identified. Model runs can then be undertaken, manipulating and changing the controlling variables as required. Sensitivity analysis can be used to determine how the model output depends on the input parameters and to examine the effect that changing different variable conditions or parameter values has on system behavior. Model results are then analyzed and interpreted. Finally, as outlined in Box 21.2 and pages 369–72, model evaluation should take place to assess the strength of the model.

## Modeling in Human Geography

Researchers in most branches of human geography question and/or reject the assumptions necessary for modeling, or simply have not adopted modeling in their work (see Chapters 3 and 5). The issues that are most commonly questioned include: the assumption that abstracting a portion of reality is legitimate and meaningful; that the model is representative of all instances it refers to; that the abstraction/model is an adequate and effective representation of reality; and that knowledge derived from the model can explain reality.

There are important reasons why some human geographers are skeptical about models and the assumptions that underpin them. While, for example, models may be useful in policy making, this utility does not come without cost. One obvious cost is that the simplifications that are necessary to quantify geographical phenomena may necessitate the removal of, or poorly express, an essential component of human geography: human behavior. Indeed, individual human behavior is often unpredictable and may be difficult to capture in mathematical terms; consequently, it does not lend itself easily to modeling. Models used in human geography often only seek to explain that portion of geographical reality that is quantifiable, and do not often attempt to explain individual behavior. This inherent limitation of models in human geography leaves a wide gap that human geographers have tried to fill in various ways. In the past, human geographers have often suggested that only one approach to the discipline was viable, sometimes criticizing other views. A recent trend in human geography suggests that integrating different conceptual approaches and analytical tools can lead to a more thorough comprehension of geographical phenomena. The integration of modeling with qualitative tools can perhaps provide an adequate response to the limitations of each approach.

Within the history of human geography, the legitimacy of modeling and the related assumptions have been fully accepted within the tradition known as **spatial science**, or more broadly quantitative human geography (Livingstone 1992). The quantitative approach has had a strong influence on contemporary geographical research, and many of the models currently used in human geography are rooted within this framework. Spatial science pervaded human geography throughout the 1950s and 1960s. While human geographical thought developed into different traditions in the following decades, developments in computing technology that began in these same decades allowed for rapid advances in mathematical modeling. For example, the early development of GIS was independent of spatial science, but many geographers later came to realize the possibilities of GIS, not just for data representation and management, but also for analysis and modeling. Over the past few decades this interaction among the various sub-fields of quantitative human geography has resulted in the development of spatial analysis for which several approaches are possible; some modelers use advanced statistical methods to explore empirical patterns among pertinent variables while others use mathematics and spatial statistics as a language to develop and test explanatory theories. Many quantitative human geographers work with the analytical methods used in spatial analysis, making the two fields somewhat indistinguishable.

Human geography is not a unified and cohesive discipline; indeed, a number of fields can be identified within its core (Fellman et al. 1997). Each sub-discipline is somewhat independent, such that its foundations are not necessarily shared by other sub-disciplines, and each one is often affiliated with other social science disciplines outside of geography; for example, economic geography can be viewed as a sub-discipline of economics. The sub-disciplines that are perhaps the most strongly rooted in the modeling tradition are economic and urban geography, and adaptations of economic geography models (such as those used to study urban land use) have become an approach that many urban geographers adopt. Similarly, medical or health geography, a sub-discipline whose importance has grown with concerns about the spread of diseases, uses quantitative models to analyze the spread of epidemics and the environmental or socio-economic determinants of disease.

The various sub-disciplines of human geography are subject to a dynamic evolution of thought (see Chapter 3). Even sub-disciplines that have traditionally accepted the use of certain models have, at times, questioned the validity of some of their early models, and disciplines that have traditionally refused the use of models may come to accept or even champion them. For example, over the last few decades dramatic changes in the world economy, such as the frequent shocks induced by large variations in oil price, the collapse of centrally planned economies in Europe, and the phenomenon known as globalization, have led many researchers to question the foundations of classical economic geography, and to reject earlier models and their conceptualizations of the world economy (Knox et al. 2003). At the same time, within the sub-discipline of urban geography, there has been a recent resurgence in the use of models, particularly with regard to the issues of land use change and urban growth. The increased use of these and other models has partly been made possible by the increased availability of data and improved processing tools. Most importantly, however, their development has been driven by an increased acceptance of models in local collaborative decision-making environments (Herold et al. 2005).

## Examples of models in human geography

While conceptual models are used in human geography to synthesize and communicate relationships among essential variables and processes associated with the phenomenon under consideration, mathematical models are in common use and are the focus of the following section. The mathematical models currently used in the various sub-disciplines of human geography can be characterized as: (a) theoretical, process-based models (some of which are referred to as "classical" models), which are explanatory and often have considerable generality, but are not always capable of adequately explaining observed phenomena; and (b) empirical models, that have a weaker theoretical background, but that nonetheless may sometimes have the capacity to predict observed phenomena.

Some of the early mathematical models are important as they stand on a theoretical ground and they share a common objective; that is, to explain geographical phenomena based on variables such as space, location, or distance. Their relative simplicity is both their greatest strength and their primary weakness; the latter because human activities cannot often be explained solely by a few key variables. To achieve their objectives, these models rely on a set of assumptions; any attempt to relax the assumptions increases the models' capability to explain observed phenomena, but weakens their power and generality.

Empirical models are calibrated to explain specific instances. Such models are most useful in a variety of applied contexts and are valuable tools for decision makers. Such models are developed using spatial analysis, within the framework of statistical analysis (and may make use of GIS) (see Chapters 18 and 22). As an example of this type of mathematical model, a researcher may collect data both on air quality and disease rates for a given region over a specified period of time to analyze the relationship between air pollution and cancer incidence in a population. These data are used to calibrate the model parameters, which will help a researcher determine whether air pollution is correlated to the incidence of cancer. Such models typically involve two or more key system variables, and may describe mutual relationships, such as a spatial interaction, or possible causal relationships among variables, which can be highlighted using regression analysis. For example, the cancer incidence model may take a multivariate form, where the disease incidence is a function of air pollution, but also of demographic and socio-economic factors recorded at specified points (e.g., census tracts). In most cases, a stochastic element is incorporated in the model, to express uncertainty in the relationship. In the case of cancer incidence, for example, even the most complete set of environmental and socio-economic factors cannot completely explain the incidence of the disease, which depends partially on variables such as genetics and diet that cannot be included in the spatial model. The stochastic element accounts for the portion of the phenomenon that the model cannot capture, and may lend itself to further investigation. Once the model parameters have been computed, the data collection and analysis can be repeated for different regions and time periods, enhancing the generality of the model. In conducting this exercise, however, researchers have come to realize that spatial data possess unique characteristics that affect parameter calibration and the reliability of the model's estimates (Bertazzon et al. 2006). Complex analytical procedures now exist to overcome these problems.

Expanding on the ideas discussed above, a specific example making use of multivariate regression is now introduced below and in Box 21.3. Many primary health concerns of our societies are spatial in nature: detection and monitoring of environmental health hazards; prompt and efficient response to epidemic outbreaks; and effective accessibility to health care services – all of these rely on spatial analysis. Spatial epidemiology – also referred to as geographical epidemiology – is concerned with the study of the spatial patterns of disease and mortality, and with the determinants of disease and their dynamic interaction in space and time (Waller and Gotway 2003). For example, heart disease is one of the leading causes of death in the developed world. In addition to non-modifiable risk factors, such as age, gender, and genetic background, the disease has been found in association with modifiable risk factors, such as stress, limited physical activity, smoking, high intake of calories, and high proportion of saturated fats (Ahlbom and Norell 1984). These modifiable risk factors, in turn, are related to demographic and socio-economic characteristics (such as age, occupation, and income), which display a specific spatial distribution and can be measured by census variables. A multivariate regression model on these variables can link the disease prevalence to these demographic and socio-economic variables (Box 21.3), providing a realistic picture of where in a given city higher disease incidence may be expected in the near future, which housing policies increase disease prevalence, and which locational decisions promote accessible health care services to the population at risk (Bertazzon and Olson 2008).

Two theoretically-based, mathematical models are now considered: (a) agricultural and urban land use (O'Sullivan 2003); and (b) spatial interaction and innovation diffusion (Bailey and Gatrell 1995). Perhaps the best-known example of a *spatial economics* model is Johann Heinrich von Thünen's (1783–1850) agricultural land use model, which was presented in his book, *The Isolated State* (1826). As a landowner, von Thünen was interested in the factors that help farmers maximize returns. His analysis focuses on the variables that determine profit: namely, an optimal use of the land and the minimization of transport costs. As transport costs in the model are a function of space and distance, the

---

**Box 21.3   Spatial Epidemiological Modeling**

An important aspect of modeling in human geography is the need for a careful interpretation of the analytical results, often supported by qualitative knowledge of the phenomena under examination. A multivariate regression model was used to examine patterns of disease in the city of Calgary, Canada (Table 21.1). Four socio-economic variables correlate significantly with the disease prevalence: "income" (positive correlation), "post-secondary, non-university education" (negative correlation), "grade 13 or lower education" (positive correlation), and "families with children" (negative correlation).

The analysis of cross-correlation among variables allows the researcher to infer richer relationships than those emerging simply from the multivariate regression coefficients. For example, the positive relationship between disease and income suggests that disease rates are higher in areas inhabited by mature professionals, possibly indicating a latent age factor. This relationship is also strongly affected by extreme income values (i.e., areas of very high and very low income: since this variable represents median family income, it is likely to display its highest values for one-person families). There are also links between disease and various categories of lonely persons, ranging from singles with very high income to single parents, or widowed persons. There is a dichotomous spatial pattern: high rates of disease prevalence are found in areas dominated by wealth and high social status, as well as in areas characterized by lower social and economic status. Nevertheless, there is a less urgent need for proactive social and health policies in areas of the city dominated by the variables "income" and "post-secondary, non-university education," which are indicators of relatively high economic and social status (although they do not guarantee well-being), than in areas of social and economic concern where the variables "grade 13 or lower education" and "families with children" dominate.

**Table 21.1** Variables incorporated in and output (correlation matrix) from a multivariate regression model used to examine patterns of disease in the city of Calgary, Canada

| Variable category | Variable name | Variable definition |
|---|---|---|
| Dependent variable | cases | Number of cardiac catheterization cases |
| Demographic variables | a45_54 | Number of residents aged between 45 and 54 years |
| | a55_64 | Number of residents aged between 55 and 64 years |
| | a65pl | Number of residents aged 65 years or older |
| Family variables | single | Number of single-person families |
| | w_d_s | Number of families with widowed, divorced, or separated persons |
| | sing_p | Number of single-parent families |
| | 2p.wchld | Number of families of two parents with children |
| | mar.claw | Number of couples married or living in common law |
| Education variables | gr13ls | Number of residents with grade 13 or lower education |
| | non_uni | Number of residents with post-secondary, non university education |
| | uni | Number of residents with university education |
| Economic variables | f.m.inc | Family median income |

| | Demographic | | | | Family | | | | | Education | | | Economic |
|---|---|---|---|---|---|---|---|---|---|---|---|---|---|
| | cases | a45_54 | a55_64 | a65pl | single | mar.claw | w_d_s | 2p_wchld | sing_p | gr13ls | non_uni | uni | f_m_inc_k |
| cases | 1.000 | 0.041 | 0.569** | 0.794** | 0.105 | −0.377** | 0.577** | −0.495** | 0.317** | 0.181* | −0.235** | −0.051 | −0.229** |
| a45_54 | 0.041 | 1.000 | 0.491** | −0.104 | −0.326** | 0.273** | −0.403** | 0.448** | −0.295** | −0.275** | −0.382** | 0.381** | 0.596** |
| a55_64 | 0.569** | 0.491** | 1.000 | 0.415** | −0.237** | 0.047 | 0.051 | −0.074 | −0.028 | −0.026 | −0.292** | 0.141 | 0.195** |
| a65pl | 0.794** | −0.104 | 0.415** | 1.000 | 0.096 | −0.416** | 0.628** | −0.555** | 0.194** | −0.098 | −0.334** | 0.216** | −0.099 |
| single | 0.105 | −0.326** | −0.237** | 0.096 | 1.000 | −0.911** | 0.550** | −0.737** | 0.469** | 0.163* | −0.046 | −0.115 | −0.642** |
| mar.claw | −0.377** | 0.273** | 0.047 | −0.416** | −0.911** | 1.000 | −0.790** | 0.819** | −0.620** | −0.224** | 0.127 | 0.132 | 0.665** |
| w_d_s | 0.577** | −0.403** | 0.051 | 0.628** | 0.550** | −0.790** | 1.000 | −0.798** | 0.702** | 0.378** | 0.035 | −0.322** | −0.639** |
| 2p_wchld | −0.495** | 0.448** | −0.074 | −0.555** | −0.737** | 0.819** | −0.798** | 1.000 | −0.450** | −0.087 | 0.044 | 0.053 | 0.572** |
| sing_p | 0.317** | −0.295** | −0.028 | 0.194** | 0.469** | −0.620** | 0.702** | −0.450** | 1.000 | 0.661** | 0.197** | −0.620** | −0.782** |
| gr13ls | 0.181* | −0.275** | −0.026 | −0.098 | 0.163* | −0.224** | 0.378** | −0.087 | 0.661** | 1.000 | 0.251** | −0.919** | −0.698** |
| non_uni | −0.235** | −0.382** | −0.292** | −0.334** | −0.046 | 0.127 | 0.035 | 0.044 | 0.197** | 0.251** | 1.000 | −0.611** | −0.294** |
| uni | −0.051 | 0.381** | 0.141 | 0.216** | −0.115 | 0.132 | −0.322** | 0.053 | −0.620** | −0.919** | −0.611** | 1.000 | 0.691** |
| f_m_inc_k | −0.229** | 0.596** | 0.195** | −0.099 | −0.642** | 0.665** | −0.639** | 0.572** | −0.782** | −0.698** | −0.294** | 0.691** | 1.000 |

*Notes*: * and ** denote the 0.95 and 0.99 significance levels, respectively
*Source*: after Bertazzon and Olson (2008)

model is based on a set of simplifying assumptions, which define his pre-industrial *isolated state*: a circular, homogeneous plain, with a single, central market, and no external (international) relationships. This completely flat, "isolated state" has a uniform soil and climate, no rivers or mountains, and is surrounded by wilderness. Other simplifying assumptions are that farmers behave rationally to maximize profits, transport their own goods directly to market in the central city using oxcarts (because there are no roads), and have good knowledge of the costs and distances they will encounter.

A mathematical formula determines the profit associated with each crop as a function of its transportation cost,

$$R = Y(p-c) - Yfm \qquad (21.1)$$

where $R$ is the land rent; $Y$ is the yield per unit of land; $c$ is the production expense per unit of commodity; $p$ is the market price per unit of commodity; $f$ is the freight rate; and $m$ is the distance to market. Thus, as the distance from the central city increases, crops of lower and lower values are produced and the model predicts patterns of agricultural activity in the form of concentric rings (Figure 21.4).

Von Thünen's model has been extended and is currently applied outside its original agricultural context. In urban geography, for example, it has been used to explain land use; typically commercial and service activities are located in the inner city (Central Business District) with manufacturing and warehousing located in the outer rings. It has also been used to analyze monocentric and polycentric cities, and the dynamics of urban land use (O'Sullivan 2003).

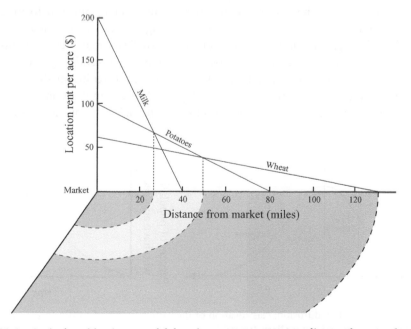

**Figure 21.4** Agricultural land use model: location rent curves versus distance from market for different commodities (after Dicken and Lloyd 1990)

Spatial interaction models, a second case examined here, focus on flows of people, goods, or ideas between places. These models apply to a variety of human activities, ranging from commuter flows between cities, to visits to a hospital or a shopping mall. Drawing an analogy with physics concepts, quantitative geographers have often used the concept of gravity, so that the expected interaction between two places is proportional to the mass of their power of attraction, which is reduced by the friction of distance. The first important step in the definition of such models is the identification of the relevant indicators, or what factors exert the attraction: typical examples are population, services, or number of jobs. The second is the identification and measurement of a meaningful distance for the phenomenon under scrutiny: examples are distance along a road network, travel time, or travel cost. The final step is the calibration of the parameters of the relationship. By estimating such parameters, the researcher comes to a deeper understanding of the mechanisms ruling the interaction. One aspect that has received much attention by geographers is the role of distance, and more specifically the importance of the separation between the two places, and how strong is the friction it exerts on their interaction.

An important class of spatial interaction models is the innovation diffusion model based on the theoretical work of Swedish geographer, Torsten Hägerstrand (1916–2004). He observed that there was a striking spatial order in the adoption of innovation, and he created a theoretical model to simulate the diffusion process and to attempt to predict it. The relevance and importance of his contribution relates to his introduction of time in the analysis of human geographic processes, and his work on space-time geography represents a fundamental innovation in quantitative human geography. To this day, however, spatiotemporal processes remain difficult to model, and much research is still devoted to this problem.

In Hägerstrand's model, a diffusion process can be dissected into a number of stages (Figure 21.5), allowing the researcher not only to understand, but also to predict the outcome of an observed process. The structure and stages of the process involve: penetration, with initial agglomerations and a few adopters (usually in small areas); expansion, by radial dissemination involving the creation of new agglomerations and a significant increase in the number of new adopters; and saturation, marked by a significant shrinkage in the number of adopters and a decreasing rate of adoption. Current applications of innovation diffusion models range from the penetration of technological innovations (such as the

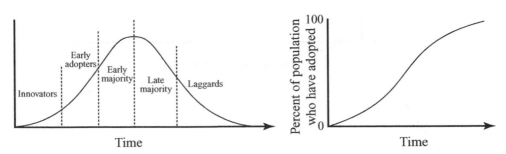

**Figure 21.5**   Innovation diffusion: classification of the various stages of a diffusion process according to Hägerstrand (after Clark 1984)

Internet or Blu-ray DVD players), to the analysis of disease transmission in the sub-discipline of medical geography and epidemiology.

Gravity and other spatial interaction models are also used in geographic and epidemiologic studies to analyze the way in which diseases are transmitted: in a contagious transmission, nearby individuals tend always to be affected because distance plays a very strong role. Diseases, however, can also appear in some places and then trickle down through the nodes of urban hierarchies and transportation networks. An example of hierarchical transmission was the 2003 SARS (Severe Acute Respiratory Syndrome) epidemic, which originated in a remote Chinese province. SARS was transmitted along the major transportation nodes of international flight routes and in this manner quickly reached distant locations, such as Toronto, Canada (Bowen and Laroe 2006).

## Model Evaluation

Model evaluation is an essential stage in any modeling exercise, whether the model is rooted in the physical geography or the human geography tradition. Without critical assessment of a model, the significance that can be attached to results is unclear. That being said, it is an issue of much debate as to what exactly constitutes a "critical test" of a particular model and its performance. The following discussion covers some of the essential characteristics that need to be considered during model evaluation.

First of all, the quality of a model is not simply equivalent to empirical agreement of model results with data observations (Hodges and Dewar 1992). Model quality should cover a range of factors, such as soundness of underlying principles, quantity and quality of input parameters, and the consistency of model results with observed data (model confirmation). Moreover, model evaluation should focus on both positive and negative aspects and outcomes of the modeling exercise (Oreskes 1998).

The actual content of the model is important to consider when evaluating its overall "quality." Flawless representations using conceptual, physical or mathematical models are not possible for most systems, and it is generally accepted that assumptions and simplifications must be made. A researcher must consider what "flaws" in a model's representation of the system must be tolerated, while still providing insights regarding essential system characteristics (Martin and Church 2004). For example, it may be possible to model some suitably "average" behavior of what may be very complex processes and changes in form; internal consistency is thus preserved, despite simplification. Other factors to consider are model complexity, the number of input variables, and the quality of data required to produce "acceptable" results. Finally, model confirmation is a key element in model evaluation. This involves comparison of model results with independent empirical data to assess the degree of consistency that is achieved. The philosophy of the science community has dedicated much effort to the question of whether hypotheses or theories can ever be demonstrated to be "correct" through model testing and empirical confirmation. As was noted in Chapter 3, Karl Popper (1902–94) put forth the notion of falsification, in which he claimed that science only progresses by conclusively demonstrating theories to be incorrect, and that they can never be proven true.

Depending on the goals of a study, the various factors outlined above may be weighted differently when evaluating model performance. However, models should not be relied on to provide unequivocal answers to explain phenomena. They should be used in conjunction with other relevant approaches and methodologies during the processes of theory development and testing.

# Conclusion

Although models represent idealizations of reality, they can be effective in understanding the operation of natural and human phenomena. Models are not utilized uniformly in all branches of geography. In particular, models are widely used in physical geography, and have been accepted in only some branches of human geography. Models can provide speculative formulations of a problem, test our understanding of a phenomenon, or enable us to see how a system responds to changes in input variables or model parameters. Although assumptions must be made and simplification is necessary, fundamental system properties must still be preserved; it is important to strike a critical balance if modeling is to be a meaningful exercise. It is the assumptions and simplifications associated with modeling that underlie many of the concerns of human geographers. Moreover, it often remains unclear how non-linear uncertainties propagate through complex models typical of environmental and human/societal systems.

Three types of models used in physical geography were discussed in this chapter. Conceptual models are used at all stages of research, and represent a way to synthesize ideas and understand system operation. Physical models are powerful tools in scientific investigation as they allow for direct and precise control of variables, which is rarely possible in field-based studies. Mathematical models likewise allow for control of input variables and parameters by systematic variation of variables during model runs. The relative simplicity of empirical models often makes them desirable, while the explanatory power of process-based models is considered a valuable attribute. While modeling is not utilized in many branches of human geography, some subfields utilize conceptual or mathematical models. Many of the models developed in human geography are rooted within a spatial science tradition, and are aimed at identifying and establishing the universality of theories regarding the behavior of spatial phenomena; a distinctive trait of these models is the central role of distance in determining spatial relationships. However, the rigid assumptions often required in such models often limit their practical value and applicability.

Model evaluation remains one of the least well understood aspects of modeling, but should encapsulate factors including soundness of underlying principles, quantity and quality of input parameters, and the consistency of model results with observed data. Models can provide a powerful methodology in the study of systems considered by geographers, providing one line of evidence within the broader range of approaches and methodologies adopted in the study of geographic phenomenon.

# References

Ahlbom, A., and Norell, S. (1984) *Introduction to Modern Epidemiology*. Chestnut Hill, MA: Epidemiology Resources Inc.

Bailey, T. C., and Gatrell, A. C. (1995) *Interactive Spatial Data Analysis (E. The Analysis of Spatial Interaction Data)*. Harlow: Longman Harlow.

Bertazzon, S., and Olson, S. (2008) Alternative distance metrics for enhanced reliability of spatial regression analysis of health data. *Proceedings of the International Conference on Computational Science and its Applications Part I. Lecture Notes in Computer Science, volume 5072*. Berlin: Springer, 361–74.

Bertazzon, S., Micheletti, C., Critto, A., and Marcomini, A. (2006) Spatial analysis of the bioaccumulation of distributed pollutants in clams *Tapes philipinarum*: the Venetian Lagoon (Italy) case study. *Computers, Environment and Urban Systems* 30: 880–904.

Bowen, J. Jr, and Laroe, C. (2006) Airline networks and the international diffusion of severe acute respiratory syndrome (SARS). *The Geographical Journal* 172: 130–44.

Clark, G. (1984) *Innovation Diffusion: Contemporary Geographical Approaches*, CATMOG 40. Norwich: Geobooks.

Dicken, P., and Lloyd, P. (1990) *Location in Space* (3rd edn). New York: Harper & Row.

Fellman, J., Getis, A., and Getis, J. (1997) *Human Geography. Landscapes of Human Activities* (5th edn). Madison, WI: Times Mirror Higher Education Group.

Herold, M., Couclelis, H., and Clarke, K. C. (2005) The role of spatial metrics in the analysis and modeling of urban land use change. *Computers, Environment and Urban Systems* 29: 369–99.

Hodges, J. S., and Dewar, J. A. (1992) *Is it You or Your Model Talking? A Framework for Model Validation*, Rand Publication Series R-4114-AF/A/OSD. Santa Monica, CA: Rand Corporation.

Knox, P., Agnew, J., and McCarthy, L. (2003) *The Geography of the World Economy* (4th edn). London: Arnold.

Livingstone, D. (1992) *The Geographical Tradition: Episodes in the History of a Contested Enterprise*. Oxford: Blackwell.

Marshall, S. J., Björnsson, H., Flowers, G. E., and Clarke, G. K. G. (2005) Simulation of Vatnajökull ice cap dynamics. *Journal of Geophysical Research* 110: F03009, doi:10.1029/2004JF000262.

Martin, Y., and Church, M. (2004) Numerical modeling of landscape evolution: Geomorphological perspectives. *Progress in Physical Geography* 28: 317–39.

Oreskes, N. (1998) Evaluation (not validation) of quantitative models. *Environmental Health Perspectives* 106: 1453–60.

O'Sullivan, A. (2003) *Urban Economics* (5th edn). Boston: McGraw-Hill.

Pyrce, R., and Ashmore, P. (2003) Particle path length distributions in meandering gravel-bed streams: results from physical models. *Earth Surface Processes and Landforms* 28: 951–66.

Smart, J. S. (1979) Determinism and randomness in fluvial geomorphology. *Eos* 60: 651–5.

Vogel, R. (1999) Stochastic and deterministic world views. *Journal of Water Resources Planning and Management* 125: 311–13.

Waller, L. A., and Gotway, C. (2003) *Applied Spatial Analysis of Public Health Data*. New York: Wiley.

## Additional Resources

## Physical geography

Chalmers, A. F. (1999) *What is This Thing Called Science?* (3rd edn). Indianapolis: Hackett Publishing Company (see Chapter 3: "Experiment"). Provides an introductory discussion on the concept of experimentation. An understanding of experimentation allows for an appreciation of how the control and manipulation of key variables can provide insights into system operation within a modeling context.

Haines-Young, R., and Petch, J. (1986) *Physical Geography: Its Nature and Methods*. London: Harper & Row (see Chapter 9: "Modeling") Provides an introduction to modeling at a level suitable for mid- to upper-level undergraduates. This chapter is a good supplementary reading for the material in the present chapter.

Mulligan, M., and Wainwright, J. (2004) Modeling and model building. In *Environmental Modeling: Finding Simplicity in Complexity* Mulligan, M. and Wainwright, J., eds., Chichester, John Wiley, 7–73. Provides an extensive overview to modeling and model building, and is a good complement to the material covered in this chapter.

Oreskes, N., Shrader-Frechette, K., and Belitz, K. (1994) Verification, validation, and confirmation of numerical models in the earth sciences. *Science* 263: 641–6. Discusses verification, validation and conformation of numerical models (similar to our mathematical models as defined herein) in the earth sciences, with a particular emphasis on the role of models in policy and decision making.

Richards, K., and Oke, T. R. (2002) Validation and results of a scale model of dew deposition in urban environments. *International Journal of Climatology* 22: 1915–33. An article that addresses the concept of formal scaling criteria for physical modeling.

## Human geography

Berry, B. J. L. (2004) Spatial analysis in retrospect and prospect. In *Spatially Integrated Social Science*, Goodchild, M. and Janelle, D., eds. Oxford: Oxford University Press, 443–5. Highlights the relevance of applied quantitative analysis in the social sciences, offering some thoughts about the roots and future of the discipline.

Fotheringham, A. S. (2006) Quantification, evidence, and positivism, In *Approaches to Human Geography*, Aitken, S. and Valentine, G., eds. London: Sage, 237–50. A paper that offers a passionate perspective from a leading quantitative geographer on theory and applications of quantitative human geography.

Longley, P. (1998) Foundations. In *Geocomputation: A Primer*, Longley, P., Brooks, S., McDonnell, R., and MacMillan, B., eds. Chichester: John Wiley, 3–15. A view of the field of geocomputation, a recent development within the quantitative geography tradition.

Poon, J. (2004) Quantitative methods: past and present. *Progress in Human Geography* 28: 807–14. *Progress in Human Geography* is a leading journal in the field; this paper offers a critical perspective on quantification in human geography.

Webber, M. (2006) Classics in human geography revisited: Brown, L.A. 1981: *Innovation Diffusion: A New Perspective*. Methuen, London. *Progress in Human Geography* 30: 487–94. A commentary on one of the fundamental texts in the discipline, this paper provides an overview of the historic development of innovation diffusion models.

## Exercise 21.1   Conceptual Modeling

### Example: Physical geography

Regolith thickness is determined by the interaction of hillslope transport processes and bedrock erosion (see Chapter 7). Develop a conceptual model, in the form of a "boxes and arrows" diagram (see Figure 21.1), that illustrates the operation, relationships and feedbacks of forms and processes associated with the development of the weathered layer. You should base your model on a particular regional setting and include all the principal hill-slope transport processes that can be expected to operate in that location.

## Example: Human geography

Obtain a map of land use in your city or town as well as of the surrounding rural area. Can you observe any evidence of von Thünen's concentric pattern? Do you observe any distortion from a circular to a more complex shape? Are any major physical elements (rivers, roads, mountains, etc.) affecting the pattern? Provide a discussion of the reasons why the observed environment does/does not conform to von Thünen's theory (relate your argument to the assumptions underpinning the model).

# Chapter 22

# Geographic Information Systems

*Michael F. Goodchild*

- Introduction
- What is GIS?
- Spatial Thinking
- The Data Supply
- GIScience
- Conclusion

Keywords

Attribute
Ecological fallacy
Field
Geographic data
Geographic information system
Georeference
Geospatial data
Layer
Modifiable Areal Unit Problem

Place
Scale
Spatial context
Spatial data
Spatial dependence
Spatial heterogeneity
Spatial thinking
Uncertainty

## Introduction

Over the past few decades **geographic information systems** (GIS) have become one of the most important components in the arsenal of tools at the disposal of the geographer, as well as one of the most important ways in which the results of geographic research can be applied to the solution of everyday problems. This chapter begins by explaining what GIS are, and then outlines how they are used in research by geographers, as well as the principles that underlie those uses. A text on GIS that discusses these topics in greater detail is Longley et al. (2005).

## What is GIS?

At the core of a GIS is a **georeferenced** database. Such databases are distinguished from all other kinds by the fact that all of their records are given a location on the Earth's surface, usually in the form of coordinates, such as latitude and longitude. For example, a database maintained by an airline that stores the locations of all of the airline's aircraft, as well as their flight numbers, numbers of passengers, and other data is georeferenced, as is a database containing the digitally scanned contents of a map, or a database containing the sort of remotely sensed data discussed in Chapter 10. Indeed, many other types of data collected by geographers may have been georeferenced before being stored in a database. A database of customers that includes each person's street address is also in effect georeferenced, because street address can easily be converted to latitude and longitude through a process known as geocoding. Data that are georeferenced are termed **geographic**, **geospatial**, or **spatial** (though the last term strictly applies to any space, not only the space of the surface and near-surface of the Earth, which is the domain of GIS). The first of these terms is perhaps most appropriate for a book designed to be read by geographers, so it will be used here, but in practice all three terms are used virtually interchangeably.

A GIS is a collection of software, normally manipulated by its user through a single interface, and designed to perform a wide range of operations on geographic data. In fact today's GIS software is capable of performing virtually any conceivable operation on geographic data. It can generate maps in a minute fraction of the time required to do so by hand; compute the shortest distances between points and generate driving directions; keep inventories of assets that are distributed in space and help manage maintenance scheduling; detect patterns and outliers; test theories and hypotheses; and execute models that predict everything from the tracks of severe storms to the growth of cities.

This versatility reflects a basic truth about the computing industry: once the foundation has been built for handling a particular type of data, it is very simple and cost-effective to add a wide range of functions to the foundation. Thus Microsoft's Excel, for example, is a set of tools geared to the manipulation of data arrayed in tables or spreadsheets, and Word is a set of tools geared to manipulating text – GIS is similarly a set of tools geared to manipulating geographic data, though as will become clear to anyone who delves into GIS in any depth, the simple title *geographic* covers a vast array of options, making GIS inherently complex. Despite this, however, GIS has become enormously popular over the past two decades since it first became available commercially. It appears to appeal to people's basic fascination with maps and geography, it offers a comparatively straightforward way to solve a wide range of problems, and it is a powerful tool for working with geographic data. In recent years a number of very attractive tools have appeared that make certain limited GIS functions much easier to learn and use.

This popularization of GIS is exemplified by a number of websites that offer simplified versions of GIS to the general public. They include sites such as MapQuest that generate driving directions; sites such as Expedia that help users to search for local hotels; and

Google Earth, which quickly captured popular attention when it was released in early 2005. Many of the features of Google Earth, including rapid zooming to greater detail, searching for locations based on street address, and overlaying other information in complex *mash-ups* are features of GIS, offered in a package that is easy for even the most inexperienced user to learn and manipulate. Brown (2006) provides an excellent survey of interesting applications and extensions of Google Earth.

The history of GIS goes back to the 1960s, when several projects began to explore the use of computers to handle geographic data (Foresman 1998). Like any computer application, however, GIS merely automates what people had always been able to do by hand, and history offers abundant evidence of the basic ideas of GIS being applied well before computers were invented. But it was always very difficult to perform any kind of map analysis by hand, and GIS was quickly recognized as a tool that would allow geographers to implement methods that had previously seemed too tedious, too inaccurate, too slow, too complex, or too expensive to do manually. Instead of laborious hand counting, areas could be measured by a simple computer program once the necessary data had been input, and the overlaying of maps also proved to be almost trivially easy in a computer. GIS was hailed enthusiastically, particularly by geographers, who compared it to the earlier impact of the invention of the telescope or microscope in the richness of the research that it would make possible.

GIS first appeared as a commercially viable software product in the late 1970s. Today, virtually any researcher whose work concerns the surface or near-surface of the Earth will have some familiarity with GIS. GIS has been widely adopted as a research tool in archaeology, criminology, epidemiology, ecology, geology, and many other disciplines. But the discipline of geography has always had a special relationship with GIS. Many of the early developments were made by geographers, many of the companies selling GIS software were started by geographers, and the majority of the courses on GIS found in universities are offered through departments of geography.

This chapter takes a broad view of the nature of research, reflecting the range of applications of GIS. These stretch from basic research in pursuit of pure knowledge and motivated largely by human curiosity, through the application of such research in public policy, planning, and development, and from such mundane everyday activities as wayfinding to the larger project of maintaining the assets of a utility company. GIS is used in all of these, and courses on GIS will often cover the full range of GIS applications.

Geographers use GIS in their research, and they offer courses on GIS to others. But in addition, geographers have staked out a research area that focuses on the technology itself – this is research *about* GIS rather than research *with* or using GIS. It forms the basis of a field that is now generally known as geographic information science, or GIScience for short. The last section of this chapter looks at research methods in GIScience, after the previous sections have discussed the methods used by geographers to exploit the power of GIS as a tool.

## Spatial Thinking

At the heart of all applications of GIS are the concepts that we collectively describe as **spatial thinking**. A person can be said to be thinking spatially when he or she draws inferences from data arrayed spatially – data presented in the form of a map, for example, or data contained in a GIS database. The process of spatial thinking can be very powerful, since the eye and brain are enormously efficient at detecting patterns, seeing movement, and detecting anomalies in data when they are presented appropriately; and a range of powerful techniques known as spatial analysis have been developed to augment the intuitive power of the researcher. Spatial thinking is explored in a report of the National Research Council (2006), with particular emphasis on its place in primary and secondary education.

We say that "a picture is worth a thousand words," implying that thoughts can be conveyed pictorially with great efficiency, and we often rely on cartoons, icons, and other simple drawings to convey ideas. Edward Tufte has discussed the importance of effective visualization in many areas of human activity (Tufte 1983; and subsequent books by the same author), and has used many geographic examples. Geographers have been trained to work with maps and other spatial presentations of data, and may as a result have more highly developed skills at spatial thinking – but on the other hand we know as a result of research in cognitive psychology that spatial thinking skills vary widely in the population. In essence, the tools of GIS are designed to enhance our spatial thinking skills, by making it possible to present data in a variety of useful ways, and to expose meaning in data that would not otherwise be apparent to an observer.

The following paragraphs identify some of the more important and commonly encountered concepts of spatial thinking, and the ways they are implemented in GIS and used in research. They begin with simpler concepts and move on to the ones that play a larger role in gaining insight and understanding through GIS. The topics mentioned are certainly not exhaustive, but they give an impression of the kinds of concepts that geographers have in mind when they use GIS in their research:

## Places

Humans divide the world into pieces for many purposes, and give the pieces names and georeferences of various kinds. A GIS user might encounter placenames, addresses, and coordinates as ways of identifying **places**, but he or she might also make use of administrative units such as counties or wards that are commonly employed by agencies to aggregate and report social statistics, and might also use the regular grids of remote sensing.

## Attributes

In addition to placenames, humans attach characteristics of various kinds to places, including its population size, the average income of its inhabitants, its rate of growth, or its elevation. Geographers use the **attributes** of places to describe them, and to draw parallels between places with similar attributes.

## Spatial objects

In the simplest possible conceptualization, the geographic world is a space littered with things of various kinds, just like a tabletop (Figure 22.1). In a GIS, each thing must have

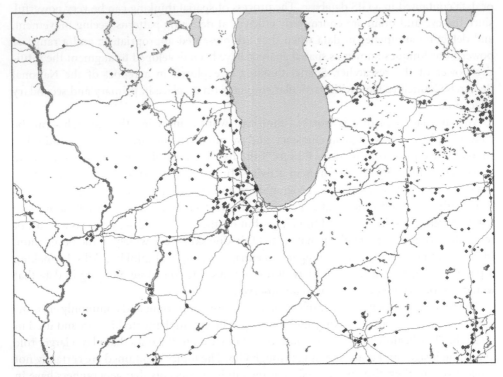

**Figure 22.1**   Geography as a collection of discrete objects littering an otherwise empty space: populated places (points), roads (lines), and major water bodies (areas) in the Chicago area

a geometric form as a point, a line, or an area. Each of these forms has associated measures such as length, area, or shape. GIS operations include the ability to evaluate such measures, and to link objects based on overlap or containment. For example, there are functions to compute the area of overlap of two given areas, or to identify whether a point lies inside or outside a given area – the latter operation might be used to determine the number of road accidents (points) falling in a given county, for example. Geographers use measures such as area to determine the amounts of land devoted to specific uses, or to evaluate changes in land use through time. Shape measures are used by landscape ecologists to determine the fragmentation of habitat, and its potential impacts on the ability of species to survive and reproduce.

## Maps

The geographic world is visualized through maps, which are planimetrically correct representations of the spatial distribution of phenomena. Maps simplify and **scale** the world, allowing the researcher to search for patterns and correlations. Maps are the primary form of graphic output of GIS, and their contents are major sources of input. Cartographers are familiar with many types of maps. Increasingly, computers are being used to visualize the

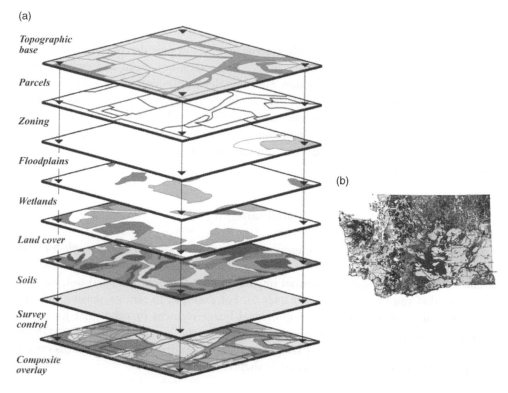

(a)

Topographic base

Parcels

Zoning

Floodplains

Wetlands

Land cover

Soils

Survey control

Composite overlay

(b)

**Figure 22.2** Geography as a series of layers, each describing the spatial variation of some property or set of properties over the area of interest (after US Geological Survey).

Earth's surface in its true near-spherical form, rather than in the flattened and distorted form that is familiar in maps and atlases. Google Earth is one example of such tools, many of which can also be found in full-blooded GIS.

## Multiple properties of places

We often conceptualize the geographic world as a series of **layers**, each layer mapping a specific property or class of properties (Figure 22.2). In developing a land-use plan, for example, a planner might consider such factors as existing land use, transportation, housing, groundwater vulnerability, or land ownership. Each factor would form one layer, and the planner's GIS would contain digital representations of each layer. The GIS would then be able to integrate the data contained in each layer, perhaps weighting each layer depending on its importance as a factor in determining appropriate land use. Layers are often compared in the search for relationships, such as between land use and soil type, or between average income and levels of educational attainment.

## Spatial heterogeneity

The geographic world is fundamentally **heterogeneous**; not only do conditions vary dramatically from one part of the Earth's surface to another, but social and even physical processes may show substantial variation. For example, county-level relationships between average income and educational attainment are likely to vary from one state to the next. Statisticians refer to such effects as *non-stationarity*. Geographers have devised a range of techniques, generally known as *place-based analysis*, for exposing these differences, mapping them, and drawing inferences about them, and this continues to be an active area of methodological research.

## Spatial sampling

Because the geographic world is so complex, for practical reasons we often sample it rather than attempt to describe it completely. There are many methods of sampling, and the differences between them have substantial impact on how research is carried out, and the inferences that can be drawn (see Chapter 6). For example, in remote sensing it is the *spatial resolution*, or the size of the individual image element or *pixel*, that limits the researcher's ability to make inferences. In climatology, the locations of weather stations impose similar limitations, while urban geographers must cope with the problems caused by the lumping of statistics into irregularly shaped reporting zones.

## Distances

Geographers often distinguish between *site*, or location specified in absolute terms, and *situation*, or location relative to other things. GIS can be used to measure distances using a variety of metrics, including straight-line distance, distance through a road or street network, or distance in a more abstract social network. One of the most popular GIS functions is the *buffer* command, which can be used to map areas within specified distances of points, lines, or areas. A researcher might buffer a major highway, for example, in an effort to understand the effects of its atmospheric pollution on nearby residents.

## Spatial context

One of the most powerful concepts in the geographer's toolkit is the notion that events at one location on the Earth's surface are determined by neighboring events, as well as by other factors present at the same location. By making a map, it is often possible to gain insight into the kinds of processes at work on the landscape, because the map or the researcher's own knowledge might link an anomaly to other possible factors in the anomaly's **spatial context**. Mapping the locations of disease incidence, for example, can often suggest potential environmental causes that can then be examined more systematically. Central to spatial context is the concept of *neighborhood*, the region of space with which an individual or organism interacts on a frequent basis.

## Spatial dependence

Numerous ways have been devised for measuring the degree to which the conditions at one location on the Earth's surface are similar to conditions nearby. The generality of this principle is often expressed as *Tobler's First Law of Geography*: that nearby things tend to be more similar than distant things (for several distinct perspectives on this concept see Sui 2004). This principle is the basis for many operations in GIS, including *spatial interpolation*, which is the process by which we make inferences about places based on data obtained at nearby sample points. For example, we use spatial interpolation in the making of weather maps, since the data on which they are based are obtained only at a limited number of points (weather stations), yet the maps show values everywhere. Spatial models, such as those discussed in some earlier chapters of this section, often exploit **spatial dependence** in making predictions based on conditions at nearby locations.

## Competition for space

Despite the previous comments about spatial dependence, in some cases social and physical processes produce vast contrasts on the Earth's surface – quite different than the kinds of smooth variation that characterize contour maps. One such process is the competition for space that occurs in retailing, or in territorial behavior, where the presence of one object makes other objects of the same type less likely in the immediate vicinity. Geographers have devised many methods for mapping the hinterlands, trade areas, or territories of such objects, and have implemented them in GIS in the form of procedures such as Thiessen polygons (Figure 22.3). GIS is also used to *design* service areas to meet specified criteria, and methods of location-allocation have been implemented in GIS to support the search for optimum locations for such central services as retail stores, hospitals, or fire stations (Ghosh and Rushton 1987).

## Directions

For most purposes it is safe to assume that the processes operating on the Earth's surface, and responsible for the patterns we find there, operate uniformly in all directions, in other words they are *isotropic*. But direction is sometimes important, for example in processes that are affected by solar radiation or by wind. GIS can be used to measure direction, to analyze directional data, and to use such data in models of non-isotropic effects. For example, a researcher might use such data in modeling the air pollution from highways subject to prevailing winds, or to predict the movement of a wildfire.

## Networks

Many geographic phenomena are limited to the nodes and links of linear networks, such as roads or rivers. Road networks are important in understanding traffic flows and

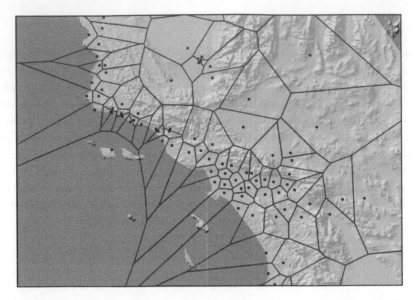

**Figure 22.3**   Thiessen polygons, defining the areas closest to each of the points in the diagram. Such methods are often used to estimate the trade areas, territories, or hinterlands of competing locations
*Source*: Longley et al. (2005)

evaluating highway development options; in planning evacuations during emergencies; and in designing optimum locations for services that make use of road networks. Hydrologists use GIS representations of river networks to model the downstream impacts of floods or chemical spills.

## Scale

The level of detail of a geographic data set is one of its most important characteristics. Any GIS database must simplify the infinite complexity of the world in order to achieve an effective representation in a computer of limited capacity, and this is often done by establishing a threshold level of detail. GIS procedures exist for changing this threshold through generalization, and also for simulating missing detail when only its broad characteristics are known. Scale is a large and complex topic, and geographers continue to explore its many facets.

## Modifiable areal units

When the fine detail described in the previous paragraph is deleted by aggregating data into larger spatial units, it is tempting to believe that the exact choice of larger units has a minimal effect on the results of any analysis. We use counties, for example, for reporting many statistics, and while the average size of counties is clearly important, we tend not to worry about the effects of the precise locations of county boundaries. Unfortunately it seems that results do depend very substantially on the choice of geographic analysis units, even at the same scale, and some early experiments using GIS showed that the effects can lead to widely varying conclusions. This has become known as the **modifiable areal unit problem** or MAUP. We normally differentiate it from the related problem of scale – the dependence of results on the level of aggregation – and the **fallacy of ecological inference**, which refers to making conclusions about individuals based only on aggregate data. The ability of GIS to aggregate data and to work with a variety of sizes and shapes of reporting zones makes it a very powerful technology to address both of these issues.

## Uncertainty

No representation of the geographic world can be exact or complete, so it is often necessary to describe the degree to which a representation leaves its users uncertain about reality. Over the past two decades uncertainty has been one of the most active areas of research related to GIS (Zhang and Goodchild 2002). Where measurement is involved, such as the measurement of position, then it is useful to know about error and accuracy. In other cases it may be impossible to identify any value as the truth, particularly where human judgment is involved. For example, it may be impossible to measure the accuracy of a soil map since there is a degree of subjectivity involved in assigning soils to classes, such that two equally qualified soil surveyors might well not agree about the appropriate classification at a given location. To handle such cases geographers have made increasing use of concepts of "fuzzy sets," and have implemented many of the associated techniques in GIS. Uncertainty is now an important component of *metadata*, or the information used to describe and document a GIS database so that it can be shared effectively.

## Fields

A previous section referred to spatial objects as one of the ways of conceptualizing the geographic world. The alternative is to think of the geographic world as a series of continuous surfaces, each representing the spatial variation of one property. This view is closely related to the concept of layers. Computers are inherently discrete machines, and conceptualizing geography as a continuum is clearly not as compatible with the digital world as thinking about it as a collection of discrete objects. But GIS has adopted several ways of handling continuous **fields**, by breaking them up into discrete points, lines, or areas. A remote-sensing scene, for example, breaks a continuous image into pixels, while a *triangulated irregular network* or TIN treats a topographic surface as a collection of tilted triangles. Many functions exist in GIS for analyzing fields, including techniques for simulating the development of drainage networks on topographic surfaces, or for computing the area visible from a given point on such a surface. Archaeologists have used such techniques to ask whether visibility was an important factor in the locations of early burial mounds in northern Europe, while engineers use GIS to determine the best locations for highways or pipelines across complex geographic landscapes.

## Density estimation

While the continuous field and discrete object conceptualizations are distinct, there are many ways of making transitions between them, and indeed any discrete object can be converted into a field simply by defining the field has having the value 1 wherever the object is present, and 0 wherever it is absent. One of the more powerful transformations is density estimation, a set of procedures for estimating a field defined by the density of objects per unit area. Density estimation is used to define the density of events such as disease occurrences, as a step in asking questions about the factors influencing density. It can also be applied to lines, for example when using street density as a basis for a definition of *urban*.

## Spatial probability

Geographers are often interested in maps of probability, which define the chance of something happening at a place. Probability maps are used in studies of risk, as in estimates of the risk of a landslide based on layers of slope, rainfall, soil type, and bedrock geology, or estimates of the range of a plant species.

As noted at the beginning of this section, this discussion of spatial concepts is far from complete, but it does give a sense of why researchers use GIS, and the kinds of inferences they make when they apply the lens of GIS to geographic data.

## The Data Supply

Traditionally most geographic data originated in various agencies of government; base mapping in the United States, for example, was provided largely by the US Geological Survey, and similar national mapping agencies exist in many other countries. NASA and other space agencies have developed into major sources of remotely sensed data. But increasingly it is possible for virtually anyone to become a supplier of geographic data, using simple global positioning system (GPS) devices, and today there is a vast network of suppliers and users, linked together by the Web and by a large number of data repositories, warehouses, and portals. The US Government's Geospatial One Stop (Figure 22.4) is one of the largest and most sophisticated of these, focused on providing a single point of entry to federal-agency data, but many others exist in both public and private sectors. In Europe the INSPIRE project is aimed at harmonizing the enormously complex situation that exists between different countries of the European Union, each with its own legacy of data and standards.

With this complexity and the great variety of data formats in existence, it has become essential to develop better standards and better methods of achieving interoperability among data sets from different sources. The Open Geospatial Consortium (OGC) is an organization of companies, agencies and universities that is dedicated to improving access to geographic data, and its standards have become enormously helpful over the past decade.

**Box 22.1** GIS Software

A modern GIS contains literally thousands of commands, procedures, and types of analysis, each of which can be applied by the user to the data contained in its database. Essentially, a GIS today is capable of performing virtually any conceivable operation on spatial data – and if one's favorite GIS lacks a particular function, it is almost certain that it can be found somewhere, with a little searching on the Web. Much thought has gone into making sense of this enormous complexity, by classifying functions into groups, or by developing cookbooks or guides that can help users to navigate within a GIS toolbox (see Exercise 22.1).

At time of writing, the market leader in the commercial GIS software industry was Environmental Systems Research Institute, of Redlands, California. Estimates of its market share vary widely, but on all obvious measures its share is the largest. ESRI software is widely used in universities and research organizations, so much so that its terminology and data formats are effectively standards. Other significant commercial products include those of MapInfo, Intergraph, and Autodesk. But while commercial products dominate, there is growing interest in open-source software. GRASS is perhaps the most conspicuous of these, and is widely used in the environmental sciences. Moreover, there is growing interest among the developers of statistical packages in providing some GIS functionality, and researchers with a strong mathematical focus will find that Matlab offers substantial power in this area. Smaller companies occupy particular niches in the marketplace. There are specialized products for three-dimensional GIS applications, for example, or for applications in transportation and other areas.

## GIScience

As noted earlier, geographers have a special relationship with GIS, especially given traditional geographic concerns with space, with the relationship between humans and their environment, and with integration across the sciences that deal with the Earth's surface. All of these themes are strongly supported by GIS tools. GIS raises many fundamental issues that are poorly understood, and over the past 15 years a research community has emerged that is focused on resolving them and moving GIS technology forward. The term GIScience is only one of a number of terms that have been adopted by this loosely knit community, though it seems to be the term of preference, at least in the United States.

The US University Consortium for Geographic Information Science (UCGIS) has devoted much of its attention to lobbying for better funding, synthesizing the curriculum of GIS courses into a Body of Knowledge (DiBiase et al. 2006), and identifying the GIScience research agenda. The agenda includes such perennial issues as scale and **uncertainty**, the need to extend methods of representation in GIS to include a more comprehensive treatment of time and the third spatial dimension, and the need for better methods of analysis and modeling. But it also addresses issues of broader social significance. Privacy is a

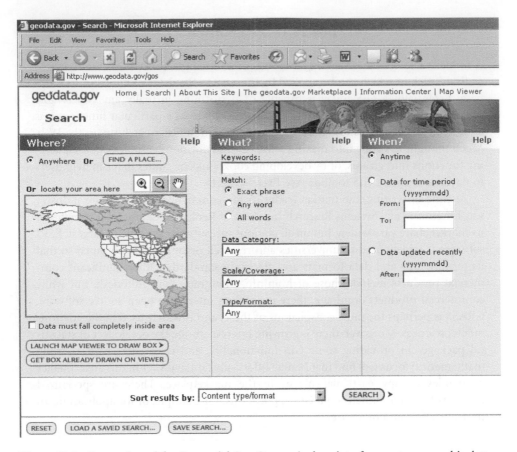

**Figure 22.4**   Screen shot of the Geospatial One Stop, a single point of access to geographic data from a multitude of sources. The screen allows a researcher to search a catalog for data covering a specified area, and also to specify theme, date, and other properties. Selected data will then be retrieved from its stored location, which might be anywhere on the Web

particularly important topic, especially given the ability of GIS to link together records based on street address, the prevalence of fine-resolution satellite imaging, and the growth of surveillance technologies such as closed-circuit television (CCTV) and GPS tracking. Another rapidly growing area of GIScience concerns public participation in the decision making process, questions about the role of GIS and geographic data in resolving community issues, and the power that GIS and geographic data give to those who possess them (Craig et al. 2002).

GIScience brings researchers together from a number of disciplines. Besides the disciplines that have traditionally concerned themselves with the production of geographic data, including surveying, geodesy, photogrammetry, remote sensing, and cartography, GIScience engages experts in spatial cognition who study the design of user interfaces and the skills of spatial thinking; computer scientists who study the design of databases and algorithms; and spatial statisticians who study spatial analysis and the modeling of uncertainty. In the center, of course, and overlapping all of these, are the geographers who have long thought of the description and understanding of the Earth's surface as uniquely their own domain.

# Conclusion

It is appropriate for GIS to be the subject of one of the last chapters of this book, given its synthesizing and integrating role. As a platform for research, GIS implements the subject matter of many of the previous chapters, and represents an increasingly important medium both for geographic research, and for encouraging a geographic perspective among researchers in other disciplines. GIS has truly been one of the most conspicuous aspects of geographic research over the past decade, and is increasingly associated with the discipline of geography in the minds of many outsiders.

At the same time, and as the previous section explained, GIS is both a tool for research and a subject of research. When critical social theorists in geography began to ask fundamental questions about the social context of GIS in the early 1990s, the initial reaction from the GIScience community was one of dismay – what would the outside world think of a discipline that seemed increasingly racked by internal debates? But in the longer term, the addition of a strong social dimension to GIScience has proven enormously beneficial, and today critical social theory ranks as one of its more important paradigms.

# References

Brown, M. C. (2006) *Hacking Google Maps and Google Earth*. Indianapolis: Wiley.

Craig, W. J., Harris, T. M., and Weiner, D., eds. (2002) *Community Participation and Geographic Information Systems*. London: Taylor and Francis.

DiBiase, D., DeMers, M., Johnson, A., Kemp, K., Luck, A. T., Plewe, B., and Wentz, E. (2006) *Geographic Information Science and Technology Body of Knowledge*. Washington, DC: Association of American Geographers.

Foresman, T. W. (1998) *The History of Geographic Information Systems: Perspectives from the Pioneers*. Upper Saddle River, NJ: Prentice Hall.

Ghosh, A., and Rushton, G. (1987) *Spatial Analysis and Location-Allocation Models*. New York: Van Nostrand Reinhold.

Longley, P. A., Goodchild, M. F., Maguire, D. J., and Rhind, D. W. (2005) *Geographic Information Systems and Science* (2nd edn). New York: Wiley.

National Research Council (2006) *Learning to Think Spatially: GIS as a Support System in the K-12 Curriculum*. Washington, DC: National Academy Press, http://books.nap.edu/catalog/11019. html [accessed March 17, 2009].

Sui, D. Z. (2004) Tobler's First Law of Geography: a big idea for a small world? *Annals of the Association of American Geographers* 94: 269–77.

Tufte, E. (1983) *The Visual Display of Quantitative Information*. Cheshire, CT: Graphics Press.

Zhang, J.-X., and Goodchild, M. F. (2002) *Uncertainty in Geographical Information*. New York: Taylor and Francis.

## Additional Resources

Burrough, P. A., and Frank, A. U. (1996) *Geographic Objects with Indeterminate Boundaries*. Abingdon: Taylor and Francis. An introduction to issues of vagueness in GIS, and to methods applicable when features on the Earth's surface are fuzzy.

Burrough, P. A., and McDonnell, R. A. (1998) *Principles of Geographical Information Systems*. New York: Oxford University Press. An excellent text emphasizing the environmental applications of GIS, with special emphasis on issues of relevance in this domain.

Knowles, A. K. (2002) *Past Time, Past Place: GIS for History*. Redlands, CA: ESRI Press. A collection of examples of the use of GIS to shed light on historical events.

Longley, P. A., Goodchild, M. F., Maguire, D. J., and Rhind, D. W. (1999) *Geographical Information Systems: Principles, Techniques, Management and Applications*. New York: Wiley. A 1999 state-of-the-art review in two volumes, including chapters solicited from leading experts on the key topics of GIS.

Maguire, D. J., Batty, M., and Goodchild, M. F. (2005) *GIS, Spatial Analysis, and Modeling*. Redlands, CA: ESRI Press. A collection of articles describing the growing applications of GIS in support of advanced modeling, including simulation of dynamic processes.

McMaster, R. B., and Usery, E. L. (2004) *A Research Agenda for Geographic Information Science*. Boca Raton, FL: CRC Press. Each of the major topics of the GIScience research agenda is addressed in detail in this review.

Mitchell, A. (2005) *The ESRI Guide to GIS Analysis Volume 2: Spatial Measurements and Statistics*. Redlands, CA: ESRI Press. Excellent reading at an introductory level on the analysis tools available in GIS, with abundant examples and illustrations.

O'Sullivan, D., and Unwin, D. J. (2003) *Geographic Information Analysis*. Hoboken, NJ: Wiley. An introduction to many of the more interesting questions that can be asked using GIS.

Pickles, J., ed. (1995) *Ground Truth: The Social Implications of Geographic Information Systems*. New York: Guilford. A widely cited compilation of the major social issues raised by GIS, from personal privacy to the marginalization of social groups who are unable to afford the technology.

Worboys, M. F., and Duckham, M. (2004) *GIS: A Computing Perspective*. Boca Raton, FL: CRC Press. An introduction to GIS from the viewpoint of computer science, emphasizing the conceptual as well as technical underpinnings of the field.

## Exercise 22.1    The National Map

GIS software can be expensive, and the exercises used in GIS courses can be elaborate and require some initial instruction. However many websites now offer substantial GIS capabilities, and give a sense of what is possible, while requiring only a simple Web browser such as Internet Explorer. The National Map, a project of the US Geological Survey, is one such site, and can be accessed at http://nationalmap.gov. You may need patience: working over the Web with a browser is not as fast as having your own GIS and database on your desktop. The concepts identified below refer to paragraphs in the chapter, and are things you might think about as you work through this exercise.

1    Open the National Map in your Web browser, read the background information, and click on the Go To Viewer link. Familiarize yourself with the map that appears on the screen, the display of latitude and longitude for the cursor position, the way the Earth

has been distorted in order to flatten it (compare Google Earth), and the capabilities for moving horizontally and changing scale. (Concepts: Maps; Directions.)

2 Zoom in to Colorado and explore some of the other available layers, such as hydrography. Zoom in to Denver and watch how the display changes its level of detail, making more and more information available as you zoom in. Try clicking the Identify button and querying some of the features shown on the map. (Concepts: Spatial objects; Scale; Networks; Fields.)

3 Zoom in to an area you are familiar with, such as your neighborhood (or if you are outside the United States, try this with a tool like Google Earth). Click the Measure button and try measuring the length of a path through the neighborhood. Explore some of the available layers at this scale. How do the contents of the National Map compare to your own knowledge of the neighborhood? Do all of the layers fit perfectly? (Concepts: Distances; Spatial context; Uncertainty.)

# Chapter 23

# Analyzing Meaning

*Deborah P. Dixon*

- Introduction
- Generating Data through Content Analysis
- Theorizing Meaning I: Critical Discourse Analysis
- Theorizing Meaning II: Deconstruction
- Conclusion

Keywords

Content analysis
Critical discourse analysis
Deconstruction
Discourse
Discursive event

Meaning
Poststructuralism
Signifier
Structures and mechanisms
Texts

## Introduction

Traditionally, when analyzing meaning, academics have focused on the obvious mediums through which information about ourselves, others, and the world around us is imparted. **Texts** such as newspapers, books, and other examples of the written word, are all designed so as to convey information to a reader. In addition to these, geographers have also expressed an interest in image-based mediums such as the map; indeed, the argument has been made that in societies saturated with high technology media such as TV, video, and the Internet it is the image, rather than the word, that provides us with a window onto the world (Adams 2009).

In recent years, however, it has become apparent that information is expressed not simply through those media designed for the purpose, but also through the vast range of objects we live and work with on an everyday basis. Consider, for example, the clothes we wear: each item sends a message to those around us regarding how we think about ourselves and our relationship to them. Even the body itself communicates information: body height and weight, color, hair adornment and so on can all be taken as indicative of our individual strengths and weaknesses, our desires and anxieties. Indeed, information is not simply the preserve of words and images, as the manner in which the body undertakes particular practices – that is, how we walk, sit, eat, greet, etc. – can also tell us much about the individual concerned (Longhurst 1995). This explains why the body itself has been likened to a "text" in that it can be "read." Words, images and practices, then, are all capable of being what are termed **signifiers**; that is, they can communicate information.

What has also become clear in recent years is that the terms "information" and "message" do not sufficiently capture the varied tone of **meaning**. This is because, first, such terms tend to imply that texts such as books, maps, and bodies are simply a channel for the conveyance of facts and figures, fiction, and opinion. Meaning, on the other hand, is very much shaped by the medium through which it is expressed. That is, it matters as to whether a particular story is reported by an academic journal as opposed to a tabloid newspaper: the former can be seen to lend legitimacy to a story, while the latter adds entertainment value. Second, such terms tend to focus attention on the way in which our perceptive and cognitive functions allow us to encode and decode information, such that we "get" or do "not get" the message. Meaning, by contrast, can invoke a more complex, emotive sense of engaging with objects; we can respond to them with feelings of hope or fear, we can ignore them, and we can even appreciate the limits of an exchange, when words, images and practices are no longer sufficient to convey how we feel. Meaning, therefore, is not something we are simply exposed to when we engage with these objects, but is something that is *made* through this very process of engagement.

How, then, do we analyze such meanings? The goal of this chapter is to point to some of the ideas that academics, and especially geographers, have brought to bear on this question. I begin by looking at a commonly used technique termed **content analysis**, which refers to the manner in which the particular meanings expressed by an object such as a book or the body can be discerned according to the pattern of signifiers (that is, words, images, or practices) perceived to be present.

I also want to show how such a technique can be given more conceptual depth by delving into the character of these meaning-laden objects: that is, how they are produced under particular conditions, how they are able to move in time and space, and how they connect with particular people in different ways. After outlining content analysis, therefore, I look to how this technique can be deployed as part and parcel of a theoretically informed methodology called **critical discourse analysis.** I also, however, want to point to a second methodology, **deconstruction**, which, while also dwelling on the relationship between signifiers, holds to a different understanding of how meaning itself works. As such, it challenges the utility of content analysis as a technique for generating data.

## Generating Data through Content Analysis

At heart, content analysis is a technique by which one can discern the way in which words, images and practices hold to a specific pattern (see Neuendorf 2002). Although any object, be it an individual or even a landscape, can be subject to content analysis, it is more usual to find that traditional texts, such as books, reports, periodicals, and interview transcripts, are assessed in this manner, such that the signifiers of interest are written and image-based.

Central to content analysis is the idea that each signifier does not exist in isolation, but is, rather, tied to a whole host of other signifiers. Consider, for example, the written word "dog." Taken purely as a sequence of symbols, d-o-g, it makes no sense to us. But, understood as that signifier which is commonly used to indicate, or mean, a four-legged canine, of which we may well have an associated mental image, it does make sense. This is because we have learnt to identify the right word, "dog," out of all the various written words, such as "cat" and "mouse." Put another way, we have understood what the written word "dog" is supposed to mean by noting how it is related to – that is, how it is different from – all of these other possibilities.

The point to be drawn from this example is that the meaning associated with a signifier can only be understood as part of a broader, relational system of meanings. In the case of written or spoken words, we refer to this system as a "language." However, it is important to bear in mind that images and practices can also be considered as part of their own relational system, as when we learn to associate the color white with purity or the practice of smiling with happiness. In order to identify a meaning, the individual must be aware of these wider systems, and the rules by which they operate.

What content analysis assumes, then, is that each series of words, images, or practices, or combination thereof, works to a set of rules, such that the meaning of each individual example can indeed be pinpointed and, significantly, that this meaning is *constant* whenever this example is used. This allows us to assume that if a particular signifier, for example a written word, is found to be present again and again within a book or article, then the meaning associated with it is of special significance to the author. More importantly, if the use of a particular signifier is compared against another factor then a pattern may well be found. If two or more words are found to be in close proximity time and again, for example, then we can conclude from this spatial pattern that the meaning of these words is also closely entwined, at least for those responsible for writing or uttering such words. When different kinds of signifiers, such as words and images, are found to be in close spatial proximity, then it may also be concluded that their meanings have been closely allied, such that one indicates the other. Moreover, if the frequency with which a word is compared against time – if the word "terrorism," say, appears more and more in our daily newspapers – then this temporal pattern indicates how prominent this topic has become for many people and may even indicate something of why this is the case.

In order to illustrate these rather abstract points regarding content analysis, let us look to an example from the academic literature. *National Geographic* is a magazine published in the US that aims to bring the rest of the world to an American readership through glossy photos and travel stories. The authors of the book, *Reading National Geographic* (1993), Catherine Lutz and Jane Collins, argue that it is important to understand the meanings that this magazine conveys because these play an important role in shaping the ways in which ordinary Americans think about and respond to the rest of the world. Their analysis has three main lines of inquiry: an assessment of the production of *National Geographic*, using a series of interviews with people who work on the magazine; a content analysis of the photographs used in the magazine; and interviews with readers of the magazine to see what kind of meanings they associate with these photographs.

To begin their content analysis, the authors looked through numerous issues of *National Geographic* before deciding on 22 signifiers, known as a coding frame, which could be used to ascertain the meanings that this magazine expresses. These included characteristics of the photos themselves, such as size and captioning, as well as the characteristics of their subjects, including age, gender, degree of nudity, ethnicity, and so on. The presence of particular objects, such as military hardware and urban backdrops, was also noted. In the subsequent analysis of 594 articles for the period 1950 to 1986, wherein the temporal and spatial incidence of all of these signifiers was noted, several patterns were found. Importantly, some of these patterns were apparent as a repeated *absence* of a signifier, rather than its *presence*.

For example, very few photos depicted military hardware. Those that did were not matched with images of dead or wounded bodies. For the authors, this pattern of absence tells us something about *National Geographic*; it has a meaning. In this case, it is that an American audience does not like to be reminded of the many conflicts present aboard, including those in which the United States itself is engaged. Indeed, American audiences are more comfortable thinking of the rest of the world as a series of other cultures as opposed to economic and political entities. To bolster this conclusion, the authors point to a particular pattern of presence, whereby many of the subjects photographed are engaged in local ceremonies or festivals. These photos portray a romanticized world that is as yet untouched by modernity, one that has kept alive a series of traditions and rituals. In similar vein, it was noted that many of the images printed in *National Geographic* deploy a close-up of a face; their subject, more often than not, is a happy, smiling child or young adult. What this particular pattern tells us, the authors conclude, is that an American readership wants to see this world as populated by mostly young, healthy people who would be happy to entertain them as visitors.

There are several "how to" points that can be usefully drawn out of this example:

1   While a qualitative overview of a text such as a magazine is important as a means of indicating which patterns might well be present, a content analysis allows the researcher to assess more carefully the extent of these patterns, as well as the degree to which they intersect with each other, via a process of quantification.
2   In choosing a coding frame – that is, which signifiers to count – it is useful to consider those which are not immediately obvious. Words and images are often the first signifiers to be noted in an analysis. And yet, one can also consider such visual signifiers as a close-up shot, a point of view shot or the use of juxtaposition (Rose 2001). In regard to the written or spoken word, one may well consider such signifiers as metaphor, assertion, and force of utterance. In this regard, and depending on the type of signifier under analysis, a content analysis would benefit from knowledge of the methodologies deployed in subjects such as literary analysis, media studies and ethnography.
3   While it is certainly possible to deploy a content analysis of one object, such as a single magazine, or even a single page, it is useful to consider how a temporal pattern can be discerned from the analysis of a series of objects. In this case, several conclusions about the changing state of American foreign policy could be drawn out from looking at subsequent issues of the *National Geographic* (also see Sharp 2000).

4   When looking at the meaning of a particular text it is useful to think of the different types of signifiers that are present and how they are combined. In the above example, a number of image-based characteristics are drawn out; in addition, however, attention was also paid to the captions attached to each photo and the use of key words therein.

5   Last but not least, a content analysis is made considerably more thorough if the signifiers that constitute an object in question can be considered at length. A book or magazine, for example, can be bought or photocopied, and its words and images subsequently analyzed again and again. The signifiers associated with another type of object, such as the practices associated with a particular body, may only be on view for a short time. It is important in this case that considerable effort is made in noting as much information as possible within this time frame; "capturing" aspects of these practices through photographing, filming, tape recording, and transcribing them would aid subsequent analysis.

There is also a more conceptual point to be made here, and that is that while such an analysis may seem to be an impartial one, reliant on quantifying patterns that are simply present in the *National Geographic*, the manner in which particular signifiers are selected for analysis, as well as the manner in which their patterns are subsequently interpreted, or given meaning, depends on the theoretical suppositions of the researcher. The danger of assuming these patterns to be simply "given" is that their production and interpretation may well be taken at face value alone; in this case, it could be concluded that the world outside of the United States *is* primarily interesting because of its exotic cultures and that warfare *is* notable by its absence. And yet, Lutz and Collins (1993) take a much more critical attitude, noting that the photographs that make up much of this publication tell us much about how Americans would *like* the world to be. In other words, they do not assume that this magazine is a mere channel for the portrayal of information about the world, but is, rather, a complex, often emotive object that is given meaning by those who produce it and those who read it.

This point is well illustrated when several texts on the same theme are examined over and against each other. While Lutz and Collins dealt only with one journal, the *National Geographic*, Deborah Martin (2003) choose to look at the texts produced by four organizations working in one neighborhood in St Paul, Minnesota, in order to compare how they gave meaning to this place (see Box 23.1). Her coding frame of words and images was chosen in order to emphasize motives for action, important problems, and proposed

---

### Box 23.1   Framing Neighborhood Identity

An article by Deborah G. Martin (2003) offers another illustration of content analysis. Her goal was to analyze the different descriptions, or "framings," of a neighborhood in St Paul, Minnesota. These are summarized in Table 23.1 (adapted from Martin 2003: 745). Her work shows just how differently a single neighborhood can be "framed" in the different "texts" of neighborhood organizations.

**Table 23.1** Neighborhood "framings"

| Neighborhood organization | Motivations for action<br>****<br>Neighborhood description | Diagnoses of problems | Proposed solutions |
|---|---|---|---|
| Thomas-Dale District Seven Planning Council | • Plan for future<br>• Clean up neighborhood<br>• Support/protect children<br>• Create community<br>****<br>• Racial, cultural, economic diversity<br>• Historic homes | • Lack of green, public space<br>• Cycle of disinvestment<br>• Broader processes/ decisions affect local conditions | • Plan for future development (industrial, socio-economic, infrastructure, long-range comprehensive plan)<br>• Clean-up days |
| Greater Frogtown Development Corporation | • Increase home ownership<br>• Improve housing stock<br>****<br>• Architecturally and historically significant housing<br>• Modestly priced housing<br>• Run-down housing | • Degraded, run-down houses<br>• City policies increase number of vacant lots<br>• Lack of investment<br>• Financial barriers to homeownership<br>• Negative images of neighborhood | • Promote neighborhood as a residential location<br>• Build and rehabilitate houses |
| Frogtown Action Alliance | • Everyone faces the same problems and should help solve them<br>****<br>• More economic activity needed in neighborhood | • Fracturing by race, ethnicity<br>• More money leaves neighborhood than is invested<br>• Neglect by the city<br>• Negative perceptions, undefined identity | • Unite individuals and Organizations (create geographic identity, foster community control)<br>• Entrepreneurship classes<br>• Business development and support |
| Thomas-Dale Block Clubs | • Keep neighborhood and homes clean<br>• Individuals responsible for Community<br>****<br>• Cultural, religious, racial diversity<br>• Children | • Garbage in streets, yards<br>• Poor attitudes, behavior, and lack of responsibility by residents | • Foster residential interactions and neighborhood pride<br>• Clean up area, plant flowers, build pocket parks<br>• Protest criminal behavior (work with police; hold property owners accountable for tenants, clean up) |

solutions. As noted in the introduction, if we wish to analyze the meaning of such texts, then we need to understand how they have been produced under particular conditions, how they are distributed in time and space, and how they connect with particular people in different ways.

What the analyses of Lutz and Collins, as well as Martin, amply demonstrate is that a content analysis provides some interesting results which can then be speculated upon. Why, for example, does a particular pattern emerge? Or, why are some patterns conspicuous by their absence? In the process, it is useful to look to how various theoretical approaches can be used to interpret these results. Below I outline two possible approaches.

## Theorizing Meaning I: Critical Discourse Analysis

Critical discourse analysis (CDA) is just one of the theoretically informed methodologies within which a content analysis, alongside other methods of data acquisition, can be applied. The term **discourse** is important here in that it indicates much more than the relational system of meanings outlined above. Whereas such a system can be considered simply a means of conveying information, discourse is what ensues when this system, for example a language, is considered to be embedded in the broader social realm, one that is, moreover, bound up with all kinds of power relations. Hence, it follows that in order to understand how language works, the character of society itself must be theorized. CDA accomplishes this by understanding society to be thoroughly enmeshed in *structures, mechanisms,* and *events*.

A social structure can be thought of as a series of routinized behaviors, carried out according to formal laws and regulations but also informal norms and expectations. So, for example, we can talk of a capitalist economy as constituted in large part from a highly complex set of behaviors wherein objects, from land parcels to automobiles, are commodified, to be sold on the marketplace for a profit. In similar vein, the relations between men and women can be thought of as highly routinized, such that, for example, it is usual practice in a given society for men to work for a wage while women maintain a house and raise a family. A political system, such as a democracy, can also be thought of as a complex social structure, routinizing the relationships between the state and its citizens in terms of duties and obligations. It is important to bear in mind that all of these structures help to determine the power relations that exist between people, in that they govern who gains access to which resources, who is allowed to act in a particular way and, last but not least, who is punishable for not following these strictures.

Mechanisms can be thought of as the particular means by such structures are "realized" as when, for example, within capitalism a wage payment system is set up and maintained. Setting up apprenticeship programs for young boys is one such example: this sort of mechanism may well reveal the presence of more than one structure, as such a program serves both a gendered labor pool and capitalism. Last but not least, events are the specific instances that together go to make up a routine, as when a specific purchase is made, or a new house cleaning product is advertised on TV to a female audience.

Importantly, the social structures noted above need to be learnt and remembered, and so depend for their existence to a great extent on systems of communication, or discourse. Hence, talking with people, dressing in an appropriate manner, reading books, writing reports, paying wages and following traffic signals are all instances of what are termed **discursive events**: by engaging in them, we maintain both the discourses and the broader, structured social realm of which they are a part.

This emphasis on discourse is key to how CDA proceeds. In addition to noting the patterns of words, images or practices revealed by a content analysis, attention is also paid to questions around (Jones and Natter 1999):

1    the *production* of a given object, such as: Who is responsible for bringing it into existence? What resources were available to them? And, why have they produced it in a particular way?
2    the *character* of that object, including: What kind of communication strategies, such as rhetoric and metaphor, are deployed? How extensive is it in time and space? And, how does it compare with other examples of its genre?
3    the *consumption* of that object, such as: Who engages with it? How have they been able to gain access to it? And, what sort of influence does it have on them?

Each of these lines of research requires additional techniques of data acquisition, such as archival research, observation, and interviewing (Fairclough 1995).

If we refer back to the example of the *National Geographic* study by Lutz and Collins (1993), then these again fairly abstract points regarding CDA can be illustrated. Content analysis was used by the authors to generate data from the photographs and associate captions in the form of patterns or "themes." The meaning of these patterns was subsequently assessed, however, via reference to the structures present in the production, distribution, and consumption of the magazine. To begin with the photos themselves, these are obviously taken by individual photographers who, moreover, are embedded in a series of broader, social structures, from the institutional context of the magazine to the capitalist system itself. Each photographer has a charge (a particular story) and a brief (a set of photos which are requested) as they go out into the field. The stories themselves do not usually come from the photographer, but from people within the magazine whose job it is to think up stories that fit in with the magazine and that will sell. Every story has a picture editor, who picks out the photographs that they like, before the story goes to the overall editor, who narrows down the pictures even more to the ones that will actually get printed. All of these practices are routinized, and can be thought of as the work-based events through which the magazine's pictures are generated.

In order to understand the magazine's distribution over time and through space (that is, where and when it is sold), it is necessary to assess the economic dynamics of the magazine industry. In order to maintain itself as a publication year after year, *National Geographic* must successfully occupy a market niche over and against its competitors. Accordingly, the meanings associated with the magazine's pictures – for instance, that there exists a basically harmless but exotic world outside of the US – are not simply an expression of how Americans would like to the world to be, but are also commodities to be bought and sold on the marketplace. Informative and entertaining, they are the central 'product' in this exchange.

Lutz and Collins were also interested to find out how readers responded to the magazine, and how it shaped their perceptions of the Third World in particular. An important point here is that people respond in different ways according to the social **structures and**

**associated mechanisms** they themselves are embedded within. For example, one person from America's Deep South says that her parents always referred to *National Geographic* as "that nigger magazine" – clearly, the magazine's attempts to ask people to take a broadly sympathetic view towards Africa had failed in this case. Another person commented that he had read *National Geographic* for a school project, but not since he became an adult; here, the magazine is seen as somewhat boring and irrelevant to everyday life, but still a significant part of one's education. So, just because the magazine seems to present a particular vision of the world, we cannot assume that the meanings taken from it are those originally intended by the magazine's photographers, writers and editors.

To sum up this section, we have seen how CDA theorizes that systems of communication such as the media are embedded in a broader social realm, characterized by structures that routinize the way in which we live. Accordingly, in order to understand the meaning that has been ascribed to signifiers within a particular medium, it is necessary to be aware of the presence of these structures in the production, distribution and consumption of a particular object. It is possible, however, as the next section shows, to theorize meaning in a manner that does not presume such structures to be a fact of life.

## Theorizing Meaning II: Deconstruction

Deconstruction is a methodology associated with non-essentialist understandings, such as **poststructuralism**, which have in common the notion that there exists not one "truthful" account of the world, but rather many different accounts, often in conflict with each other and constantly undergoing change (Dixon and Jones 2004). Indeed, what is at stake here is the notion of truth itself. If we refer back to CDA, it was assumed that there exists a social realm replete with structures. This account is considered to capture the "reality" of our world in that it holds true regardless of any one person's experiences and beliefs. Hence, it follows that any other account can be dismissed as false or simply mistaken.

Non-essentialist understandings, by contrast, do not presume that there is one single reality that exists regardless of what most people think and believe. Instead, it is argued that reality is very much what we make of it. In other words, while there is indeed "stuff" that makes up ourselves and the world around us, it has no inherent meaning outside of that given to it by people. What is more, the belief that there is indeed a single reality outside of our individual perceptions is actually symptomatic of the human desire for there to be some kind of external order to the world, whether that be God's divine purpose or a series of social structures. It is important to note that in this theorization, the role of meaning in relation to the social realm has been inverted. No longer is meaning a constituent part of the social realm; rather, the social realm is the product of meaning. In other words, we believe that "society" exists – that it forms our reality – because we have grown used to using this term to indicate, or mean, the complex of relationships that hold people together.

Within this theoretical context, the project of deconstruction is to analyze how meaning is given to the world, and with what effect. This can be accomplished by focusing on the pervasive discourses through which meaning is imparted and ascribed. Discourse is here understood quite differently than within CDA, in that it refers to a particular way of thinking about and engaging with the world at large – a worldview – as manifest in the words people use and the manner in which they behave. In other words, a discourse provides one with an ontological stance in regard to that world, delineating the nature of people and things as well as the relations between them (see Chapter 2). Though there are a variety of lines of inquiry that can be followed – for example, one may look to the emergence or transformation of a discourse – geographers have, for the most part, focused on questioning "taken for granted" discourses that are manifest in our words and actions but are rarely critically reflected upon simply because we mistake them for truthful knowledge about what the world is actually like. And so, for example, a Medieval, European Earth-centered view of the Universe, wherein all objects revolve around the Earth, was once taken as truth. Now, it is viewed as a discourse that underpinned all manner of religious and scientific beliefs and their associated practices.

In part, this focus on the taken for granted has ensued because one of the foremost proponents of deconstruction, the French philosopher Jacques Derrida, operated in this manner by drawing attention to how Western thought, understood as a particularly complex but highly pervasive discourse, has uncritically centered time and again on "essential" terms, such as God, cause, origin, and structure, which are thought to be the fulcrum around which truthful explanations of reality can be built. For Derrida, these terms do not capture a fundamental truth of the world but are in fact highly contingent terms, the meanings of which have developed and changed as they have been handed down through generations of philosophers and scientists. If the meaning of such terms is so dependent on the time and place within which they are used, Derrida asks, then how could they be *the* key to understanding how the world works?

For a number of geographers following in these footsteps, deconstruction offers a way of discerning how other examples of discourse have similarly become "taken for granted" as truthful ways of describing and explaining the world (Dixon and Jones 2005). More often than not, it is argued (and this is shown in Chapter 2 of this book in the section on ontology), this has been accomplished by means of what is called an either/or type of discourse, wherein one side of the binary is presented as "good" or "necessary," while the other is dismissed as "bad" or "irrelevant." In binaries of this sort, phenomena are neatly split into this group or that, are of this kind or of that. To illustrate this, consider the way in which we think and talk about gender; for the most part, it is simply assumed that gender exists as a binary. It is surely a basic fact of life that individuals are *either* male *or* female? And yet, when we attempt to define male/female gender according to culture, biology, or a combination thereof, we find that it is not so simple, and in fact that many individuals resist any such categorization. In a similar vein, an either/or discourse has often been used in relation to geopolitics, with states being described as either "good" or "evil." Again, when we attempt to pin down the exact differences between the two, the situation becomes much messier.

Why, however, is it important to point out the fact that we tend to rely uncritically on such either/or discourses to describe ourselves and the world around us? Why would we wish to undertake a deconstruction of these? A first answer to this is that if we assume that such discourses are an accurate and truthful picture of the world, then it severely limits the kinds of questions we can ask about it: indeed, these questions can only tell us about ourselves and our desire for convenient categories. Second, such an analysis directs our attention to the issue of power: simply put, who gains from the prevalence of particular discourses and who loses?

**Box 23.2**   Deconstructing US Farm Policy

An article by Deborah Dixon and Holly Hapke (2003) provides a useful example of how political documents, as well as debates, can be "deconstructed" in order to draw out the discourses underpinning how people and place are described and afforded significance. They argue that a series of either/or binaries are at work in the 1996 farm legislation, two of which – free versus fettered market, investment versus welfare – are noted in Table 23.2 along with illustrative quotes from the Senate and House debates. Importantly, these are termed "discursive geographies" insofar as they are an attempt to delineate how objects of debate – such as people and place – are to be demarcated and placed in relation to each other. Dixon and Hapke also trace the "geographies of discourse" that have led to the emergence of the legislation

**Table 23.2**   1996 farm legislation binaries

| *Free market* | *Fettered market* |
| --- | --- |
| Senator Lugar (R-IN): The [Freedom to Farm] bill we consider today offers a straightforward, common sense policy … With this bill, agriculture has done its part to help balance the federal budget in seven years … Farmers will have full planting freedom – thus the label given to this act, the "Freedom to Farm," the ability to manage your land, to make decisions for the market. | Senator Boxer (D-CA): This bill is being sold to the agriculture community as the best vehicle to guarantee an income safety net to farmers … [but] Under this bill, one farmer could be receiving windfall gains while another hardworking farmer could go bankrupt in a bad year because of lack of assistance. Furthermore, the owners of the corporate farming enterprises do not even have to produce to make money – they can become absentee landlords. |
| Senator Grassley (R-IA): We give increased flexibility to the farmers to make planting decisions. We take that decision-making out of the hands of Washington bureaucrats and public servants. It will be in the mind and office of every farmer to decide how many acres of corn or how many acres of soybeans to plant. Presently, those decisions are made, to the greatest extent, by people in Washington, far removed from the reality of farming, ignoring the marketplace and trying to insert their judgment upon the people on the spot. Full flexibility means plant what you want to plant, not what some Washington bureaucrat says. | Senator Dorgan (D-ND): Today we are considering what will prove to be one of the most important and most disastrous pieces of legislation affecting … family farms across this Nation…. We are going to see the small fail and the large and powerful prevail … The big landowner should love the Freedom to Farm bill because their farms are going to get bigger, their farms are going to become richer, their farms are going to produce more and more, and the small family farmers are going to be there with less and less…. This takes a safety net we have had for fifty years and yanks it right out from under family farmers. |

**Table 23.2**  *Continued*

| *Investment* | *Welfare* |
| --- | --- |
| Senator Bond (R-MO): Farmers can manage a predictable seven-year income stream ... just as well or better than Washington can do it on their behalf ... I know there are some who may call this welfare ... Farmers know it is not welfare and most senators do not consider the existing program welfare.<br><br>Senator Grassley (R-IA): The legislation that is before us will guarantee an investment ... in rural America ... an investment in rural America at a time when there is a tremendous transition from the agriculture of the last half of the 20th century to the more free market, international trade-oriented agriculture of the 21st century ... Some people have said on the floor of this body that we are giving welfare to farmers ... How ironic ... | House Speaker Newt Gingrich (R-GA): Obviously, it is unfair for one part of agriculture to block reform in its programs as we are pursuing change across the rest of American agriculture. |

at this particular time and place, including the role of "discursive sites" (such as research centers, media organizations, policy forums, and religious centers) where ideas and concepts are brought together in the formulation of knowledge concerning the world, and from which such knowledge is then disseminated via various means, such as the media.

Notice here that while individual quotes can be used to illustrate one particular binary divide – say, unfettered versus free markets – they also presume other divides such as *family:farm*, or *rural:urban*. This means that, unlike Table 23.1, which neatly placed differing views in relation to each other, a deconstruction draws out a much more complex picture of how binaries constantly cut across each other in diverse ways. In regard to the farm bill, moreover, binaries do not fall evenly into political lines (e.g., Republican versus Democrat), thus revealing more real world complexity in regard to how discourses are manifest than is apparent in content analysis.

To help illustrate these points, let us consider how deconstruction can be applied to an object of analysis we have used previously, the *National Geographic*. Within the pages of the magazine, both words and images are used to portray a world saturated with race. And yet, if we look more closely at how this is accomplished, we can discern an either/or binary at work; in this case, it is that while blacks, Hispanics, Indians and so on are all considered to be "exotic" in some fashion, those people with white skins are not. While this binary may appear at first sight to associate the category non-white with positive values, such as youth and vitality, there is actually a more worrying process at work here. That is, non-whites are being linked with cultural, largely ceremonial activities, as opposed to political and economic ones. This actually associates them with notions of primitivism. By default, on the other side of the binary, whites have become associated with urban-based development and progress. They are the ideal to which others aspire and, accordingly, they are the measure against which non-whites are judged to be lacking.

Such an assessment might well appear to be based on a qualitative form of content analysis, such that, for example, images of non-whites are found to be matched with words such as primitive and undeveloped. And yet, this is to misunderstand how deconstruction considers meaning to "work." Recall that in content analysis it is assumed that we learn to associate each signifier with a particular meaning. Moreover, it is assumed that this meaning stays the same whenever a particular signifier is used. Indeed, it is only when meaning is assumed to be constant that we can talk of a pattern emerging.

Within deconstruction, by contrast, the relationship between signifiers and meaning is understood to be much more messy and complicated. This is because the meaning we give to words, images and practices actually varies over space and through time. The meaning that we associate with the color pink, for example, has changed considerably over the years. At the turn of the century, pink was associated with maleness, and so became a key motif in American college fraternity symbols. Today, pink is almost exclusively considered to be a female color, as manifest in girl-orientated toys and clothing. If the color that is pink has changed meaning so dramatically, think of the transformative potential of much more complex terms such as democracy, gender, class, and race.

What this emphasis on change and transformation means is that we cannot assume that a signifier such as skin color conveys the same thing across several decades of the *National Geographic*, nor even across the same issue. Instead, deconstruction suggests that we look to the particular "context" within which meaning is presented. This requires that we treat each appearance of skin color as a whole new object of interest. In some photos, for example, whiteness may be present in clean and orderly surroundings, while in others non-whiteness may be pictured near to nature and alongside low technology. Though skin color is surrounded by a different array of signifiers each time, we can yet conclude, however, that there is a general tendency within the magazine for skin color to appear as an indicator of a person's level of progress and development, with whiteness, moreover, occupying the top level.

To take another example, consider the work of Deborah Dixon and Holly Hapke on the 1996 farm legislation in the United States. Theirs is a deconstruction of the discourses manifest in the US House and Senate debates leading up to this legislation, as well as the text of the bill itself. As can be seen in Box 23.2, such debates were predicated time and again upon particular binaries, such as us/them, rural/urban and US/rest-of-the-world, binaries which attempted to sort out a messy, chaotic real world into something that could be effectively changed via the implementation of particular political policies.

## Conclusion

In concluding I want to make three brief, observational points concerning the ideas under discussion. First, the recognition that we are not exposed to meaning so much as we make it through our engagement with other people and objects, has meant that analyzing texts has become a much more complex, theoretically-laden process. For some, an understanding of how the social realm works is crucial to understanding one of its key, constituent parts: discourse. For others, any such explanation is itself based on a series of meanings that we have imparted to the world around us. Second, and following on from the above, we cannot consider techniques such as content analysis as simple "tools" for generating data, ready and waiting to be adopted within any such theoretical framework. As we have seen, while content analysis can be utilized within CDA, its underlying assumptions regarding the role of signifiers mean that it is incompatible with deconstruction. And third, we must bear in mind that our concepts and ideas regarding how the world works are themselves constantly undergoing transformation; whatever the methodology we choose to use, we can be sure that it will itself be subject to change across space and through time.

## References

Adams, P. (2009) *Geographies of Media and Communication*. Oxford: Wiley-Blackwell.

Dixon, D. P., and Hapke, H. (2003) Cultivating discourse: the social construction of agricultural legislation. *Annals of the Association of American Geographers* 93(1): 142–64.

Dixon, D. P., and Jones III, J. P. (2004) Poststructuralism. In *A Companion to Cultural Geography*, Duncan, J. D., Johnson, N., and Schein, R., eds. Oxford: Wiley-Blackwell, 79–107.

Dixon, D. P., and Jones III, J. P. (2005) Derridian geographies. *Antipode* 37: 242–5.

Fairclough, N. (1995) *Critical Discourse Analysis*. London: Longman.

Jones III, J. P., and Natter, W. (1999) Space "and" representation. In *Text and Image: Social Construction of Regional Knowledges*, Buttimer, A., Brunn, S. D., and Wardenga, U., eds. Leipzig, Germany: Selbstverlag Institut für Länderkunde, 239–47.

Longhurst, R. (1995) The body and geography. *Gender, Place and Culture* 2: 97–105.

Lutz, C., and Collins J. (1993) *Reading National Geographic*. Chicago: University of Chicago Press.

Martin, D. G. (2003) "Place-framing" as place-making: constituting a neighborhood for organizing and activism. *Annals of the Association of American Geographers* 93: 730–50.

Neuendorf, K. A. (2002) *The Content Analysis Guidebook*. London: Sage.

Rose, G. (2001) *Visual Methodologies*. London: Sage.

Sharp, J. P. (2000) *Condensing the Cold War: Reader's Digest and American Identity*. Minneapolis: University of Minnesota Press.

## Additional Resources

Aitkin, S. (1997) Analysis of texts: armchair theory and couch-potato geography. In *Methods in Human Geography – A Guide for Students Doing a Research Project*, Flowerdew, R., and Martin, D., eds. Essex: Longman, 197–212. A good introduction to the analysis of texts in geography.

Barnett, C. (1998) Impure and worldly geography: the Africanist discourse of the Royal Geographical Society, 1831–73. *Transactions of the Institute of British Geographers* 23: 239–51. An application of discourse analysis and deconstruction to the Royal Geographical Society's nineteenth-century explorations of Africa. Barnett shows how these expeditions were represented as "scientific" by hiding behind a racially unmarked (i.e., "white") colonial narrator.

Chouliaraki, L., and Fairclough, N. (1999) *Discourse in Late Modernity: Rethinking Critical Discourse Analysis*. Edinburgh: Edinburgh University Press. Excellent but advanced text laying out the theoretical and methodological dimensions of critical discourse analysis.

Del Casino, V., and Hanna, S. (2000) Representations and identities in tourism map spaces. *Progress in Human Geography* 24: 23–46. Textual analysis applied to geography's most important form of representation: the map. Includes discussion of the relationship between identities, space, and their representations. The empirical example centers on reading tourism maps from Bangkok, Thailand, including several associated with sex workers.

Harley, J. B. (1989) Deconstructing the map. *Cartographica* 26: 1–20. Important early theoretical article calling for applications of discourse analysis and deconstruction to the subfield of cartography.

Wodak, R., and Meyer, M., eds. (2001) *Methods of Critical Discourse Analysis*. London: Sage. Edited volume that engages with the theoretical aspects of different forms of discourse analysis.

## Exercise 23.1   Analyzing Corporate Responsibility

Corporate social responsibility is an important part of the modern business world. In fact, social responsibility programs, especially those in large international organizations, are themselves big business, drawing on and contributing to both the financial and human capital resources of corporations. In this exercise you are asked, first, to develop a coding frame for the comparative analysis of two large, multi-national corporations. These firms can be in any sector of the economy, from industrial (e.g., John Deere, Hyundai) and service (e.g., Carnival, Enterprise Rent a Car) to financial (e.g., American Express, HSBC), real estate (e.g., Century 21, ReMax), or transportation (e.g., British Airlines, Delta Airlines). Take a few minutes to jot down some key aspects of corporate responsibility "lingo" and imagery – for example, "sustainability," "green," and "environment," "protection," and other terms are often used to signify that a corporation is environmentally responsible. Other terms can indicate a corporation's effort to project themselves as culturally responsible in a diverse world, as philanthropically minded, or as committed to the communities in which they have far-flung operations. After you have developed some potential categories, select two corporations for further analysis. Analyze the contents of their documents or websites in terms of your set-upon coding frame. After conducting a content analysis for key terms, revise your coding frame and re-work your analysis, until you feel satisfied that you have developed a full picture of the words and images put forth by the corporation. Then answer these questions:

1   *Content analysis:* What does your research say about the corporation's intended message with respect to these geographically significant categories: (a) environmental sustainability; (b) cultural preservation; (c) global citizenship; (d) and community engagement? How do the two corporations compare to one another on these terms, and why do you think they are similar and different? Are there any patterns of association that have emerged? For example, is environmental sustainability always represented

textually and visually in terms that avoid discussions or depictions of modern technology?

2 *Critical Discourse Analysis:* Now turn to the methods of CDA, and ask how the various contents you identified and analyzed are themselves embedded in wider structures and mechanisms. These should be analyzed not simply with respect to the contents of the texts themselves (see 1, above), but with the *contexts* of their production and distribution, namely: (a) the fact that it is a corporation, perhaps in alliance with a public relations firm, that has put together the materials, and (b) that there is an intended audience for the materials, for example, potential investors, public officials, members of non-governmental organizations, concerned citizens. As you think about these wider contexts of production and reception, keep in mind that both producers and the audience are embedded in wider social structures of capitalism, patriarchy, and racism. Can you find instances in which contents seem to be informed by these structures? For example, how do the corporations represent their global responsibilities relative to those of the communities within which they work? Are representations that invoke gender and "race" tied to emotive appeals, or to nature, ceremony, and the like?

3 *Deconstruction:* The first step in this process is to conduct some thought experiments. To begin, filter some of the relevant passages or visuals from your content analysis through the following pairs of binary relations: realism:idealism; individual:society; nature:culture; danger:safety; local:global; feminine:masculine; public:private; chaotic:orderly; western:non-western; subjective:objective; science:art; and modern:traditional. What binaries are being tapped by the corporation? Can you discern any underlying biases, or preferences, among these binary pairings in the pages of the responsibility statements? Are the underlying binaries linked to one another in a series (e.g., nature/traditional/non-western/local/chaotic)? How do commonsensical notions of individual responsibility play out on the larger corporate stage?

# Part IV
# Obligations

# Chapter 24

# The Politics and Ethics of Research

*David M. Smith*

Keywords

| | |
|---|---|
| Advocacy | Normative |
| Action research | Partiality |
| Confidentiality | Plagiarism |
| Consulting | Politics |
| Deception | Positionality |
| Ethics | Privacy |
| Expedition | Relevance |
| Morality | Values |
| Neutrality | |

## Introduction

**Moral** values are defining characteristics of being human. We have notions as to what is right or wrong, good or bad, what we ought or ought not to do in various circumstances. Moral **values** impinge on research in numerous ways. For the purpose of this chapter they may be grouped under two broad headings, albeit with some overlap and interconnections. The political concerns choice of priorities with respect to subject matter along with broad strategies of investigation, with power in various forms deployed to achieve particular ends.

The ethical concerns specific aspects of how inquiry is conducted, as a moral project. Both involve the **normative** dimension, concerned with how we should behave in a research context.

There was a time when geographers were not greatly concerned with **politics** and **ethics**. During the so-called quantitative revolution, which greatly influenced research in the 1960s and subsequently, it was quite widely believed that geography could be value free or **neutral**. Measuring characteristics of spatial organization in numerical form, as systems of nodes, networks and surfaces, for example, could be conveyed as purely "scientific," as could the adoption of mathematical models purporting to explain what was observed. Indeed, detachment from what might be conveyed as subjective opinions was regarded as a virtue of the new geography as location analysis or "spatial science." Largely unrecognized was the fact that the methods adopted, as well as the subject matter to which they were usually applied themselves, rested on "opinions." Regarding human beings as automatically responding to the mechanical imperatives of the gravity model of spatial interaction, for example, is a **partial** and rather peculiar view of human motivation, influenced as this is by all manner of other circumstances, or "random variables." Similarly, the spatial organization of settlement patterns and the location of industry, to which the measurement techniques and models were disproportionately applied, were surely no more important than some other aspects of the human condition ignored by the quantifiers.

Reaction set in during the late 1960s and the 1970s, in the form of a more behavioral approach. This promoted a more realistic view of human behavior, though still subject to analysis via numerical techniques, among other methods. What was perhaps more important was the reorientation of subject matter away from a rather narrowly defined economic and urban geography towards conditions which had hitherto not featured prominently if at all in analysis of the spatial arrangement of human life. Under the categories of "socially relevant" and "radical" geography (Peet 1977), these included poverty, hunger, crime, health, social disorganization, the situation of racial and ethnic minorities, environmental pollution, and the quality of life more generally as a spatially variable condition. As inequality came into focus, values came to the fore (Buttimer 1974). Radical geography began to shed new light on such familiar subjects as housing, location theory, the environment, resources, and spatial planning, under the rubric of uneven development. Today, the discipline of geography has a wide ranging and vibrant subfield that examines issues of morality, ethics, and social concern (Smith 2003, 2004).

## The Politics of Research

The politics of research concerns choice of research subject, topic or problem, along with some aspects of the broad strategies of investigation. It also concerns who may gain (or lose) from the research: from its conduct and from the availability and application of its findings. Research agendas are influenced by institutional politics and structures of power. For example, in Britain the Economic and Social Research Council and the Natural Environment Research Council set priorities for research funding which are to some extent influenced by the central government. The utility and status associated with large research grants is easier to obtain in some fields than others. Individual universities can encourage certain kinds of research, for example by providing good laboratories or computing facilities. The preferences of journal editors and book publishers perform a similar role. As priorities take shape and fashion effects set in, broad movement for consolidation or change take place. Thus it was during the quantitative revolution, with the very term "revolution" underlining its political character. Radical geography and the social relevance movement which followed proceeded in a similar way, responding to particularly influential and powerful scholars as well as institutions.

The radical geography movement shifted not only the research agenda with respect to subject matter, but also the possible beneficiaries. There was intended to be a shift from narrowly academic beneficiaries of research to the population at large. Furthermore, if such subjects as poverty, hunger, crime, ill health, environmental pollution and racial disadvantage became prominent fields for investigation, with a view to their amelioration, then the beneficiaries would most likely be the poor population rather than the affluent. An underlying assumption, of course, was that the poor would gain from geographical research on poverty, for example, entering and influencing the realm of public policy, an expectation that was not necessarily realized in practice.

As radical geography broadened to a general concern for social **relevance** and responsibility in geographical research, then particular issues came to the fore, especially race and gender (with the rise of feminist and anti-racist perspectives), along with a focus on further disadvantaged groups of "others," such as the disabled, post-colonial subjects, and gay/lesbian persons. Qualitative research methods began to augment and displace the quantitative. By the beginning of the twenty-first century there was even talk of "dissident geographies" – named so as to stress a departure from conventional political orientations. A book on the subject identified the following dissident approaches: anarchism; Marxism; feminism; sexual orientation; and post-colonial perspectives (Blunt and Wills 2000). Each of these posed significant challenges to the political positions usually adopted in the mainstream of geographical research.

If social relevance involves practical concern with societal problems, then there may be more or less effective strategies of engaging in it. A distinction is often made between **advocacy** and **consulting**. Advocacy involves working for a particular cause that benefits from geographical research, such as identifying places poorly served by health care or other services, or sources of pollution of residential neighborhoods. Such research may be conducted in association with local people, in a collaborative participatory capacity, helping to ensure that the project really does proceed with their interests in mind. The term **action research** is sometimes used in cases where there in a strong collaborative element in a project with a clear advocacy objective. Consulting differs in the sense that the researcher(s) usually work in a paid capacity for a particular client, who, by virtue of the pecuniary relationship, is in a position to dictate the research agenda. The client may be in the private or public sector, the latter often generating expectations of contribution to the general good not necessarily promoted by private interests.

Both advocacy and consulting may have the advantage of opening up research problems and sources of data that might not otherwise be accessible. Disadvantages include working to relatively short-term objectives when longer-term research might be more useful, and being tied to the agenda of local communities, public authorities or corporations. In both cases there may be an element of uncertainty over whether the supposed benefits of the research will really materialize, and which population groups in the community or society at large might benefit. Those contemplating advocacy or consulting should weigh the relative advantages and disadvantages before embarking on such a course – rather like the cost-benefit analyses which may be a prelude to or part of the project itself. In the background of such decisions will be the fact that the poor are not usually in a position to commission research on the open market, after the fashion of a private corporation or wealthy foundation, which could tip the scales in favor of working for the public sector or in a private advocacy capacity for disadvantaged populations. The choice between the two is sometimes couched in terms of working within the "corridors of power" represented by local or national government, as opposed to working within and with local communities who may feel badly served by politicians at any level.

An early and still famous example of advocacy, or participatory action research, is the Geographical Expedition in Detroit headed by William Bunge at the height of the radical geography movement at the end of the 1960s and beginning of the 1970s (Merrifield 1995). The idea was to explore the relatively unknown world of the "black inner city," setting up what was referred to as a base camp there. The designation of **expedition** was a deliberate subversion of those of the nineteenth century, which took an earlier generation into exotic lands overseas: the inner city was viewed a similar *terra incognita*. The purpose was to demonstrate the practical significance, or social relevance, of geography and its techniques outside the theory-dominated world of the university, in understanding and improving the lives of deprived populations. Thus, geographers brought their knowledge to the poor, with whom they worked in a collaborative association not marked by the usual intellectual hierarchy. The Detroit expedition facilitated the creation of a critical vantage point from which the inner city could be seen from the perspective of those living there. And conditions were very poor: Bunge showed that traveling from the affluent suburbs to the inner city was like going from the First to the Third World in terms of such an indicator as infant mortality. Bunge's politics was overtly expressed in a commitment to socialism, with strong anti-capitalist and anti-racist sentiments. The point of view of local people was given priority in building up geographical knowledge, very much in the spirit of situated, positioned and representative approaches applauded today by some post-modern skeptics of the Enlightenment notion of a dominant and supposed objective external perspective, or the "view from nowhere."

The Detroit Geographical Expedition and its Institute went on to develop a program of community research and education for local African Americans, as well as urban planning services. Thus a geography department at Wayne State University and the University of Michigan was linked to the inner city. Research focused on such issues as educational resources, political districting, money flows, transportation problems, and the geography of child deaths in a particularly hazardous environment where playspace competed with automobiles on the streets. These were issues around which the community could mobilize, armed with relevant information, in an attempt to effect change. In their turn, Bunge and his colleagues had access to research data that would not otherwise have been available: he published a book on the neighborhood of Fitzgerald (Bunge 1971). This study was criticized by some contemporary reviewers for its polemic content, departing from prevailing academic convention, but was defended by Bunge as a justified reaction to dreadful living conditions. His work nevertheless involved both traditional and modern geographical research techniques, including maps of social conditions and flows of money (from inner city to suburb). As well as contributing to the research, poor people brought their experiences to the university classroom, enlivening teaching and the experience of students. However much it might reflect the politics of a particular time and place, this expedition remains a model that could be adopted for geographical research elsewhere.

A contemporary example is provided by The East London Communities Organization (TELCO). This is a broadly based peoples' organization in the east end of London (UK), composed of almost 40 independent grassroots institutions, in particular churches, mosques, trade union branches and schools. Committed to working for the common good, TELCO is based on the shared values of justice, dignity and self-respect. It provides local organizations across East London with a way of acting together as citizens and communities to improve life. One of their activities is the Living Wage for London campaign, launched in 2001 in response to growing concern about declining wages and deteriorating working conditions. The gap between the official minimum wage and what is actually paid is particularly pronounced among those employed by the private sector to provide such services as care, hospitality, portering, catering, car parking and security. The main thrust of the campaign has been to persuade local publicly-funded institutions to introduce living wage clauses into contracting procedures for services. The campaign uses academic research to highlight the scope of the problem.

Links have been established between TELCO and the Geography Department at Queen Mary, University of London. The idea is to enable students to undertake a community action research project of value to TELCO. The aim is to provide the students with knowledge about citizens organizing in the local community, to offer practical experience of questionnaire surveys, interviews and focus groups, to demonstrate the value of research to local community groups and illustrate ways in which research can be put to practical use, to highlight the politics of the research process and the ways in which action research might be used as part of a campaign, and to demonstrate the need for flexibility in implementing research projects by responding to new issues and problems as they arise. All this is part of the students' training and preparation for their own final year research project. Just as students work in the locality with TELCO, so too do representatives of TELCO come to the college to give talks to students. Like the expedition, this model is capable of adaptation to similar situations elsewhere.

The notion of local, situated knowledge raises the broader political (and ethical) issue of who has the entitlement, or authority, to represent the lives of particular people to a wider audience. The conventional position is that this is the role of the academic research worker, characterized by the dispassionate objectivity of the detached external observer. It is these characteristics that give academic research its particular authority. The alternative view is that knowledge built up by, or with, local people themselves has an authenticity lacking in purely academic research. Furthermore, the dissemination of such knowledge gives a voice to various groups of hitherto marginalized "others" lacking the privilege of access to institutes of higher education. Politics is involved in the struggle of one form of knowledge to dominate claims to truth. The academic often has the advantage, not least in the form of access to publishers of books and journals, over the credentials held by community activists, for example. Clearly, there is much in favor of the two models summarized here, in which local people or students are encouraged and trained to conduct work consistent with accepted technical standards of research design and implementation, while keeping in mind the research needs of the particular causes for which campaigns may be mounted. Such was social relevance at the time of the Detroit expedition, and such it is in East London and elsewhere today.

The models outlined above by no means exhaust the scope of politically charged research, which can involve the full range of topics, from social segregation to soil erosion, and techniques from the qualitative, through the quantitative, to the laboratory. While the focus thus far has been on human geography, political considerations can also arise on the physical side. For example, in spite of a wealth of scientific evidence regarding climate change, its acceptance as a fact, as well as what should be appropriate responses, is to some extent a matter of politics. On another note, research on environmental degradation can also be highly political, as the sources of atmospheric, water or land pollution are causally linked to particular noxious facilities. A book on contested environments invites a distinctly political stance on such issues as the production of food, the provision of parks, energy and water supply, and values in environmental decision-making (Bingham et al. 2003). It also covers the environmental justice movement, which is an example of an issue in which politics and ethics are closely intertwined (Cutter 1995). So the choice of where to undertake research, nationally, regionally and locally, is a matter dealt with in the next section.

## The Ethics of Research

The politics of research merge with its ethics. Both are concerned with the normative, and areas of overlap are obvious. Here we concentrate on questions of what is good (or bad) research from an ethical or moral point of view, recognizing that there is an element of politics in some of this. It is important to stress at the outset the distinction between what is good (or bad) research from an ethical as opposed to technical point of view. Research may be done well from a technical point of view but in an ethically objectionable manner. Similarly research may be accomplished to a high ethical standard but with faulty technical skill. Good research in its most general sense will be good from both points of view. Indeed, for research to be good technically it may require proper concern with its ethics, while part of undertaking ethically sound research may be technical competence in the sense of making sure that the researcher is aware of relevant techniques and how to apply them, and indeed that high technical standards have been achieved.

One of the problems entering the realm of ethics, as with politics, is that there are more questions than answers. Some things are surely wrong from an ethical point of view, but more often a question poses ethical dilemmas to which there may be no definitive answer, as much will depend on the context or prevailing circumstances. So, let us begin with things that are surely wrong. It is wrong to falsify data, by inventing it or changing actual data to make a case more persuasive. It is also wrong to falsify results, like a statistical test, to suggest a better fit to some hypothesis than is actually the case. The unethical nature of these kinds of falsification is too obvious to require further elaboration.

Another thing that is wrong is to deny others credit for their research findings, say by failing to acknowledge them or properly reference their work. This raises the issue of **plagiarism**. While some definitions of plagiarism stress deliberation in a process of **deception**, such as making an explicit or implicit claim to someone else's research, there are broader notions that attribute plagiarism to poor scholarship and careless attention to referencing (see Box 24.1 and Exercise 24.1).

**Box 24.1** Definition of Plagiarism

Plagiarism is the presentation of statements, usually from another's work, in your own written work (whether an essay or thesis or examination script), without citation or any indication that the statement is a quotation (viz. a verbatim transcription of the writing of another person). The use of quotations or data from the work of others is acceptable, provided that the source of the quotation or data is given. Failure to provide a source or to put quotations marks around material that is directly copied from somewhere else gives the appearance that the comments are your own. Similarly, direct quotations from an earlier piece of your own work, if unattributed, suggests that the work is original, when in fact it is not. The direct copying of one's own writings qualifies as plagiarism if the fact that the work has been or is to be presented elsewhere is not acknowledged. You must note that even paraphrasing, when the original statement is still identifiable and has no citation, is plagiarism. It is not acceptable to put together unacknowledged passages from the same or different sources, linking these together with a few words or sentences of your own and changing a few words from the original text (this is regarded as over-dependence on other sources, which is a form of plagiarism). All material that is copied from another source must be acknowledged.

*Source:* Adopted by the Geography Department, Queen Mary, University of London.

It is worth quoting this definition at length because of the ease with which plagiarism (including self-plagiarism) is possible with the use of word processing and the Internet. Plagiarism is widely regarded as one of the most blatant and blameworthy forms of unethical behavior in research and scholarship.

Plagiarism leads to another issue, that of intellectual property rights. The correct attribution of credit or responsibility for research, published or otherwise, can be associated with proprietary claims (Corry 1991). Hence the importance not only of accurate citation but also of giving co-workers proper credit in the case of collaborative research. This is particularly important when there may be status differences, for example between a research supervisor and a graduate student. Gaining personal, departmental or institutional credit for research is becoming increasingly important in these days of formal performance assessment, in which money to support research may be allocated in accordance with some measure of past achievement. And of course in circumstances where research has commercial implications, intellectual property rights take on special significance.

Whoever undertakes research comes to the task with what is referred to as **positionality** (Rose 1997). In other words, they may be influenced in one way or another by personal characteristics such as social origin, "race" and gender. In terms of research, these could be an advantage as well as a disadvantage, enabling greater understanding (of what it means to be disabled, for example), than is possible for those not members of the group in question. Given this, there are arguments for including a biography of the researcher in any project, as a matter of background information as well as of ethics. Then, something of

**Box 24.2**   Ethical Issues in Research

Prior to data collection

1   Gaining access to research population: voluntarism
2   Obtaining cooperation from subjects – informed consent vs. coercion
3   Privacy
4   Reactive effects

During data collection

1   Humane and decent treatment – freedom to disengage
2   Irreversible effects on participants – harm vs. benefits
3   Use of deception

After data collection

1   Debriefing
2   Data confidentiality
3   Publication of results

*Source*: Mitchell and Draper (1982: 50).

the worker's possible sources of bias or special insight can be made known. The importance of personal biographies on career trajectories is increasingly recognized by students of ethics in research.

Box 24.2 lists a range of ethical issues that can arise at different stages in the research process, and for which there may not be a clear a resolution.

In the case of interviewing of various kinds, the research population should usually participate in a voluntary manner. However, there may be situations of compulsion, like the requirement that all persons should submit national census returns. Voluntarism may not always be practical, as in the case of less direct forms of observation – including participant observation – the purpose of which may be undermined if those observed are aware that they are the subject of research. Any research may invade the **privacy** of the subjects, which underlines the importance of informed consent. Reactive effects require anticipating how people respond to the research process. Involving research subjects in the project as co-workers may be considered, especially if they might provide special insight into the problem being investigated, such as the impact of disability which may be illuminated by the disabled themselves. During data collection the research subjects should be treated with respect, both individually and with reference to their culture, and allowed to disengage from the process if they find it onerous or otherwise objectionable. Given that there will be some cost to the research subjects, if only their time, thought should be given to ways in which they might benefit from the research, for example by contributing to the betterment of their lives or by ensuring that taking part is as pleasurable an experience as possible. The issue of deception can arise in a variety of ways, including not revealing that research is going on (e.g., in participant observation) or not revealing the true purpose of the research if this might prejudice data collection. After the data has been collected there may be a case for sharing it with the research subjects in the form of debriefing. Then there is the issue of **confidentiality**, which may require disguising the respondents and the location of the project. Finally, if the research is published (and there are ethical reasons which it should be, so as to make the findings widely available), those who provided data might be given copies so that they can see the use to which their information has been put, as well as to provide reassurance if confidentiality has been an issue. There are ethical issues involved in the process of writing up, such as whether the style should always be dispassionately academic prose or whether commitment, passion or even anger might be conveyed (Keith 1992).

Some of these issues require a careful balancing of advantages and disadvantages. For example, is it always possible or even necessary to respect one's research subjects, even if they may be involved in some dubious business practices or racism, and what are the implications for the research of ceasing to give them respect? In the case of covert methods of data collection (such as using a hidden tape recorder or unrevealed participant observation), these may be the only means of obtaining data: does the purpose of the research justify the deception? Is it right to conduct research in places where the local population stands to gain nothing, or may even lose if the results of the research are used to their disadvantage? Is it right to undertake research on a poor population when the major objective might be to gain another research grant or add another book or research paper to one's resume or CV?

An important geographical question that raises political as well as ethical issues is where to conduct research. The point just raised, about whether local residents might gain or lose from research, has implications for work in the impoverished inner city, for example. It could be asserted that there is a potentially exploitative relationship between a university and its local (often poor) neighborhood, if the area is frequently used to provide research subjects who gain nothing from the process. This underlies the ethical aspects of expeditions and similar arrangements, as discussed in the previous section. At a wider scale, it is common for geographers from the United States and Western European countries to undertake research in the underdeveloped world (Sidaway 1992), with the risk of similar exploitative relationships if nothing is offered in return. At the least, it is sometimes asserted, those working in such places should involve local geographers in a collaborative role, to assist with capacity building and otherwise pass on some of the benefits of coming from well-endowed universities in affluent countries where conditions are the envy of many others elsewhere. This is part of the wider ethical issue of what we in the more fortunate parts of the world may owe to "distant strangers" (Smith 1994).

Working in some particular places raises special ethical and political issues. For example, there was a partial academic boycott of South Africa during the era of apartheid, with the approval of the Africa National Congress (ANC). It was argued that for people from other countries to undertake research in South Africa helped to bestow legitimacy on the apartheid regime, and that some of the research might even help the Nationalist government's purpose. Against this it was argued that doing research in South Africa helped to expose and explain the important geographical dimension to the racial discrimination and exploitation characteristic of apartheid, and that to stay away from the country deprived progressive academics who opposed apartheid with important sources of external support (Lemon 1988; Smith 1988). Similar arguments are being made today over academic links with Israel. Some see them giving implicit encouragement to Israel's domination of the Palestinian population, while others consider research by outsiders important in revealing geographical dimensions of Jewish settlement of the occupied territories.

It is increasingly the case that applications for research funding and other sources of support require reference to ethical issues and how it is proposed to deal with them. Ethical review panels are the norm for any research involving human or animal populations in the United States, the United Kingdom, Australia, and New Zealand. Research is subject to increased scrutiny if the research subjects belong to vulnerable groups, such as children, the elderly, the disabled, or prisoners. A fruitful exercise would be to take a research paper or set of related papers and seek to expose their ethics. Of one thing we can be sure: that ethics is very much part of contemporary research policy and practice, with all the opportunities and demands that this entails.

## Professional Ethics

These and other political and ethical issues raise the question as to whether there might be some overarching code of conduct by means of which members of an academic profession should be governed. After all, if there is such a thing as medical ethics governing the conduct of physicians, why not a geographical ethics covering what it means to be a good geographer? Geographers in many countries lack such a code, but the Council of the Association of American Geographers (AAG) adopted a statement on professional ethics in 1998 (see Box 24.3).

This list of headings covers much of what has been raised in this chapter. It could apply to almost any academic profession. But one feature is worth noting as special to geography if not entirely distinctive of our discipline: ethical behavior during field research. Geographical research has a virtually unrivalled capacity to affect people, places and things in those parts of the world where we chose to work, and this carries an enormous ethical responsibility. However eloquent may be a profession's written code of conduct, it is the attitudes, values and behavior of individual scholars, in the laboratory and classroom as well as in the field, which makes the difference between informed moral responsibility and indifference or worse.

One of the AAG's concluding paragraphs provides an apt conclusion to this chapter:

> The concept of well-being that underlies the statement is not to be understood as the product of any particular personal or political agenda. Instead it is inspired by a concern with individual, social, and environmental "health." What constitutes "health" will always be a matter of debate that can and should be informed by a diversity of perspectives. And geographers will differ regarding its ends and means. Some will emphasize the well-being of animals, humans and/or the natural environment, focusing, for example, on the rights of sentient animals, oppressed minorities, or endangered species and ecosystems. Others will emphasize the role of human rights, social justice, or ethics of care in the pursuit of well-being. For still others, well-being may exist as an unarticulated commitment, or as the central focus of research. This diversity of views is to be welcomed because an ongoing conversation, conducted with respect, can deepen personal and shared insights into moral relations between humans and the world in which they live and work (AGG 1998: 9).

**Box 24.3**   Association of American Geographers Statement
on Professional Ethics

I        Preamble
II       Professional relations with one another
         A   Avoiding discrimination and harassment
         B   Sustaining community
         C   Promoting fairness in hiring
III      Relations with larger scholarly community
         A   Attributing scholarship
         B   Evaluating scholarship
         C   Self-plagiarism
IV       Relations with students
         A   Instructional content
         B   Pedagogical competence
         C   Training students with funded research
         D   Confidentiality
V        Relations with people, places and things
         A   Project design and development
         B   Ethical behavior during field research
         C   Reporting and distributing results
VI       Relations with institutions and foundations that support research
         A   Funding research
         B   The use of results from funded research
VII      Relations with governments
         A   Government research support
         B   Government employment
VIII     Conclusion – ethical debates in geography

*Source*: AAG *Newsletter* (1998), Vol. 33, No. 8: 6–9.

It is in undertaking research very much in the spirit of these words that we will become better geographers, and do better geography.

## References

*AAG Newsletter* (1998), Vol. 33, No. 8: 6–9.

Bingham, N., Blowers, A., and Belshaw, C. (2003) *Contested Environments*. Chichester: John Wiley.

Blunt, A., and Wills, J. (2000) *Dissident Geographies: An Introduction to Radical Ideas and Practices*. London: Prentice Hall.

Bunge, W. (1971) *Fitzgerald: The Geography of a Revolution*. Cambridge, MA: Schlenkman.

Buttimer, A. (1974) *Values in Geography*. Resource Paper 24, Association of American Geographers, Commission on College Geography, Washington D.C.

Corry, M. R. (1991) On the possibility of ethics in geography: writing, citing and the construction of intellectual property. *Progress in Human Geography* 15: 125–47.

Cutter, S. L. (1995) Race, class and environmental justice. *Progress in Human Geography* 19: 111–22.

Keith, M. (1992) Angry writing: (re)presenting the unethical world of the ethnographer. *Environment and Planning D: Society and Space* 10: 551–68.

Lemon, A. (1988) "Reason from whatever quarter": reflections on the academic boycott. *Area* 20: 353–7.

Merrifield, A. (1995) Situated knowledge through exploration: reflections on Bunge's "geographical expeditions." *Antipode* 27: 49–70.

Mitchell, B., and Draper, D. (1982) *Relevance and Ethics in Geography*. London: Longman

Peet, R., ed. (1977) *Radical Geography: Alternative Viewpoints on Contemporary Social Issues*. London: Methuen.

Rose, G. (1997) Situating knowledge: positionality, reflexivities and other tactics. *Area* 26: 305–20.

Sidaway, J. D. (1992) In other worlds: on the politics of research by "First World" geographers in the "Third World." *Area* 24: 403–8.

Smith, D. M. (1988): Academic links with South Africa: is ignorance a greater sin? *Area* 20: 357–9.

Smith, D. M. (1994) On professional responsibility to distant others. *Area* 26: 359–67.

Smith, D. M. (2003) Geographers, ethics and social concern. In *A Century of British Geography*, Johnston, R. J., and Williams, M., eds. Oxford: Oxford University Press for The British Academy, 625–44.

Smith, D. M. (2004) Morality, ethics and social justice. In *Envisioning Human Geographies*, Cloke, P., Crang, P., and Goodwin, M., eds. London: Arnold, 195–209.

## Additional Resources

Cloke, P. (2002) Deliver us from evil? Prospects for living ethically and acting politically in human geography. *Progress in Human Geography* 26: 587–604. This paper takes up the challenge of bridging the divide between theorizing ethics and politics and practicing it in one's everyday life. Living ethically and politically means being sensitive to others and collectively and responsibly addressing organized "evil."

Crampton, J. (1995) The ethics of GIS. *Cartography and Cartographic Information Systems* 22: 84–9. Proposes that ethics in GIS cannot be simply or easily achieved by looking within the GIS community; the standard for ethical GIS requires external engagements. Offers a four-fold standard for ethical practice in GIS.

Hay, I. (1998) Making moral imaginings: professional research ethics, pedagogy, and professional human geography. *Ethics, Place and Environment* 1: 55–76. This is an early call for geographers to consider the role of ethics and morality in their field, and to think about how professional organizations can be involved in the process.

Hay, I. (2003) Ethical practice in geographical research. In *Key Methods in Geography*, Clifford, N., and Valentine, G., eds. London: Sage, 37–54. A good introduction to the topic of ethics for beginning geographers.

Kirby, A. (1991) On ethics and power in higher education. *Journal of Geography in Higher Education* 15: 75–7 (and the papers that follow). If you're interested in the twists and turns of politics within geography departments, this is a fascinating collection of papers.

Lee, R., and Smith, D. M., eds. (2004) *Geographies and Moralities: International Perspectives on Justice, Development and Place*. Oxford: Blackwell. An edited volume focusing on case studies of morality from around the world, with case studies of inequality, human rights, and territorial conflict. Argues that there is a geography to morality.

Proctor, J. D., and Smith, D. M., eds. (1999) *Geography and Ethics: Journeys in a Moral Terrain.* London: Routledge. A wide ranging theoretical collection with an emphasis on geography and ethics. Separate sections are devoted to ethics and space, place, nature, and knowledge.

Smith, D. M. (2000) *Moral Geographies: Ethics in a World of Difference.* Edinburgh: Edinburgh University Press. Addresses some of the big questions in moral and ethical geographies, including care for distant others, absolute versus relative moral positions, and the possibility of social justice. Organized by geographic concepts (territory, nature, development), there are examples drawn from a wide range of places.

Valentine, G. (2005) Geography and ethics: moral geographies? Ethical commitment in research and teaching. *Progress in Human Geography* 29: 483–7. A progress report on geography's engagement with ethics, with a call to increased attention to ethics in the academy, including the classroom.

Winchester, H. P. M. (1996) Ethical issues in interviewing as a research method. *Australian Geographer* 27: 117–31. A helpful piece on uneven power and gender relations as they affect the interview process.

## Exercise 24.1   Plagiarism Court

All universities and major professional societies have honor codes that safeguard the research enterprise against plagiarism. Plagiarism, however, is complicated, and students do not always come to university knowing what constitutes plagiarism, or how to avoid it. Fortunately, the DiMenna-Nyselius Library of Fairfield University, in Fairfield, Connecticut (USA) maintains a "Plagiarism Court" – a wealth of knowledge about: what constitutes plagiarism; the consequences of plagiarism; proper documentation; note taking tips; quoting caveats; citation styles; and a quiz you can take to assess your knowledge. Fairfield's Plagiarism Court website site can be found at: http://data.fairfield.edu/library/lib_pcobjectives.html. The flash version is found at: http://data.fairfield.edu/library/documents/Library/plagicourt.swf. Take some time to work through the materials and then take the flash version of the plagiarism quiz!

# Chapter 25

# Writing It Up

## *Dydia DeLyser*

- ■ Introduction
- ■ Getting Started
- ■ Telling your Tale
- ■ Working Outlines
- ■ Style and Voice
- ■ Anecdotes
- ■ Arguments
- ■ How to End
- ■ Revisions
- ■ Final Words

Keywords

## Introduction

Though for some of us **writing** itself is a pleasure, the simple fact of the matter remains that while both the process and the results can be rewarding, writing well about our research is hard work. And that work is not just hard, it is also work for which many of us are but poorly prepared: though many colleges and universities offer advanced courses in research methods and courses that help students learn how to conduct successful research, far fewer offer writing classes for those in the sciences and social sciences that go beyond introductory composition. Further, while most of us survived, or perhaps even enjoyed our first-year writing courses ("Freshman Comp."), such courses are generally not geared specifically to the writing of major research papers. Thus, even an excellent undergraduate education can often leave a gap: students may be ill-prepared for the serious writing work of research papers (DeLyser 2003).

This chapter hopes to help fill that gap. In it I will offer advice based on my own experiences as a writer of research papers and also as an instructor who has taught writing to both graduate and undergraduate geography students. This chapter also presents advice culled from some of the best books available to help struggling writers with their craft; at the end of the chapter I provide references to some of these texts for those interested in further developing their skills. Throughout the chapter I refer to exercises that draw out points in the text designed to help you focus and train your skills. But even for those who do not wish to seek help from books, many colleges and universities have excellent Writing Centers where students can go for free help with their papers and other writing projects. Students seeking help with their writing have only to reach out: help is available, but most likely you will have to be the one to take the first step.

## Getting Started

After working your way through this text on research methods, by the time you have reached this chapter, you will be familiar with the basics of conducting research in geography. If we imagine the research for your own project is now finished (whether because you really find you are finished or because your paper is due tomorrow), you will probably find yourself "ready to write." Often this is a daunting moment, where we sit staring at a blank piece of paper, or, more likely, a blank computer screen.

In order to help us face this task, before we sit down to write many of us wish first to have everything "just right" – the right time of day (or night), the right music on, the right caffeinated beverages at hand, even perhaps the right color paper or the right position in a room. To sociologist Howard Becker such habits add up to a series of "magic rituals," rituals not unlike those used at other times to encourage rain, or a successful catch of fish: each is designed to help us with something over which we feel we have little control. Of course, the rituals do little beyond providing comfort – they don't really affect the outcome (of the day's catch or the evening's writing). But the prevalence among writers of such ritual behavior suggests that our writing fears (of the blank screen, or of writing poorly) are widely shared (Becker 1986).

Further, many of us have heard, about other writers, "He's really talented," or "She's really good!" All of which implies that some people simply *are* good writers, while others among us, therefore, are *not*. This sort of thinking encourages us to see writing as an innate talent, something we are born with – or born without. While it is possible that this is the case for some of us, it is much more productive, much more positive, and much nearer the truth to think of writing instead as a skill, or a craft. Writing, like carpentry, gymnastics, or drawing, is only partly talent-determined. Like the other three, writing can and indeed must be learned and practiced, our skills honed and sharpened. Thought of in this way, each of us must begin as an apprentice, learning tools and techniques, training, and hopefully eventually even perfecting our abilities.

But in order to do so we must first begin – a process so difficult for some writers that the staff at the Harvard Writing Center found the best tool for the writers they advised was a bucket of glue: "First you spread some on your chair, and then you sit down" (Bolker 1998: 33). Avoiding glue, it may be possible to approach the writing process and its difficulties in a more encouraging way. If we hold ourselves to a notion that whatever we first write, whatever first words appear on the page, will remain that way, then we are indeed putting a great deal of pressure on those first few words. But if, on the other hand, we allow our first words merely to suggest a direction, we may find that blank screen far less daunting. What I am suggesting is what teachers of writing have called "the **sloppy first draft**" or even "the **zero draft**" (meaning the one *before* the first draft; see Becker 1986 and Bolker 1998). Allow yourself, in the first stages of a writing project, to throw out sentences and ideas without care that they be "perfect" just get some words onto the page. Once over that hump, most writers find they can proceed much more easily – and then if need be **revise** the material in the sloppy first draft. You can use this technique to yield a beginning for your paper, but you can also use it to generate a draft of the entire work.

For some of us, beginning on page one remains difficult – especially at first, when we often don't yet exactly know what we're going to say. In that case, why not start with a later section of the paper? Try beginning somewhere in the middle, with a part of your **argument** that you are confident of or excited about. Often starting with the "easiest" or "most fun" part of a paper (never mind if it turns out to be in the middle of page seven when the paper is finished – word-processing software has eliminated any concerns about that) may give us the confidence to work both forward and backward into the more difficult material.

When all else fails, sometimes writing under pressure is just what we need: with the minutes ticking away until the paper is due you'll find that you *have to* write something. While this is not the best strategy for creating truly well-crafted work, nearly every writer has experienced its effectiveness. Even the most seasoned scholar who begins her work long before a deadline is near, may find herself with much left to be written as the deadline looms. While truly excellent work takes much more time to craft, a finished paper is better than no paper at all and, even under pressure, other techniques of good writing can still be applied.

## Telling your Tale

At its most basic, each research paper tells a story, so, although we may feel that our academic writing is, in some ways, less "creative" than say fiction writing, there is still a great deal of flexibility, and indeed creativity in how academic research papers are written. One problem for academic prose is that, in our commitment to "telling the facts," we may get lost in a less-than-creative (or even boring!) way of presenting those "facts." Yet, no matter how clear your evidence may be, each research paper does much more than simply present the "facts" about a particular topic. In fact, long before we sit down to write, even before we began doing research on a given topic, we had ideas about what that topic was about, and even maybe ideas about which "side" of an argument we wanted to pursue (Becker 1986). In ways like this, our research papers are always clouded – but also enlightened – by such issues as how we began thinking about our topics, how we actually proceeded with the research, and ultimately also by *who we are*. In other words, though research papers do tell facts, those facts are always shaded (to different degrees) by the circumstances and the person doing the research and writing. Each research paper then, by its very nature, grants its author the opportunity "tell a story" in a myriad of ways.

## Working Outlines

Settling on which story to tell and how you wish to tell it is one of the challenges of writing about your research – but it's one of the fun parts too. Though most students new to research fear they have little flexibility in the writing of their papers, it doesn't have to be that way. One of the most important tools in taking charge of the story you are about to tell is writing a working **outline.** Such an outline – one written for yourself, as a series of notes and reminders, not one written for your teacher – needn't be formal. In fact, my outline for this chapter consisted only of eighteen short words or phrases designed to remind me of the things I wanted to write about, and to suggest an order to those topics (one which I ended up following only a little bit – that's part of the creativity of writing and the flexibility of a working outline: you can change *everything* as you go!). But whether you stick to your outline or not, creating it, and then having it there as a reminder, will help you to explore the possibilities for telling your tale. It will help you to see how emphasizing one point instead of another, or telling a different part of the story first, may lead you also to different conclusions. So, while writing an outline may at first feel restrictive, as if the outline hindered your creativity, you can also use an outline as a flexible tool to help you to better expose and then express that creativity. Once you begin to play with different possible outlines you'll see that, with so many ways to tell the story your research presents, each paper is a truly creative endeavor (see Exercise 25.1).

## Style and Voice

In order to engage your creativity, and to recognize it, for you as you write and for your readers as they read, academic writers use many of the same techniques that journalists and fiction writers do. Some use metaphors and descriptive language to help evoke the feeling of a scene, a place, or a person. What is most important here is that you write in a style that is comfortable to you. Many writers of research papers attempt to imitate the often stodgy style of much academic writing, choosing words that sound "harder" or more complex (and are often longer) rather than using the "regular" words of everyday language. They write sentences in the passive voice rather than assigning agency or claiming credit by using the active voice (for example, "the gun was fired" instead of "Pat fired the gun" and "data were gathered" instead of "I gathered data") (Williams 1995). New researchers often mistakenly take, as our examples, not the best, but the worst prose of our field (Williams 1995). Sorting the wheat from the chaff in this case can be difficult, so a cautious writer will stick to a simple and direct style, one that will make your ideas clear to readers and runs not the risk of imitating an embarrassing example of scholarly turgidity. Read your prose carefully back to yourself, and listen for your own voice in it. Strive to recognize yourself, not someone else, in your own writing.

# Anecdotes

One of the best ways to enliven a research paper is by using **anecdotes**, interesting pieces of data told as an illuminating story (instead of being presented simply lumped together with other "evidence"). Granted, good anecdotes can be difficult to find because they must shed light *directly* on what you are writing about, but when they do, they draw the reader in, and they raise interesting questions that the paper promises to answer.

Here's an example of an anecdote and how to use it in a research paper: I once wrote an article about how I did the research for my PhD dissertation. Now that's a topic that on one level probably really only interests me, after all, why would anybody else care? But I realized that because I had studied a community of which I was a member, I faced challenges that "insider researchers" face all the time, so, I reasoned, if I told the story right, then others *would* be interested (and hopefully even able to learn something from my challenges – and mistakes). My research involved an ethnographic study of a popular ghost town in California where tens of thousands of tourists come each year, many of them hoping to glimpse the "Wild West." But I also lived and worked in the town (as a member of its State Park staff). So, in order to draw readers in to my situation (and the article), I began with the following anecdote.

> One summer Saturday, I was sitting at the living-room table in a run-down old house in the gold-mining ghost town of Bodie, a California State Historic Park located in the high-altitude desert east of the Sierra Nevada. Signs on the outside walls of the house identified it as an "Employees' residence." A nearby number post linked the building to the Park's self-guided-tour brochure, which described it as "The Gregory House," and detailed the lives of the home's historic inhabitants. I was busy writing when small running footsteps approached: children, some of the 200,000 or so annual visitors to Bodie. A brown-haired girl of about eight and her tow-headed younger brother strained to pierce the relative darkness inside the house. What they saw was me. Turning away from the window, the girl hollered to her parents, "There's a guy in there! And he's dead! He died writing!"
>
> Being taken for dead – and for a man – may seem shocking to some, but this was not the only time that I was seen as a ghost – or as a man – during the 14 summers that I worked and did field work in Bodie. But experiences like this one led me to contemplate the interactions between my physical presence and my role as insider in the public place that I was trying to study.
>
> As a researcher I was interested in how visitors and staff understood Bodie's past and made room for it in their present, in how they made meaning in and from the landscape. But as a staff member and part of the Bodie community, I myself was part of that process. An important part of my work became understanding how I was a part of my own research, and negotiating the challenges that being an "insider" presented. (DeLyser 2001: 441)

The anecdote itself ends with the girl's exclamation; the rest is my analysis of what the story means to this chapter – an essential part of any anecdote used in an academic setting.

The anecdote works for several reasons. First, it really did come from the data I gathered for my research. But, to readers, it's also relatively interesting (particularly compared to an **introduction** that begins, "This paper is about how researchers are a part of the communities they work with"), and possibly it's a bit surprising. It works to draw readers in. And critically, in these three short paragraphs, the anecdote (with its analysis as a "back up") serves to outline pretty much everything that the whole article is about.

Taken on its own, one story about my research is not enough to "prove" much of anything. To use it as "evidence" I would have to support it with a number of other observations. But used as an anecdote it tells readers what the article will be about and does so in an interesting way. While it is not always possible (or necessary) to begin (or end) a paper with an anecdote, thinking about writing up your findings in that way may help you to enjoy the writing process more, and, when an anecdote works, it may also help your readers to enjoy better what you've written (see Exercise 25.2).

## Arguments

When not thinking about writing, the word "argument" conjures the image of a fight or a dispute. In scholarly writing however, arguments are the key to a successful paper, each research paper makes an argument: it states a case, and presents evidence to support that case. Though some academic arguments (polemics) are forceful, and may actually feel like a fight (they may, for example, take on the work of another scholar, refuting it as incorrect), others can be far more subtle, or complex. A good research paper may even present an argument showing multiple sides of an issue – the argument may be that each side is valid, that each side has its merits.

Whichever type of argument you choose, it should be summed up in that always-difficult-to-write-or-to-say single sentence that answers the question, "What is your paper about?" If, at first, you find no "argument" in a paper about, say, "the effects of transportation developments on the American city," (which you could more simply call "Transportation and the American City" or even "Transporting the American City") think about how you will tell your tale: which developments will you focus on? Which ones will you leave out from that story and why? What other factors that influence the city, or transportation, played a role? Were transportation developments the same everywhere, or did particular conditions or particular individuals influence them in particular places? In deciding what (and *who*) to focus on, we make decisions about our arguments that may not be at first apparent when we mention the title or topic of our papers. But by thinking harder about what exactly the argument will be, we simultaneously uncover more of what that paper will really be "about."

So too, each argument made in a geographical research paper seeks to engage themes, issues and/or ideas of relevance to geographers. In a field as broad as geography it can be daunting, at first, to ask yourself, "How is my paper geographical?" It may help to remind yourself of prominent themes from lectures – and it matters whether you're writing about physical geography (perhaps about coastal dune erosion), about human geography (perhaps about the interplay between landscape and memory), and which geographical techniques you seek to engage (perhaps using satellite imagery to analyze tropical deforestation, GIS to document urban land-use changes, or ethnography to understand the meanings of migration). While each geographer typically studies a particular empirical, "real-world" situation in a particular place or setting, we each also engage broader themes that can be suggested by key geographic terms. In human geography, for example, these might include region, landscape, place, space, network, or mobility. Each of those, in turn, is cross-cut by (many different) understandings of class, gender, "race," ethnicity, sexuality, and ability (among other things). All of them are linked to broader literatures within geography, to the works of others who have attempted to think through similar issues, even if their work took place in different locations. Thus, identifying key terms and themes relevant to your paper helps you to understand how your paper is geographical, and links you to a rich geographical literature that can, in turn, provide support for your argument.

Further, these decisions about arguments are critical to the rest of the paper because everything in the paper must in some way contribute to the argument. We muster our evidence in support of whichever argument we have chosen, whether using statistics and tables, references to published works, quotes from interviews, photographs, or all of the above. Of course, we cannot simply exclude evidence that may seem to contradict our arguments. In fact, by including and addressing such claims, we may actually be able to make our arguments stronger.

Whether the information you gather supports or refutes your point, the work you do will rest upon the works of many who came before you (whether they presented their findings in print, at a conference, in a lecture, or on the Internet), and often as well upon the cooperation of other individuals (whether, for example, as interview participants, or advisors on your work). Whatever the evidence is and whatever participation you've had from others, your research paper must carefully **cite** all its sources so that readers will be able to decide for themselves that the evidence is credible, that appropriate prior scholars have been acknowledged, that participants have been credited for their contributions, and that the argument is therefore potentially viable. While we each strive to have original ideas and to make original contributions, we support rather than undermine that originality in our generosity to other researchers and to research participants. By citing each source, and carefully acknowledging the paths along which our ideas were reached, we allow future scholars to build with confidence upon the work we have done.

As you compose your argument, and work to engage geographical themes, keep also your paper's assigned length in mind. Having a well-defined argument will help you to structure your paper to the desired length from the outset (see Exercise 25.3). Working with your outline, you can estimate the number of pages needed for each portion of your argument, and then pace the argument to that number of pages. Pacing your argument is a powerful tool, not just for page- or word-length, but also because such pacing encourages writers to keep focused: research papers with vague outlines and ill-defined arguments can easily drift off point, meandering away from their stated aims.

Still, in making sincere efforts to explain and engage all you need to, you may find that your argument has at times branched away from its focus. If that's the case, consider moving the branching portions of the argument to one or more endnotes – there they can be developed briefly but fully, and need not distract from the point of your paper. On the other hand, you may find your paper coming up short of the assigned length. Here too, a well-defined working outline, and a well-constructed argument, can help you to target the areas you can readily expand, or identify sections where more research is needed.

## How to End

Though at first the most daunting part of writing an academic research paper is that blank screen, once we've begun, many beginning writers (as well as those more experienced) acknowledge that a bigger challenge lies in writing the **conclusion**. Here we must sum up our findings: careful not to repeat the exact words we have already used, we nevertheless must, in a sense, say what we already have said. In your conclusion, an academic reader will expect to see your findings summarized, and your main points spelled out clearly and concisely.

But if your conclusion stops as a summary, you're missing an opportunity to show how original and important your own work is. The best conclusions do summarize, but they also move at least one step farther, suggesting, for example, the broader implications of the work in question ("geographers studying A should pay attention to B and C"). Others may call for more work in the same area ("geographers should conduct more work on X"). Others still suggestively point to issues not directly covered by the paper's narrow scope of evidence ("my research on Y may have important implications for Z").

Be careful, however, to avoid dropping entirely new ideas and entirely new evidence into your paper's conclusion. The conclusion is not the place to build a new argument – it's the place to wind up the old one. Information that builds a new argument should be worked into the body of the text, or left for another paper.

Still, the possibilities are broad. The type of conclusion you choose will depend upon the strength of your data and evidence, as well as the confidence you feel in your argument. Either way, much of the value of your paper rests on its conclusion.

In fact, for better or for worse, many hurried academics struggling to keep up with all the latest research in their field simply skip to an article's conclusion – they only read the rest of the article if the conclusion is provocative enough to suggest a further time investment. Understood in this way, whatever goes into the conclusion is clearly critical (see Exercise 25.4).

## Revisions

Professional writers, whether academics, journalists, fiction writers, poets, or playwrights, spend a great deal of time revising their written work. Some very famous writers even make changes after their work is first published, altering what they have written before their already-published work appears in a new edition (see, for example, a Henry James manuscript revised after publication and reprinted in Bolker 1997). For students and beginning writers this may seem a rather odd situation. Since most of us begin as writers with relatively short essays or papers on topics we can easily hold in our minds at one time, and since most of us (as beginners at least) tend not to start writing until the last minute (often literally the last few hours!), we tend to save very little time for revisions. Indeed, even when we begin with the best intentions, deadlines have a way of leaping up upon us, and even the most prepared writer may find himself rushing as the clock ticks down. When you find yourself in that situation you will find too that you have no time left for serious revision. If that is the case, you can still do something to make your manuscript *seem* more polished, you can still do something to try to hide your hurry.

Proofread every word on every page closely and carefully, literally listening to each word in turn. (It often helps in catching small mistakes to actually read the paper aloud to yourself.) Compared to the irritation your reader will feel at sentences whose verbs and subjects don't agree, the small amount of time it takes to attentively proofread will be well worth your while. And then, once that is finished, run your spellchecker carefully one more time – this will usually catch any errors you introduced later in the writing/editing process. Together hopefully those two techniques will at least leave you with a paper that *appears* polished.

Real revisions, however, involve much more than running a spellchecker. They are, quite frankly, more a state of mind or even a way of life. But because the papers we wrote before we entered university, and many of those we write as undergraduates, are relatively short, and because the suggestions of teachers – usually written in red ink – make us feel that we have got something "wrong," we tend to believe that a good research paper must hatch fully formed, emerging as a first draft that serves as final draft as well. For experienced writers of research papers this is not the case: the process of making revisions may in fact be far more time consuming than was the writing of the initial draft. Getting to that point, to the point where revising your own work becomes as important a part of the creative process as writing the first draft, involves a change in attitude, and a change in mind-set – a change in your understanding of the writing process.

True, some revisions are geared to correcting mistakes. But the real point of revision is not just correction but *improvement*. Even when all you do is fix spelling you are making your paper better. But when you restructure your argument, add more evidence or anecdotes, take out redundancies, reorganize your points, or write a new conclusion, you have taken the writing process dramatically into your own hands, crafting your paper as a skilled creation rather than as something that "just happened."

To begin to recognize where revisions are needed, or where revisions can improve the paper, is a challenging task. It involves not just very close and careful reading of the work, but also the ability to "step back" or "away" from the work you know so well, to imagine yourself a reader unfamiliar with your topic. How will your reader feel? Have you presented enough background information? Does your argument make logical sense, or does it jump around? Are your conclusions well supported with evidence? Do you repeat yourself unnecessarily? Does the writing itself "flow" nicely? Will your paper hold your reader's interest with its style?

The answers to such questions can be difficult to discover, but a few strategies can help. If you are working alone (and perhaps under a tight deadline), one of the best ways to make sure your argument is logically structured and that your writing flows well is to dissect the paper into smaller pieces. Look first at any subheadings. Line them up in a row. Do they make sense in that order? If not, then consider either changing the subheadings or moving the sections around.

Look next at the paragraphs themselves. Each paragraph should develop only one or two different points or ideas. Read each paragraph individually and, in the margin, write that paragraph's main topic or point. Then read down the list of topics. Do they build, or follow in a logical sequence, or do they skip around all over the place? If they skip around you can simply move paragraphs (or sentences within paragraphs) to remedy that. Even revisions that seem major, can be carried out simply, effectively, and with little risk. Save your paper as a new draft before beginning the revisions; if you don't like the changes you've made, you can easily go back.

And finally, within each paragraph do the sentences follow each other in a sequence that will be clear to the reader? Each sentence has a grammatical subject, but it also has a felt or perceived "topic," a "what-the-sentence-is-about." Those topics should be closely related for sentences in the same paragraph, and should not jump around (see Williams 1995). If they do, reorganize sentences or rewrite them (see Exercise 25.5).

Often however, it is difficult to get enough "distance" from your own work to know when your topics are confusing to readers, or when your evidence has not been clearly presented. But the best solution may also be the scariest: show a draft of your work to somebody else. I say "scary" because it may *feel* scary. After all, if somebody else reads your draft paper and it is "bad," they may conclude that you do bad work or that you write poorly (Becker 1986). But experienced readers do not expect "perfection" from a *draft*. They know that making comments on a draft paper is done not to expose the writer's flaws but rather to make suggestions that can help improve the paper.

Showing a draft of your work to a teacher, friend, or fellow student may indeed point out some "weaknesses," or places, for example, where your argument was not clear. But rather than thinking of that as exposing a flaw, you can also see your friend's comment(s) as helping you to make your paper better. After all, it will be your name at the top, not hers. Future readers will know only the (improved) finished product.

This kind of good critical feedback is one of the most important parts of scholarly work – in fact, most research articles must be reviewed and approved of by several other scholars before they can be published, and most often those other scholars will make suggestions on how to improve the work. Learning to share your work with others is thus an important part, not just of making your writing better, but also of making your work part of a broader research endeavor, part of a richer scholarly discussion.

## Final Words

In looking back over this chapter perhaps you, too, will notice that the title, "Writing it Up," now appears a bit misleading. Indeed, while many (at least initially) consider writing merely a mechanical process, where data goes in and a finished paper (with a seemingly natural and singular solution) flops out, as I hope I have demonstrated, writing is really not at all like this. The writing process itself is *formative*: we actually come up with, construct, and develop our ideas *while we are writing*. Our ideas occur to us and change literally as we write, and those ideas (along with the ultimate finished paper) are a product not just of the data that is "out there," but also of how we found that data, how we approached the topic, and even who we are. Thus, the process of writing a research paper needs to be seen as integral to the research itself (see also Mansvelt and Berg 2005).

Further, writing is a skill and a craft that takes time to develop, to polish, to perfect. Writing takes practice, and good writing takes lots of it. Some of what we do as writers is straightforward and even mechanical (like running a spellchecker, or carefully tracing out the construction of an argument). And some of the forms we write in, as scholars, may seem restrictive. Certainly, the need to closely adhere to a pre-assigned page- or word-length, the requirements of engaging existing literature, the strictures of always citing sources, and even the very format of a research paper may seem, at first glance, to constrict creativity. Viewed from a broader perspective, however, nearly all forms of writing – from sonnets to screenplays – impose similar boundaries upon writers. In fact, such boundaries can serve not as restrictive limits to creativity, but helpful guides to focus that aid writers in channeling their creative efforts into forms that readers can more readily accept.

Finally, if we still find ourselves working at the last minute into the wee hours against the pressure of a deadline, well that too is not necessarily bad: the creative and formative nature of the writing process works 24-hours a day. But if, even as our deadline looms, we make the effort and take the time to fully engage with the process of writing itself – crafting quick outlines, experimenting with anecdotes, carefully constructing a conclusion, and leaving at least a little bit of time (if not a lot!) for revision – we can begin the process that turns data into a research paper, thoughts into text, and a pile of notes and papers into a compelling argument. We then will *be* writers.

## References

Becker, H. S. (1986) *Writing for Social Scientists: How to Start and Finish Your Thesis, Book, or Article*. Chicago: University of Chicago Press.

Bolker, J., ed. (1997) *The Writer's Home Companion: An Anthology of the World's Best Writing Advice, from Keats to Kunitz*. New York: H. Holt.

Bolker, J. (1998) *Writing Your Dissertation in Fifteen Minutes a Day: A Guide to Starting, Revising, and Finishing Your Doctoral Thesis*. New York: H. Holt.

DeLyser, D. (2001) "Do you really live here?" Thoughts on insider research. *Geographical Review* 91(1 & 2): 441–53.

DeLyser, D. (2003) Teaching graduate students how to write: a seminar for thesis and dissertation writers. *Journal of Geography in Higher Education* 27: 169–81.

Mansvelt, J., and Berg, L. D. (2005) Writing qualitative geographies, constructing geographical knowledges. In *Qualitative Research Methods in Human Geography* (2nd edn), Hay, I., ed. Melbourne: Oxford University Press, 248–65.

Williams, J. (1995) *Style: Toward Clarity and Grace*. Chicago: University of Chicago Press.

## Additional resources

Ballenger, B. P. (2008) *The Curious Researcher: A Guide to Writing Research Papers* (6th edn). New York: Longman. Shows how good research and interesting writing work together in compelling research papers, offered in a step-by-step approach to focus a topic and get the paper finished.

Cook, C. K. (1985) *Line By Line: How to Edit Your Own Writing*. New York: Houghton Mifflin Co. Rich examples teach editing techniques and strategies designed to help writers see the weaknesses they may otherwise miss in their own prose.

DeLyser, D., and Pawson, E. (2010) From personal to public: communicating qualitative research for public consumption. In *Qualitative Research Methods in Human Geography* (3rd edn), Hay, I., ed. Melbourne: Oxford University Press. Offers strategies and techniques for communicating research to different audiences in different forms – in writing, in verbal presentations, on web-sites, etc.

Lester, J. D. (2009) *Writing Research Papers: A Complete Guide* (13th edn). New York: Longman. A comprehensive guide to traditional- and electronic-based research and presentation, this text also navigates the subtleties of different citation styles and methods of documentation, and answers questions about grammar and usage.

Turabian, K. L. (2007) *Student's Guide for Writing College Papers* (7th edn). Chicago: University of Chicago Press. A classic text offering both a style guide and a research guide, now updated to include extensive material covering on-line sources.

Veit, R. (2004) *Research: The Student's Guide to Writing Research Papers* (4th edn). New York: Longman Publishers. Uses examples from actual research papers to help students maintain a focus on the goals of writing research papers, explaining the research and writing processes in a step-by-step manner, and showing along the way how conventional writing rules and formats help writers achieve those goals efficiently.

Williams, J. (2006) *Style: Lessons in Clarity and Grace* (9th edn). New York: Longman. Compelling is not a customary term for a book about writing, but Williams' text is just that. It teaches writers how to craft, diagnose, and then revise their prose to make it both appealing and con-vincing to readers, introducing as well the ethical issues that such persuasion raises.

## Exercise 25.1    Working Creatively with an Outline

Prepare a skeletal, working outline for the paper you would like to write. Include the main points you'll make in your introduction and each point (in turn) that will help you develop your argument. Identify at least some of the geographical literature you will engage with, and jot down (at least tentative) ideas for your conclusion.

Then, rework that outline to put forward your argument in a different way, by re-ordering your points, or emphasizing different evidence, or engaging different literature.

Keep the outline you find most effective – but avoid sticking to it if it's not working. As you write your ideas will develop and your outline may need to change again.

## Exercise 25.2    Writing an Anecdote

Look over the information you've gathered for your paper. Think carefully about the main point you are trying to make, and how that point is demonstrated by information you collected. Then look again at that information, searching for a story that well exemplifies the point you're trying to make (as in the example above). Write a brief (two paragraph) anecdote-and-analysis passage based on this story. Consider using that in your paper's introduction or conclusion.

## Exercise 25.3    Drafting an Argument

Write one sentence (even an awkward or imperfect one) describing what your paper is about. How strongly do you feel about this topic? What sort of argument would you like to make? A polemic? A balanced argument? Then look over the research materials you've amassed and make notes about how each source contributes to that topic. As you do so, keep detailed track of your sources so that you can appropriately cite the ideas, works, and words of others (whether from print, web, interview, or other sources). Examine the materials in turn, grouping them to build a logical argument that convincingly presents your topic. Look for weak points where you'll need to find more information, as well as counter-claims you'll need to address. Examine too what it is that makes your research geographical. What key terms are you engaging, and which other geographers have written about them before you? Then re-examine the tone you took at the beginning of this exercise and check that your data, and the works of other geographers, can support the type and strength of argument you decided to make. You may need to revise your topic statement and the tone of your argument accordingly – but that is more easily done than trying to build an argument on a weak foundation.

## Exercise 25.4    Constructing a Conclusion

Look back to the topic statement you wrote when you constructed your argument, and to the opening anecdote you wrote. Begin your conclusion by finding a new way to state that point, using different words and/or a different emphasis to address your topic. Next, examine the argument you built up, and draft a summary of that. Such a summary will help you to emphasize the most significant points in your work. Then look back at the research done by others on your topic. How is your work different? What new possibilities can your work offer? What can the things you've learned from your research offer others interested in the topic? Finally, keeping the tone and intensity of your argument, draft a sentence to suggest a response to the work of others, a way forward for other geographers, or the value of the insight you've gained through your research.

## Exercise 25.5    Stringing the Pieces Together

Since the interlinked processes of thinking and of writing can be well engaged by revising what we've already written, look at the exercises you completed above to see how much of a paper is already there, and how you can now revise what you have to make a powerful paper. Using your favorite outline, place the anecdote you wrote at the paper's beginning. Follow that with the statement about your argument, and add the list of notes from your outline about how exactly you plan to develop that argument (eventually you'll expand each of these into one or more paragraphs). Then use the conclusion you wrote as a temporary ending. Check carefully that the opening anecdote-and-analysis passage fits with the argument and with your claims in the conclusion. Make sure your argument builds logically, from one point to another, for readers not familiar with your topic. Remind yourself of the works of other scholars, and make sure to address these along the way. Once this is done, you'll find that your whole paper is nearly done: you can now knit the pieces together into a polished and finished product. Importantly too, you will have written a paper in manageable pieces, and proven to yourself that you need not necessarily write without stopping from beginning to end. By doing so, you build your own skills as a writer, and harness the writing and revision process to help, rather than hinder, your own creative abilities.

# Glossary

**Abduction**  The process of inferring a cause (A) from an effect (B) and a general law relating cause and effect. For example, given B and the law "If A then B," one infers that A might have occurred (but since B might also be produced by a cause other than A, this type of reasoning is less certain that deduction).

**Absorption**  Interception of EMR by matter.

**Action research**  Research conducted for a practical purpose, often in association and collaboration with the expected beneficiaries.

**Advocacy**  The promotion of particular interests in research, by individuals or groups such as communities.

**Amenities**  Site-specific or region-wide characteristics that can make a place especially attractive for living and working. Amenities are natural (e.g., beautiful landscape) or human-made (e.g., good infrastructure, skilled workforce), and unattractive features like pollution and high crime rates are called disamenities.

**Anecdote**  A story, told in compressed form, taken directly from the information gathered for a particular research paper, whose point clearly indicates the direction of the paper as a whole. In academic papers, the story itself is followed by a brief explanation of how that story relates to the topic of the paper.

**Angle of friction**  The angle of a slope on which a potentially mobile cohesionless body is in a critical state with respect to movement.

That is, either the angle at which it comes to rest (dynamic friction), or at which motion is initiated (static friction) represented by the ratio of shear stress to normal stress (tangent of the angle of friction) at the critical state.

**Angle of repose**  The steepest angle that cohesionless material can attain and still remain stable.

**Anomaly**  Deviation or departure from the long term average. The period of time for which the average was calculated must be stated explicitly.

**Antipodal points**  Points at opposite ends of a diameter of the Earth sphere.

**AOGCM**  Abbreviation for "coupled Atmosphere-Ocean General Circulation Model." AOGCMs couple coarse-scale atmosphere and oceans models. More complex versions of AOGCMs may also include biosphere, carbon cycle, and atmosphere chemistry models. AOGCMs are used to study climate sensitivity, variability, and change.

**Archive**  Refers to a collection of information on a topic, the physical site where that information is stored, or both. These days an archive might be a library bookshelf, or a digital collection housed on somebody's server.

**Argument**  In an academic paper the point of that paper – and not necessarily anything to fight over. All information in the paper contributes to the argument – to the author's ability to convey her/his point, and convince her/his audience.

**Assignment algorithm for simple closed curves** Procedure to determine if two arbitrarily chosen points lie on the same or on different sides of a simple closed curve.

**Attribute** Characteristics distinguishing a point, line, or area on the Earth's surface.

**Authorship** This is a textual metaphor that helps us to realize that cultural landscapes are created by human action, intentional or not. A landscape author originally referred to someone who was able to effect earth-writing, or who had created or produced a cultural landscape or landscape feature. More recently, it also can refer to the landscape interpreter, or reader, who also "author's" a landscape's meaning through her interpretation.

**Autoethnography** Ethnographic research based on often highly personal experiences, written up by researchers and/or the people with whom they do research.

**Basic employment** Export-oriented industries, sometimes called city-forming or traded. These are the various local industries that sell to the industries, households, or governments of other regions. In many small community economies basic employment constitutes more than one-half of total employment.

**Becoming** The philosophical view put forward by Gilles Deleuze that human and non-human forces are always influencing each other and creating hybrid and uncertain productions.

**Binary** A distinct opposition between two terms, for example between discrete and relational.

**Biomass** The amount of above ground and below ground plant living matter in an area.

**Boundary shear stress** The force per unit area exerted on the channel or sediment as a result of the weight and velocity of flowing water.

**Cartogram** A value-by-area map. The areas of geographic polygons, such as countries, are scaled by the values of some attribute, such as population.

**Cartography** The discipline dealing with the conception, production, dissemination and study of maps.

**Catena** A linked topographic sequence of features, forms and conditions which differ because of variations in altitude, relief, and drainage (usually but not exclusively applied to soils).

**Census** A complete enumeration of the population.

**Central tendency** Measures for determining the middle of a distribution (e.g., mode, median, mean).

**Citation** A reference to the work of another that gives acknowledgement to that work. Citations may refer to books, articles, conference papers, interviews, observations, or websites and must be present whenever an author builds upon the works and ideas of another, even if they are not directly quoted.

**Climate change** As defined by the IPCC, climate change refers to any change in climate over time, whether due to natural variability or as a result of human activity. Other definitions limit climate change to only those changes attributed directly or indirectly attributable to human activity.

**Climate classification** A means of organizing climate information.

**Climate model** A quantitative way of representing the interactions of the atmosphere, oceans, land surface, and ice. Models can range from relatively simple to quite comprehensive.

**Coding** The systematic process through which researchers look for patterns in interview transcripts and participant observation notes, to properly identify their findings.

**Coefficient of determination** The degree of variation in a dependent variable explained in a regression analysis. Any variable's variation is composed of its total variability (with respect to its mean), its "explained" variability, and the unexplained variability. The coefficient of determination is the ratio of explained variability to the total.

**Communication model** An addressor makes a message (text, picture, etc.) that is received by an addressee. From this simple model all kinds of questions unfold. How should we theorize the addressor? The addressee? Or the message? One thing we know: there's a geography at every level.

**Complex behavior** A situation where outputs of a geomorphic system are not directly proportional to inputs across the entire range of inputs, and where multiple outcomes may arise from the same input, depending on the condition of the receiving system.

**Composite number** Positive integer with factors in addition to 1 and itself; all non-primes.

**Conceptual model** A type of model used to explore, summarize and reveal new insights about system functioning. These models are often narrative or visual summaries in which system components (processes and forms), and interactions among them, are described.

**Conclusion** The end of an academic paper where the writer reviews the paper's argument and main point(s), leading to a statement of the paper's (and the research's) purpose. The conclusion affords an opportunity to call for new research, point to new ideas, and draw attention to the significance of the paper.

**Confidentiality** Declining to reveal information or its source, for example in not identifying survey respondents or the location of a research project.

**Conformal map projection** Preserves angular relationships at points transferred from a 3-dimensional globe to a 2-dimensional map.

**Consulting** Acting or undertaking research services for a particular client, usually in a paid capacity.

**Content analysis** A method that seeks to scan texts for meaning by focusing on signifiers and the pattern of either their presence of absence. It assumes that the meaning of such signifiers remains constant regardless of context.

**Co-variation** A process in which two variables deviate systematically from their respective means.

**Critical discourse analysis** A method that seeks to explain the production, dissemina-

tion, and consumption of texts by virtue of their embeddedness within social structures. It assumes that these factors explain the meanings of a particular text.

**Critical realism** The realist position that there exists an objective world, yet this world is essentially contingent, embedded, and relational, depending upon higher order casual mechanisms and structural relations.

**Cultural ecology** A field of research that applies ecological concepts (e.g., system, adaptation, feedback) to explain human culture, behavior, or action.

**Cycle of erosion** A largely abandoned theory of landscape evolution developed by William Morris Davis at the end of the nineteenth century.

**Data classification** The categorization of a large number of individual data values into a smaller number of groups to facilitate explanation and interpretation.

**Data matrices** In geography, these are usually partitioned with spatial units as rows, variables as columns. At the row and column intersection is that observation's value on that variable. Some people reject the way they carve up and classify the world, but for others a data matrix is a beautiful thing.

**Data mining** A set of statistical techniques for analysis of large databases that seeks to discover the underlying patterns in data. Includes spatial data mining.

**Database querying** A set of techniques that retrieves the data from databases using structured query language.

**Deception** Deliberately concealing the existence, nature or purpose of a research project from potential or actual participants or survey respondents.

**Deconstruction** A method that seeks to draw out the discourses, generally in the form of either/or binaries, that underpin which words and/or images are used and the manner in which they are used. It privileges context in determining the meaning of discourses and the binaries they depend upon.

**Deduction** The process of inferring an effect (B) from a cause (A) and a general law relating cause and effect; e.g., Given A and the law "If A then B", one can deduce that B must occur.

**Degree of variability**  The amount of variation present in the population.

**Dendrochronology**  Tree analysis of the annual growth rings of trees in order to date the age of the tree and reconstruct past environmental conditions.

**Density**  Number of individuals per unit area.

**Denudation**  The overall lowering of the Earth's surface by the removal of material from the land, usually expressed in units such as mm/year.

**Dependent/independent variable**  The dependent variable is the one whose variation you are trying to explain. The independent variables are those hypothesized to statistically explain (or account for) variation in the dependent variable.

**Desertification**  A now largely discredited and problematic thesis postulating the global or regional expansion of deserts that directed scientific research, government activity, and political control of land management; the concept has been replaced by a more modest notion of "arid land degradation."

**Determinism**  A deterministic world view encompasses the notion that there is a direct cause and effect for initial conditions and final conditions. This implies that if all conditions are known, then system behavior can be predicted. It is often difficult to treat all model components deterministically, hence in some cases stochastic elements must be incorporated into models.

**Dialectics**  The exchange or conflict of opposing ideas and forces that produces change or synthesis.

**Diameter at breast height**  The diameter of a tree at approximately 1.3 m above the ground.

**Digital elevation model**  Topographic data stored in an equally spaced gridded network. Each grid cell, representing a square on the Earth's surface, is assigned a single elevation value.

**Discourse**  A set of commonly shared meanings or even common sense assumptions about the world which serve as the medium through which we communicate, negotiate, and challenge everyday practices.

**Discourse analysis**  A procedure that treats texts as socially constructed accounts that bear the mark of social power, including the power to determine important differences, such as truth and non-truth.

**Discursive event**  A particular instance wherein a word or image is used to convey meaning. This event is made possible by an underpinning discourse; it also has the capacity to transform that discourse.

**Distributive Law**  Algebraic law linking operations of addition and multiplication: $x(y + z) = xy + xz$.

**Dualism**  The ontological belief that existence is split between physical and non-physical substances, commonly referred to as the "mind-body" problem.

**Dummy variable**  A variable included in a regression analysis to capture categorical effects. The coding is binary (0 = not included in the category; 1 = included in the category).

**Dynamic equilibrium**  A state where changes to the system's overall general trend of change are compensated by restoring changes (negative feed back), so that the general trend is able to continue.

**Ecological fallacy**  An erroneous assumption that all individuals in a group share the average characteristics of that group.

**Economic base**  Sometimes called the export base, this is the sum of all basic or export-oriented activities found in the regional economy. In industrial society the economic base was dominated by primary, secondary, and (sometimes) government industries but, in today's postindustrial society, many service and trade industries are now included.

**Empirical**  Obtained from experiment and observation, rather than from theory.

**Empirical model**  A type of mathematical model that does not account for underlying processes, but rather data are used to define functional relationships that capture the main system trends. Empirical models are often used to predict behavior when detailed process data and/or understanding do not exist, or when data are available for only certain restricted variables.

**Endogenic processes** Processes deriving their energy from within Earth (e.g., tectonic processes such as mountain building and earthquakes).

**Ensemble forecasts** An ensemble weather forecast is a collection of forecasts. Often the ensemble is created by running a numerical prediction model multiple times but slightly perturbing the initial conditions for each run to represent uncertainty in atmospheric observations. The spread of the ensemble runs provides an indication of the uncertainty surrounding the forecast.

**ENSO (El Niño – Southern Oscillation)** A complex interaction of the tropical Pacific Ocean and the global atmosphere. El Niño is a warming of the surface waters of the eastern equatorial Pacific, and the Southern Oscillation is a large-scale fluctuation in sea-level pressure between Indonesia/North Australia and the southeast Pacific. ENSO is a natural part of the climate system and contributes to the interannual variability in weather and climate.

**Environmental determinism** A perspective in geography that advocates individual human beliefs and actions are determined or controlled by the natural physical environment.

**Environmental gradient** A continuous, gradual change in conditions and vegetation.

**Environmental perception** An approach to exploring human experience and behavior in the environment focusing on how individual cognition and understanding acts to filter information and assemble ideas about people's surroundings.

**Epistemology** The frameworks of knowledge that we use to understand the world (see Chapter 2 for an elaboration).

**Equal area map projection** Preserves areas in correct proportions when transferred from a 3-dimensional globe to a 2-dimensional map.

**Equator** Great circle bisecting the distance between the poles of the Earth sphere.

**Eratosthenes's measurement of the Earth** Classical method for calculating circum-ference of Earth sphere using Euclidean geometry.

**Ergodic transfer** The substitution of space for time, as a way of determining evolutionary changes to geomorphic systems. For example, fault scarps of different ages in an area may be examined from the youngest to oldest, in an attempt to establish the process of slope evolution that might have occurred through time.

**Ethics** The study of the moral or normative dimension of life, otherwise known as moral philosophy.

**Ethnography** Research whose central method is participant observation, but also involves other methods appropriate for addressing the issues being studied.

**Euclid's Parallel Postulate** Given a point $P$ not on a line $m$: through $P$ there passes exactly one line that does not intersect $m$ (is parallel to $m$).

**Exogenic processes** Those processes deriving their energy from or outside Earth's surface: (e.g., processes driven by gravity and climate variables).

**Expedition** An exploratory excursion into unknown territory, usually in some distant land but possibly close by such as the inner city.

**Extensive research** This requires a large sample of observations and seeks to identify patterns in the phenomenon of interest, often through the use of inferential statistics or numerical analysis.

**Factorization** Decomposition of a number or an algebraic expression into a product of expressions. The factorization of $x^2 - y^2$ is $(x - y)(x + y)$.

**Field** A conceptualization of spatial variation as continuous over the Earth's surface.

**Floristic provinces** Regions that have similar growing conditions.

**Four color theorem** All maps in the plane can be colored using four of fewer colors. Distinctions between adjacent regions will be clear.

**Frequency distribution/frequency polygon** A descriptive portrait of data after collecting observations into a number of classes. Frequency distributions are tabular; polygons are graphical (also histograms).

**Functional specialization** This is the notion that regional economies often specialize in very different functions or economic activities. Mulligan and Vias have shown that regional economies with different specializations in their economic bases can have remarkably different inter-industry transactions and very different multiplier effects.

**Fundamental Theorem of Arithmetic** Every positive integer can be expressed uniquely as a product of powers of primes.

**Generalization** The process of selecting and organizing the geographic information to be represented on the map and transforming that data into map form.

**Geodesic** Path along which the shortest distance between two points is measured.

**Geographic data** Data that is georeferenced.

**Geographic information system** Software designed to perform operations on geographic data.

**Geographic visualization** The use of visual representations to make spatial patterns and problems visible. Often used to reveal unknown geographies.

**Georeference** Information serving to locate information on the surface of the Earth.

**Geo-referenced** Geographically referenced; pertains to the representation of geo-spatial data (including remotely sensed images) by geographic coordinates, normally through geometric transformation to a specific earth coordinate system.

**Geospatial data** See geographic data.

**Geovisualization** Computer-based multiple and interactive displays of geo-spatial information.

**Great circle** Cross-section of a sphere cut by a plane that passes through the center of the sphere.

**Greenhouse effect** The warming of the Earth's atmosphere due to the absorption of longwave radiation by gases in the atmosphere.

**Greenhouse gases** Gases that are transparent to solar radiation but opaque to Earth's long-wave radiation, contributing to the atmospheric greenhouse effect. Some greenhouse gases occur naturally while others result from human activities. The main greenhouse gases are water vapor, methane, carbon dioxide, and ozone.

**Hermeneutics** The study of meaning.

**Homoscedasticity** A violation of the assumption of ordinary least squares regression. It occurs when the variability of residuals is non-constant across different values of an independent variable. In simple regression cases, the condition can be detected on an initial scatter diagram of the dependent and independent variables.

**Human dimensions of global change** A field of research that examines human responses to, and influence upon, worldwide environmental transformations, especially climatic change.

**Human-environment geography** The study of the mutual influences of humans and non-humans on one another, the impact of people on the environment, and the categorical and conceptual apparatus through which humans understand the non-human world.

**Humanism** A perspective in human geography that seeks to understand the world by reference to human consciousness, spatial experience, and everyday life.

**Humanistic geography** The phenomenological interpretation of people's sense of place, as well as the understanding of space as a repository of meaning and lived experiences.

**Image interpretation** The act of analyzing remotely sensed images and extracting useful information.

***In situ*** "In place" conventional weather observations are referred to as *in situ* observations, as they are taken at a particular location and refer to a specified level(s) of the atmosphere.

**Induction** In induction, predictions about future events are based upon observation of the regular patterns identified from past events. However, with this approach we cannot be certain that the next observation will be consistent with previous ones; therefore this approach to knowledge production often incorporates inferential statistics and probability to quantify how certain we are about our conclusions.

**Infrared** The portion of the EMR spectrum consisting of wavelengths that are just longer than visible light; near infrared = 0.7–1.4 μ; shortwave infrared 1.4–3.0 μ; thermal infrared 3.0–14 μ.

**Inherent error** The error associated with the data collection process.

**Inhomogeneity** Heterogeneous, not homogeneous. Inhomogeneities in climate records can be caused by a number of factors including changes in instrumentation, station location, and observing practices.

**Intensive research** The objective of intensive research is to determine how processes operate to produce an observed pattern. It relies on relatively small samples, such as case studies, and may involve qualitative methods to characterize social processes or detailed measurements of specific physical processes.

**Intergovernmental Panel on Climate Change (IPCC)** Established by World Meteorological Organization and the United Nations Environmental Program "to assess scientific, technical and socio-economic information relevant for the understanding of climate change, its potential impacts and options for adaptation and mitigation."

**Introduction (opening)** In an academic paper, the first section of that paper. Here the author outlines the issue, question, or problem that will be addressed, and points to the conclusions that will be reached. Different from fiction writing, in an academic paper the main point of the conclusion is usually indicated in the introduction. But the introduction is also the paper's opening – the first words encountered by the reader, and can therefore also be crafted to draw readers' attention. (See also anecdote.)

**Inventory** A quantitative accounting of land surface features, often achieved by means of statistical sampling.

**Irradiance** Amount of EMR per unit time and area incident or coming into a surface.

**Jordan Curve Theorem** A simple closed curve partitions the plane into two mutually exclusive domains.

**Kantianism** An approach to understanding based upon the ideas of seventeenth-century German philosopher Immanuel Kant. Kant argued, essentially, that any order we see in the world is imposed on it by our minds. This approach is also known as *idealism*.

**Kurtosis** A measure of the extent to which a distribution is flat or peaked. Flat distributions are platykurtic; peaked distributions are leptokurtic; normal-curve shaped distributions are mesokurtic.

**Landslide** The outward or downward movement of a defined mass of rock or soil, under the influence of gravity.

**Large scale map** Representative fraction of the map is larger than those of maps to which it is being compared.

**Latitude** Angular measure of displacement of a parallel from the equator of the Earth sphere.

**Layer** A digital representation of the variation of one or more related properties over an area.

**Leibnizianism** An approach to understanding in which space and time are constituted through the interactions between objects (the "space does not exist prior to the social/natural relationships that brought it into being" view).

**Levels of measurement** These fall into four categories: nominal (categorical, with no implied ranking, such as sex or region of origin); ordinal (categorical, with implied ranking but of unknown metric, such as poor, fair, and good); interval (quantitative measure without a known zero point, e.g., time or temperature); and ratio (quantitative measure with a known zero point, e.g., income or rainfall).

**Life worlds** The sum of organized individual experiences in the human imagination that is identified and studied by phenomenologists.

**Livelihood** The strategies, knowledges, assets, and actions required for surviving and making a living.

**Local knowledge** Also sometimes called "indigenous knowledge," it is the knowledge that communities have developed and continue to innovate from their experience and context, often tested over centuries of use, adapted to local conditions, though continuously dynamic and changing.

**Logical positivism** A philosophy of science that maintains that all knowledge is grounded in observation, especially scientific observations of objective, theory-neutral facts.

**Longitude** Angular measure of the displacement of a meridian from the Prime Meridian of the Earth sphere.

**Macrofossils** Preserved organic remains large enough to be visible without a microscope.

**Magnitude–frequency relationships** Magnitude refers to the "size" of events or features measured by work done, energy release, or degree of landform change. Frequency, expressed spatially, refers to the density of features (number/unit area) or temporally by the number of events per unit of time. Magnitude can be measured, for example by volume (landslides), weight (sediment), intensity (rainfall depth/unit time), height (waves), velocity (wind gusts) etc. Frequency–magnitude distributions usually exhibit a log-normal relationship – indicating that large events (features) are greatly outnumbered by smaller ones.

**Map** A symbolized image of geographical reality, representing selected features or characteristics, resulting from the creative effort of its author's execution of choices, and is designed for use when spatial relationships are of primary relevance.

**Map projection** A systematic method of transforming a spherical coordinate system, and the locations on the Earth's surface it describes, into a flat or planar coordinate system.

**Map scale** A mathematical measurement of how much smaller the map is than the portion of the world it represents.

**Mapping** Graphically representing land surface features and/or phenomena in a spatially explicit and geographically referenced manner.

**Marxist geography** The study of space as both a product and active participant in unequal social systems.

**Marxist theory** A perspective in human geography that borrows from the writings of Karl Marx to critically examine the dynamics and evolution of human societies, especially capitalist ones, across space.

**Mathematical model** This type of model consists of a series of quantitative relations among variables that represent the underlying theory of the system being studied.

**MAUP (Modifiable areal unit problem)** posits that data analysis is scale-dependent.

**Meaning** A piece of information or way of feeling that is conveyed through the use of signifiers.

**Meridian** Half of a great circle, going from pole to pole of the Earth sphere.

**Metadata** Data about data. Metadata for climate observations usually includes information about the station location, instrumentation, and observational practices (such as time of observation).

**Methodology** The tools and approaches used to conceptualize and investigate specific objects of analysis.

**Misspecification** A class of errors in the formulation or estimation of regression models; may include fitting the wrong function, missing variables, multicollinearity, etc.

**Mixed methods** The trend in geography and other social sciences to use multiple methods in the assessment of research questions. Often involves pairing quantitative and qualitative approaches. Works from the assumption that two or more approaches are usually better than one, particularly when the issues under study are complicated.

**Modifiable Areal Unit Problem** The tendency for the results of analysis to depend on the spatial objects on which the data for the analysis are based.

**Monism** The ontological belief that there is only one substance in the world, from which each being or object is an expression or component.

**Monitoring** Making time-sequential observations of land surface areas, normally with an interest in assessing and/or quantifying changes.

**Morality** Normative beliefs or practices concerning individual actions, kinds of actions, dispositions or characters, or principles of behavior.

**Multicollinearity** When two or more independent variables are co-varying, it can be difficult to sort out their independent effects on a dependent variable. The condition is known as multicollinearity.

**Multipliers** The very important idea that local spending expands or extends the income that is first brought into the region by its exports. Local spending is visualized as passing through various rounds of expenditures until all of the new income leaks out. In the simplest version of economic base analysis one aggregate multiplier is estimated for the entire economy, but this can be adjusted for various factors like non-earnings income. The sizes of multipliers depend upon various factors, including the size, specialization, and location of the regional economy.

**Multivariate statistics** Statistical procedures simultaneously involving a number of independent variables.

**NAO (North Atlantic Oscillation)** The large-scale fluctuation in sea-level pressure between the subtropical high near the Azores and the polar low near Iceland. The NAO is the dominant large-scale circulation mode in the North Atlantic region and is associated with wintertime variability of weather and climate.

**Narrative analysis** A method of analysis that takes as its object of investigation the first person accounts of respondents of their experiences (i.e., stories).

**Neutrality** Declining to convey moral or political values involved in research or a research project.

**Newtonianism** An approach to understanding based upon the ideas of seventeenth-century British scientist Isaac Newton. Newton argued that time and space are absolutes, and objects are simply moved around in them (the "space is simply a container for social/ natural objects" view).

**Nonbasic employment** Locally-oriented activities, sometimes called city-serving, residentiary, maintenance, or non-traded. These are various industries that sell directly to the (other) local industries, households, or governments of the same region. Such transactions locally circulate the money that was first brought into the region by its basic industries.

**Non-earnings income** Sometimes called non-employment income, this designates all personal income flowing into the regional economy that is not a reward for current labor. This income stream has two distinct components, one that is private (i.e., dividends, interest, and rent) and one that is public (including various types of transfer payments from different levels of government).

**Non-Euclidean geometry** Geometry that does not obey Euclid's Parallel Postulate.

**Normative** Having to do with what is good or bad, better or worse, or what should or should not be done.

**Observation** The human act of gathering information about the world.

**On site recruiting** The recruitment of study participants by approaching individuals at particular sites.

**Ontology** Theories about what the world is like (see Chapter 2 for an elaboration).

**Operational error** Error derived from the analysis of the data.

**Optical** Lenses and mirrors are used by a sensor to collect EMR.

**Outline** A skeleton of a paper that can be used to help chart the paper's course both before and while writing. Outlines can be very detailed and designed for others to see, but they may also serve as evolving, working reminders for the writer of the work in progress.

**Overt/covert** If you tell people you're researching their lives, that's "overt" research. If you don't, it's "covert."

**Palimpsest** Originally referred to re-using a writing surface, so that it might bear the traces of several authors at several different time periods and thus appear, at first glance, as chaotic and indecipherable. In cultural landscape study it metaphorically refers to the similar appearance of the landscape, as the result of many years of human earth-writing.

**Palynology** Scientific study of pollen and spores often used to reconstruct past vegetation.

**Paradigm** An assemblage of beliefs, visions, and values that organize and structure prominent ways to see the earth and its places and to conduct research to understand it. Thomas Kuhn popularized the word and gave it its contemporary meaning in his book, *The Structure of Scientific Revolutions*, published in 1962.

**Parallel** Small circle determined by a plane that is parallel to the equatorial plane of the Earth sphere.

**Parametric/non-parametric** Parametric statistics and their associated tests are used to assess significance in ratio and interval data; non-parametric statistics are used in cases of nominal and ordinal data.

**Partiality** The explicit or implicit association of a position or project with particular agents or values.

**Participant observation** Research in which you write notes describing your participation in, and observation of, other people's lives in particular settings.

**Passive** Capturing naturally reflected or emitted EMR from the Earth's surface or atmosphere by a sensor on an aircraft or satellite.

**Phenomenology** The study of objects or events based on our conscious experience of those phenomena. Developed by Edward Husserl, this approach assumes that "reality" consists of objects and events as they are perceived in human consciousness, such that they cannot be understood independent of human consciousness.

**Photogrammetry** Subfield of remote sensing involving precise measurements and topographic mapping based on stereo imagery.

**Physical model** A type of model involving the construction of an analog for a real-world environmental system. The physical representation of the target system, which is usually reduced in size and complexity, must maintain essential attributes and relationships to provide meaningful results.

**Place** A location on the Earth's surface.

**Plagiarism** Conveying the work of others as one's own, by deliberate deception or careless scholarship such as inadequate referencing.

**Platform** Aircraft or spacecraft on which sensors are mounted to collect data remotely.

**Political ecology** A field of research that examines socio-environmental change through the lens of political economy, environmental power/knowledge, and resource control/ access.

**Politics** The power-driven process of competitive struggle over ends and means of achieving them, for example in the realm of economic and social policy.

**Population** The total number of individuals of interest for a study.

**Positionality** All interpreters bring this condition with them to the research table; no one sits outside their history, sociality, or geography.

**Positivism** An approach to research that argues that the only valid statements a researcher can make about the world derive from direct observation.

**Postindustrial economy** One increasingly dominated by advanced service and highly specialized trade activities. The idea was popularized by the sociologist Daniel Bell. More recent urban analysts like Edward Glaeser and Richard Florida now talk about "consumption amenities" and "technology, talent, and tolerance" as being important for attracting and retaining highly-skilled workers.

**Poststructuralism** A loose collection of schools of thought that focus on the indeterminacy of meaning, the non-essentialist character of identity, and, in human geography, the fluidity and productivity of space.

**Poststructuralist geography** The interrogation of spatial discourses that legitimize and naturalize power regimes and regulate difference.

**Prime meridian** Meridian passing through Greenwich, England; vertical axis of coordinate system of the Earth sphere.

**Prime number** Positive integer whose only integral factors are 1 and itself.

**Privacy** The capacity to protect or control access to information, for example about potential survey respondents.

**Process-based model** A type of mathematical model that attempts to explain and provide understanding of process operation by defining equations for governing principles. Such models often have high explanatory power in comparison to empirical models. Many models that are largely process-based, or strive to be so, still involve some degree of empiricism in certain model parameters, as complete explanation is not always possible.

**Purposive sampling** Researchers use their judgment to select a sample they believe is representative of the target population.

**Qualitative data** Reflect differences in kind or type of phenomenon.

**Quantitative data** Measure differences in quantity or degree of a phenomenon.

**Radiometric** Pertaining to the magnitude of EMR captured by a remote sensor.

**Random sample** A sample where every individual has an equal chance of being selected.

**Rapid ecological assessment** A combination of techniques for surveying the condition of flora and fauna, which uses a step-by-step technique for sampling, surveying, recording, and analyzing the condition of any plant and animal community.

**Realism** The view that the world exists independently of human cognition of it and that some of our beliefs about the world, especially those derived from scientific inquiry, are accurate and true.

**Reconnaissance** Conducting an initial survey of a landscape, often to gain appreciation of the geographic setting or context.

**Reflectance** A property of a material that is defined as the amount of EMR reflected by the material divided by the amount of EMR incident upon it.

**Regionalism** A perspective in human geography that views the Earth's surface as being composed of a mosaic of distinctive homogeneous places of human settlement and activity.

**Regolith** The layer of (weathered) material that mantles the bedrock.

**Relational database** A database structured by record (row) and allows for connection with other databases based upon a common field.

**Relativism** The view that knowledge and knowledge claims are relative to a particular group, community, or theoretical viewpoint, i.e., no objective epistemological standards exist.

**Relaxation time** The time required for a system to restore original conditions after it has changed in response to a given impact.

**Relevance** Having practical or theoretical significance to societal problems or policy in the private or public arena.

**Reliability** A term reflecting the degree to which variables measure the same things under changing conditions.

**Remote sensing** Collection of data about features or phenomena on the Earth's surface or atmosphere, without directly being in contact with such features/phenomena.

**Remotely-sensed observations** Gathering and recording information without coming in actual contact with the object or area being investigated.

**René Descartes** A seventeenth-century French philosopher who argued that because our senses may deceive us, the only statement we can make about the world without fear of contradiction is that we are conscious – hence his famous maxim "Cogito ergo sum" ("I think, therefore I am"). Descartes believed that the world could be understood through the process of rational thought and logic. Research following his approach is often referred to as Cartesian.

**Representative fraction** Ratio showing units on the map in relation to the same units on the ground, as in 1 inch (cm) represents 50,000 inches (cm).

**Research ethics** Following formal rules about "not doing harm" to research participants, etc. but also involving everyday concerns about "doing the right thing."

**Response rates** The ratio of all of those who participate in a study to all of those who were invited to participate.

**Resultant** Vector sum of multiple vectors. The sum of the wind vectors for multiple observation times at a location is referred to as the resultant wind vector.

**Return period (recurrence interval)** The average time interval between events of a given magnitude.

**Revision** More than running a spell checker, revision – the act of editing, reworking, and rethinking a written work – is an essential part of the writing process. Writers continue to think, to refine their ideas, and to polish their prose as they (we) write.

**Sample** A selected subset that represents the population for statistical purposes (see Chapter 6).

**Sample size** The number of samples to be measured from the population.

**Sampling unit** Individual components of the sample.

**Scalar** A quantity that has magnitude but no direction. Temperature is an example of a scalar quantity.

**Scale** A measure of the geographic extent or resolution; narrowly in cartography, the ratio of distance on a map to distance on the ground.

**Scattering** Redirection of EMR upon interaction with matter (e.g., gas molecules and particles in the atmosphere).

**Scientific name** The genus and species name of an organism.

**Sensor** In the remote sensing context, the device used to collect EMR, commonly in a two-dimensional, imaging manner.

**Serendipity** Entails making accidental discoveries; findings that you were not looking for in the first place (but which nevertheless might be interesting).

**Shear strength** The maximum shear stress (force per unit area) that may be sustained by a given material before it ruptures (shears).

**Signifier** A word, image, behavior, or thing that is associated with a particular meaning such that its presence or absence can be "read" as indicating that meaning.

**Situated knowledge** All knowledge is shaped by how we are equipped – personally, culturally, technologically – to find things out and make judgments.

**Skewness** The degree of symmetry or left/right bias in a frequency polygon or histogram.

**Sloppy first draft** see zero draft

**Small circle** Cross-section of a sphere cut by a plane that does not pass through the center of the sphere.

**Snowball technique** A method for creating a non-random sample. Initial contacts within a population are asked to identify other potential participants, who in turn are asked to identify others.

**Spatial analysis** A perspective in human geography that seeks to understand the making and evolution of landscapes and places by reference to universal and generalizable principles.

**Spatial autocorrelation** When spatial units in a data set lack independence with one another, they are autocorrelated. Most spatial data is autocorrelated; regression assumes that residuals are not autocorrelated, and this has led to numerous tests and technical efforts to model and remove spatial autocorrelation from spatial data.

**Spatial context** The characteristics of a place and its immediate surroundings.

**Spatial data** See geographic data.

**Spatial dependence** The tendency for the characteristics of a place to be similar to the characteristics of neighboring places.

**Spatial diffusion** Term refering to processes involving movement, migration, and transmission over geographical space. Research into spatial diffusion, and specifically into the diffusion of innovation, was pioneered by Swedish geographer Torsten Hägerstrand.

**Spatial exploratory data analysis** The search for new spatial patterns in large databases using a variety of statistical and geovisualization means.

**Spatial heterogeneity** The tendency of properties and processes to vary over the Earth's surface.

**Spatial science** An approach to human geography that emerged and developed throughout the 1950s and 1960s. Spatial science is generally associated with the positivist tradition and relies mainly on quantitative methods. Its fundamental idea is that the distribution of human activies is governed by some regularity, and can be explained by quantitative models.

**Spatial thinking** The practice of reasoning from geographic data.

**Species abundance** Relative abundance of a species in a region.

**Species richness** Number of species per unit area.

**Spectral** Pertaining to the wavelength of EMR.

**Stereographic projection** Transformation of the sphere to the plane using the North Pole as a center from which to carry the sphere's surface to a plane tangent at the South Pole.

**Stochasticism** A stochastic world view considers randomness to be inherent to the system. Even if one's world view does not subscribe to this philosophy, stochastic elements are often incorporated into models to treat the apparently random elements.

**Stratified random sample** A sample in which the population is divided up into groups, and samples are made randomly within each group.

**Structuralism** A perspective in human geography that posits the power of unseen, humanly-bounding forces as central to understanding the earth and its places.

**Structures and mechanisms** Social structures are manifest as a series of mechanisms. Where the structure can be thought of as the underlying rules of the game that provide roles and

set goals, mechanisms are the particular "how to" instructions that allow people to take part.

**Survey** A standardized set of questions administered to a number of respondents.

**Symbolization** The transformation of geographic information into map symbols.

**Systematic sample** A sample where a constant (e.g., every 10 meters) defines the selection criteria.

**Texts** Traditionally, texts are word and/or image-based artifacts, such as books or newspapers, that are mobile vessels for signifiers. More recently, such phenomena as bodies and buildings have been included within this category.

**The scientific method** A largely positivistic approach to the study of phenomena that involves observation and description, formulation of hypotheses to explain and predict, and controlled experimentation to test and refine those hypotheses.

**Theory-determined** An approach to research that argues that data can only be interpreted in the context of the theoretical frameworks brought to bear on them. This means that researchers adopting different theoretical frameworks would not be able to legitimately evaluate each other's research.

**Theory-laden** The notion that scientific terms, methods, practices and judgments all depend to some extent on specific scientific theories, on theoretical traditions or on theoretical perspectives.

**Theory-neutral** An approach to research that argues that data can be interpreted in the same way, regardless of the theoretical frameworks brought to bear on them. It supposes that there is a difference between "fact" and "value."

**Threshold** The value or condition identifiable in a varying system parameter (or combination of parameters) that signifies an abrupt change in behavior from that characteristic of one domain to a distinctly different behavior, characteristic of another domain. An intrinsic threshold is a threshold within a system brought about by changes that occur within the system (such as progressive weathering of a cliff face, reducing rock strength below prevailing shear stress to produce a landslide), whereas an extrinsic threshold is a threshold brought about by forces external to the system (such as an increase in stream velocity to a point where sediment becomes entrained).

**Transformation** Systematic mathematical process for transferring elements from one point set to another.

**Troposphere** The layer of the atmosphere closest to Earth's surface characterized by a decrease of temperature with height. The troposphere contains approximately 95 percent of the mass of the atmosphere and extends from the surface to 8–15 km depending on location and season.

**Uncertainty** The extent to which geographic data leaves one uncertain about the true nature of the Earth's surface.

**Validity** A term reflecting the degree to which your variables align with your constructs and theoretical abstractions. Is GPA a good measure of intelligence?

**Values** Attitudes or opinions associated with normative aspects of life, or morality.

**Variability** The degree to which observations on a variable (an attribute that varies) deviate from a mean. If the world didn't vary, it wouldn't be very interesting, and spatial scientists (and everyone else for that matter) would be out of a job.

**Variation** Measures for determining the degree of dispersion of a variable about a central point. Common measures include range, variance, standard deviation, and coefficient of variation.

**Vector** A quantity that has both magnitude and direction. Wind, which has a zonal (west-east) and a meridional (north-south) component, is an example of a vector quantity. The sum of the wind vectors for multiple observation times at a location is referred to as the resultant wind vector.

**Vegetation cover** The percent of vegetation covering a specified area.

**Vernacular** In landscape study, refers to the everyday, common landscape; as opposed to landscapes created by professional designers (such as landscape architects).

**Verstehen** German for "understanding," this term is used to describe an approach to knowledge production in which the researcher attempts to develop empathic understandings of the perspectives of others so as to understand their behavior on their own terms (unlike positivism, which argues that direct observation is the only valid basis for knowledge production). This approach was largely developed by the sociologist Max Weber.

**Whiggish or teleological** Histories in which evidence from the past is marshaled in an argument that appears to lead indisputably and unproblematically to one's own present position.

**Writing** A formative and creative process integral to research itself as well as the primary means of communicating research findings in academia. Both a skill and a craft, writing must be developed, through practice.

**Zero draft** A draft before the first draft used to help the writer initiate the writing process and develop ideas (also called sloppy first draft).

**Z-scores** A measure for an observation that locates it in terms of its position in standard deviational units away from the mean.

# Index

Printed and bound by CPI Group (UK) Ltd, Croydon, CR0 4YY

27/10/2024

14580392-0005